Introduction to
Principles of Plant Pathology

Fourth Edition

Introduction to
Principles of Plant Pathology
Fourth Edition

RS Singh

Oxford & IBH Publishing Co. Pvt. Ltd.
New Delhi

(*A Unit of* CBS Publishers & Distributors Pvt Ltd)

CBSPD

CBS Publishers & Distributors Pvt Ltd

New Delhi • Bengaluru • Chennai • Kochi • Kolkata • Lucknow • Mumbai
Hyderabad • Jharkhand • Nagpur • Patna • Pune • Uttarakhand

Introduction to Principles of Plant Pathology
Fourth Edition

ISBN-13: 978-81-204-1551-5
ISBN-10: 81-204-1551-5

© 2002, RS Singh
Reprint: 2017, 2018, 2019, 2021, 2023

OXFORD & IBH
New Delhi
(A Unit of CBS Publishers & Distributors Pvt Ltd)

Published by **Satish Kumar Jain** and produced by **Varun Jain** for
CBS Publishers & Distributors Pvt Ltd
4819/XI Prahlad Street, 24 Ansari Road, Daryaganj, New Delhi 110 002, India
Ph: 011-23289259, 23266861   Website: www.cbspd.com
                                                       e-mail: delhi@cbspd.com

Corporate Office: 204 FIE, Industrial Area, Patparganj, Delhi 110 092, India
Ph: 011-4934 4934   Fax: C11-4934 4935   e-mail: publishing@cbspd.com;
                                                                                    publicity@cbspd.com

Branches

- **Bengaluru:** Seema House 2975, 17th Cross, KR Road, Banasankari 2nd Stage, Bengaluru 560 070, Karnataka, India
  Ph: +91-80-26771678/79   Fax: +91-80-26771680   e-mail: bangalore@cbspd.com
- **Chennai:** 7, Subbaraya Street, Shenoy Nagar, Chennai 600 030, Tamil Nadu, India
  Ph: +91-44-26680620, 26681266   Fax: +91-44-42032115   e-mail: chennai@cbspd.com
- **Kochi:** 42/1325, 1326, Power House Road, Opp KSEB, Power House, Ernakulum Kochi 682 018, Kerala, India
  Ph: +91-484-4059061-65,67   Fax: +91-484-4059065   e-mail: kochi@cbspd.com
- **Kolkata:** 147, Hind Ceramics Compound, 1st Floor, Nilgunj Road, Belghoria, Kolkata-700056, West Bengal, India
  Ph: +033-25633055, 033-25633056   e-mail: kolkata@cbspd.com
- **Lucknow:** Basement, Khushnuma Complex, 7 Meerabai Marg (Behind Jawahar Bhawan),Lucknow-226001, UP, India
  Ph: +91-522-4000032   e-mail: tiwari.lucknow@cbspd.com
- **Mumbai:** PWD Shed, Gala no 25/26, Ramchandra Bhati Marg, Next to JJ Hospital Gate no. 2, Opp. Union Bank of India Noorbaug, Mumbai-400009, Maharashtra, India
  Ph: 022-66661880/89   e-mail: mumbai@cbspd.com

Representatives

• Hyderabad	0-9885175004	• Jharkhand	0-9811541605	• Nagpur	0-9421945513
• Patna	0-9334159340	• Pune	0-9923910676	• Uttarakhand	0-9716462459

Printed at Chaman Enterprises, Daryaganj, New Delhi, India

Preface to the Fourth Edition

The third edition of this book was published in 1985. I am encouraged by the response to the older editions of the book. This has prompted me to update the contents for the fourth edition. Principles do not change frequently. But new ideas and new examples are added to the existing principles. In this edition there are significant additions to most chapters especially molecular and genetic basis of host parasite interaction. The last few chapters in the older editions that covered the area of disease management have been condensed since now they form subject matter of the new book Plant Disease Management. However, information in the condensed form is updated. References prior to 1960 have been mostly avoided. But references covering the period up to 2000 are listed.

With new informations and changes incorporated I am sure the book will be more useful to undergradute and postgraduate students and teachers of plant pathology.

July, 2001

R.S. Singh

Preface to the First Edition

This book has been written for an advance undergraduate course in Plant Pathology. So far, the curricula of Indian universities awarding Degree in Agriculture have been lacking the teaching of Principles of Plant Pathology. In most of the universities in India the undergraduate syllabus in Plant Pathology has remained static, there being no change in contents taught during the last 30 years or more. The teaching of plant pathology still includes only the symptoms, morphology or life-cycle of the pathogen, and its control base on old knowledge. The essential principles for understanding these aspects of plant pathology are kept away from the students at the undergraduate level. As a result the student fails to make practical use of his earned knowledge when required to do so.

Dynamism of any syllabus is unavoidable. A syllabus framed on the basis of today's knowledge needs revision after a few years since during the intervening period discoveries of new facts and development of new concepts have changed the face of existing knowledge. The syllabus must be based on and conform to the latest trends in the knowledge of technical subjects. One of the objectives of the presents book is to introduce the principles of plant pathology to students studying for their B.Sc. degree with specialisation in plant pathology or plant protection. Attempt has been made to present up-to-date but brief introduction to principles underlying various stages in the development of plant diseases with special emphasis on the control of such diseases.

Description of plant diseases, organisation of structure of pathogens, their physiology, reproduction, etc. do not fall under the scope of this book and have been avoided. In an earlier book "Plant Diseases" the author has given information on these aspects for undergraduate students. The present book is expected to be used as supplementary reading to the same.

Pantnagar
August 1975

R.S. Singh

Contents

1

Introduction

Dependence of man on plants for food dates back to more than 12000 years. Other animals and plants existed on this planet before the arrival of man and these later became the security against starvation of man. The animals domesticated by man almost wholly depend on plants. In addition to food, plants and animals are the primary sources of materials for other necessities of life such as clothing, house building materials, furniture, drugs and fuel etc. Coal and petroleum products are also derived from the remains of plants buried deep in soil. The congenial climate, the rains, etc are made and modified by the plants.

The availability of food to human population in reasonable quantity is governed by the population density needing food, cultivated land area available for food production and production of food per unit land area. Among these determinants availability of land is almost fixed. Production per unit land area does not rise fast enough while the human populations is a fast rising phenomenon. If the increase in food production can be matched with increase in the number of mouths to be fed, probably the problem will not be acute. However, the rise in human population is much faster than the rise in food production. Although total foodgrain production in the world has been rising and about 1400 million hectares of land surface (12% of the total) is being cultivated for some food crop, the rise in production has not matched with the rise in consumer numbers. For instance, the total food production in India has shown an overall increase of about 65% since 1950 while the population during the same period showed an increase of about 175%.

In the year 1940 the human population on this planet was around 2140 million which rose to 3280 million in 1965, to 3950 million in 1975 and to 4900 million in 1985. In the year 2000 it stood at around 6800 million. Overall rise in world population during the 60 years was about 218%. During the 25 years since 1940 the average rise per year was around 2%. It has shown only slight increase in the average percentage but the picture is different when these figures are considered on regional basis. Two third or more of the total world population is in the countries of Asia, Africa and Latin America. In India alone the population expanded from 360 million in 1940s to 550 million in 1971, to 920 million in 1996 and to about 8.5 million more than a billion by mid-2000. A report of the FAO in 1994 had estimated that the population of India in the year 2005 will be around 1600 million.

In the rising total food production, some selected crops have shown better increase than others. For example the productivity of wheat has gone up from 1.2 tonnes per hectare (in 1965) to 2.4 tonnes per hectare (in 1989). In India the total foodgrain production has gone up from about 130 million tonnes in 1950-51 to 208.6 million tonnes in 1999-2000. The total wheat production rose from 6.46 million tonnes in 1950-51 to 54.52 million tonnes in 1990-91. The corresponding figures for Punjab, the main wheat producing state, are 1.06 and 12.16. In 1999-2000, wheat production in India was 75.5 million tonnes, nearly 11-fold increase since 1950. One adverse effect associated with rise in production of specific grains has been the fall in production of other crops such as oilseeds and pulses. Area from such coarse grains as millets has been withdrawn in favour of cereals. Diversity has been narrowed and it has its own adverse effects on all crops. Probably the type of change in food production in India has not occurred in other food deficient countries. The spectre of famine, that has scourged humanity from the beginning of the time, and the shortage of food remains as great as ever, or even more than in the past in many regions of the world. The number suffering from hunger or malnutrition runs into billions. It is in this context that the need for fighting the competing microorganisms and other agencies that cause loss of productivity of plants has been felt. Plant diseases have been considered as stubborn barriers to the rapid progress in food production.

The plants that sustain us on this earth evolved much earlier than the first man. However, before the origin of plants, microbes had already developed in the organic evolution. Thus, the association of microbes with plants is much older than association of man with plants. This has significance in the light of adjustment or disadjustment between plants and microbes during their coevolution. In the beginning of this relationship, the microbes started living on dead organic substrates left by plants since they had no faculty to synthesize their own food. With further evolution, these microorganisms gave rise to parasitic forms which are even to-day attacking our crops and disturbing our crop production programmes.

In the pre-agriculture days when there was no systematic cultivation of specific plant types by man diversity of plant populations was great, the microorganisms had less opportunity to thrive and cause damage because of less chances of easy contact with the desired host plant. The introduction of selective culture of plant types by man gave the microorganisms better opportunity for survival, multiplication and consequent destruction. The development of scientific agriculture by man and development of plant diseases by pathogens have been on parallel lines since the very beginning of crop evolution. The attack on plants by these microorganisms changed the appearance and productivity of the crop and this observed change was called a disease.

Protection of plants against their enemies is one of the many approaches for increasing food production. In persuing their justifiable efforts the specialists in crop production emphasize the use of new cultivated plants, better high yielding crop varieties, more quick acting mineral fertilizers, ample irrigation and maximum use of land by not permitting any fallow period. Judged from plant pathological viewpoint these recommendations need careful consideration. Maximization of yields can be achieved only if along with ensured fertilizers, irrigation and seed of high yielding varieties there is also provision for protection of the crop against its enemies. It may involve modified cultural practices, encouragement of biocontrol agencies, induction of resistance and, if necessary, use of toxicants (fungicides, antibiotics, nematicides, etc.).

Introduction of new cultivated plants may involve phytopathological risks. It may cause introduction of potentially dangerous pathogens through seed and planting materials or a pre-existing pathogen may became dangerous because of the introduction of the new plant type. High yielding varieties are not necessarily disease resistant. Even if they are, the resistance is against one or two diseases while the crop is liable to attack by many more diseases. Often, when one or two diseases are checked, some minor diseases become major diseases in the crap. The uniformity of genetic material in the field, heavy plant population and presence of dense crop canopy almost throughout the year favour survival, reproducion and spread of pathogens. Thus, the measures being adopted for increasing productivity are favouring the pathogens also. Danger from diseases will continue to increase in proportion to the developments in technology of production without protection.

However, the modern technology had come to stay and atleast in India it has proved its worth by making the country almost self-sufficient in food production in 40–45 years. In other countries facing deficit modern technology is unavoidable if food is to be produced for the rising population. The compromise between the two situations lies in the recognition of the importance of plant protection so that whatever is sown in the field grows into a crop, whatever is grown is enabled to achieve its maximum production potential and whatever is produced is safely stored and finally reaches the consumer. There are diseases in which the loss caused to the crop is more than the deficit of that particular crop produce and if the diseases are checked the deficit can be reduced to a minimum.

Before the advent of high yielding crop varieties in India the varieties that were grown had not very high yield potential and plant protection was not paid much attention because the expenditure on plant protection measures was more than the income. The situation changed with the introduction and acceptance of high yielding varieties of wheat, rice, maize, many pulses and oilseeds. Attempts are made to take more than two crops in a year from the same field. From commercial point of view the value of the crop is much higher than in the past. Therefore, the loss of even a quintal of foodgrain is felt by the farmer and he has a genuine desire to stop this loss. He has realised that enemies of crops cause losses and these losses must be prevented by fighting these enemies.

In developed countries (mostly cold climate regions including North America, Europe, Japan, New Zealand, Australia, Israel) only about 11.5% of the population is engaged in agriculture while in the Asian and African countries agriculture engages 57.5% of the population. Ironically, it is the areas that engage the maximum population in agriculture, the population density is highest, crop productivity is low, and hunger most prevalent. Crop losses are much less in developed countries where the population is low then in the developing countries where crop losses are high and more food is required. Estimates differ but on a global basis, about 34% of crops is lost annually due to diseases, insect pests and weeds. Of this annual crop loss, 12% is due to diseases other than nematodes, 11% due to nematodes, 7% due to insect pests and 3% due to weeds. In cereals the annual loss is estimated to be about 18% in the developed countries compared to 46% in Asian and African countries. Corresponding figures for sugarcane are 34 and 56%. Percentages of all produce lost to diseases, insects and weeds are 25 in Europe, 29 in North and Central America, 30 in the Russian Federation of States and China, 33 in South America and 42-43 in Africa and South and South-East Asia.

In 1963, it was estimated that India loses 10% of foodgrains every year due to plant diseases. In some of the essential food crops this loss was more than the actual deficit in the supply of the produce to the consumer. In an estimate published in 1975, the total loss due to pests, diseases and weeds was approximately 18% of our total production. Of this loss the share of different agencies was estimated as follows: weeds 33%, diseases 26%, insect pests 20%, rats 6%, storage losses 7%.

The effect of these losses is not only economic. Many political and social problems originate from food shortage. Therefore, plant protection occupies a very important position in the strategy for making the countries self-sufficient in food and allied materials.

WHAT IS A PLANT DISEASE?

When the plant is under some sort of stress and is not developing and functioning in a manner expected from it, we call it diseased. This does not define the term "disease". Often the symptoms appearing as a result of a disease, the cause of the disease and the injuries caused to the plant have been considered synonymous. However, they signify only the condition of the plant due to disease or the cause of the disease.

In 1858, Julius Kuhn, in Germany, had defined plant disease as abnormal changes in the physiological processes which disturb the normal activity of plant organs. A similar definition was given by H.M. Ward in 1896 who defined disease as a condition in which the functions of the organism are improperly discharged or, in other words, it is a state which is physiologically abnormal and threatens the life of the being or the organ. In 1918, E.J. Butler had defined disease as variation from normal physiological activity which is sufficiently permanent or extensive to check the performance of natural functions by the plant or completion of its development. Similarly, according to the American Phytopathological Society (*Phytopathology* 30 : 361-368, 1940) disease is a deviation from normal functioning of physiological processes of sufficient duration or intensity to cause disturbance or cessation of vital activities. The British Mycological Society (*Trans. Brit. Mycol. Soc.* 33 : 154-160, 1950) also defined disease as a harmful deviation from the normal functioning of process. Wheeler (1975) broadly considered plant disease as all the malfunctions which result in unsatisfactory performance of the plant or which reduce ability of the plant to survive and maintain its ecological niche. Thus, the major thrust in defining a disease since the time of Julius Kuhn was on the deviation from the "normal", a condition which may be variable.

An analytical approach to definition of the term "disease" was made by Horsfall and Dimond (1959) who clarified many misconceptions. According to them disease (i) is not a pathogen, it is caused by a pathogen, (ii) is not the symptoms or effects seen on the plant, symptoms result from disease, (iii) is not a condition as the condition results from the disease, (iv) is not any injury which results from disease as well as from any traumatic cause and (v) cannot be catching or infectious, it is actually the pathogen which is catching or trasmitted. They defined disease as a malfunctioning process in the plant body due to cotinuous irritation which results in some suffering. It is a pathological process in plants and animals (including man).

Singh *et al* (1989) put up a definition of disease based on views expressed by the above and other scientists. In their opinion disease is "a sum total of the altered and induced biochemical reactions in a system of the plant or plant part brought about by any biotic or abiotic factor(s) or by a virus leading to malfunctioning of its physiological processes and ultimately manifesting gradually at cellular and/or morphological level. All these alterations should be of such magnitude that they become a threat to the normal growth and reproduction of the plant".

The pathogens bring about the irritating processes resulting in a diseased conditions of the plant through different but interrelated pathways:
1) By utilizing the host cell contents.
2) By causing death of cells or by interfering with their metabolic activities through enzymes, toxins and growth regulators.
3) By weakening of tissues due to continuous loss of nutrients.
4) By interfering with translocation of food, minerals and water.

SOCIO-ECONOMIC SIGNIFICANCE OF PLANT DISEASES

Agrios (1988, p. 20) had listed 26 diseases that have caused severe losses in the past. These include cereal rusts, potato blight, grape mildew, rice leaf spot, coffee rust, banana leaf spot, citrus tristeza, citrus canker, fire blight of pome fruits, sugarbeet cyst nematode and root knot nematodes. In the history of mankind the plant diseases have been connected with a number of important events. Some of these events are mentioned below:

The late blight of potato (*Phytophthora infestans*) is a famous example of what a plant disease can do to change the course of history. The disease coexisted with potato in the highlands of Central America. So long as the transportation across the equator was by slow moving boats, the fungus in potato tubers brought from there to Europe and North America was destroyed by the equatorial heat. With the introduction of steam boats and use of ice for storage on the boats the time taken for crossing the heat zone was shortened and the fungus could survive in the cold stores. This enabled introduction of the viable inoculum into Europe and North America. In 1845 the disease destroyed the potato crop of Ireland where potato was the staple diet of the population in the villages. The disease had started in Ireland, England and parts of the continental Europe as early as 1830 and was every year causing some damage resulting in food shortage. In England, which ruled Ireland at that time, free trade and import of foodgrains was not permitted. When the late blight epidemic destroyed the potato crop in 1845 there was famine in Ireland. The demographic data are highly variable (*cf.* Hampson, 1992) but it was reported that in 1840 the population of Ireland was 8 million which was reduced to 4 million after the famine. Hundreds of thousand perished from hunger and disease. There was large scale migration of the population to other countries including the north American continent where 1.6 million are reported to have migrated between 1847 and 1854. There are opinions contradicting the belief that the entire catastrophe in Ireland was due to potato late blight. Socio-political conditions including failure of the government to manage the situation were also equally, if not more, responsible. But the fact remains that this single disease forced man to realize the importance of plant diseases. As a result, scientific investigations were taken up, the cause of the disease was identified, concept of fungus as a cause of plant disease was finally established and extensive use of chemicals for plant disease control came into existence. The late blight epidemic not only brought the science of plant pathology to limelight, it caused many social and political changes in the affected countries. Free trade in England was permitted and import of foodgrains and other foodstuff was allowed. In order to protect shipping the country had to strengthen its navy which remained the strongest navy in the world for a long time. People who had migrated settled in other continents and helped in the development of new nations.

The history of devastation caused by cereal rusts is much older than that of late blight. The rust had been feared by farmers even thousands of years ago. In absence of scientific knowledge, the disease was considered curse of gods and rituals propitiating the rust god Robigus became routine among farmers. The annual festival of Robigalia was started in 700 BC and continued in some form even into the Christian era. The black or stem rust (*Puccinia graminis tritici*) had been appearing in epidemic form in many countries and in some areas such as the central Europe the farmers abandoned wheat cultivation and resorted to cultivation of rye as a cereal because it was less damaged by the rust. Similarly in Mexico and neighbouring areas wheat rust prompted the farmers to grow more corn (maize). These changes in cropping systems resulted in change in food habits of the population.

In the last year of the Second World War (1943) Bengal state in India had to face a serious famine. One of the reasons to which this famine has been attributed was the loss in yield of the rice crop (major diet of the population) due to attack of Helminthosporium leaf spot which had been damaging the crop for the last several years. The situation was similar to the Irish potate famine but not so catastrophic. In this case also many reasons other than loss of the rice crop are listed. The important reason was the rise in the price of rice due to short supply resulting from low production and demand by army engaged in the battle front near Bengal. The population, mostly poor villagers, could not, thus, pay for the food.

In the middle of the nineteenth century coffee and tea were equally consumed in England because these were available in plenty from such occupied countries as India, Sri Lanka and Malaysia. Sri Lanka (Ceylon) used to produce maximum coffee in the world. In 1867 coffee rust attacked the plantations in Sri Lanka and by 1893 the export of coffee from Sri Lanka had declined by 93 per cent. The economic crisis forced the planters to cut down coffee plants and take to tea planting. Export of tea revived the economy to some extent and at the same time consumption of tea increased in England. When coffee rust was spreading in Sri Lanka the science of plant pathology was just developing and control measures for the disease were not known. Tea was also attacked by a blight but by that time chemical control measures were known and the situation did not deteriorate. The system of monoculture in coffee plantations of Sri Lanka was considered a contributory factor in devastations caused by the coffee rust which was not prevalent in coffee growing countries of South America. Coffee rust was first seen in the Western Hemisphere in 1979, in Brazil, where it is now spreading. In Brazil coffee trees are often surrounded by other kinds of trees. The decline of coffee cultivation in Sri Lanka gave a boost to coffee industry in Brazil which became a major coffee exporting country in the world. Nearness of this country to USA could be one reason for popularity of this beverage in USA. These instances of plant disease epidemics are worth mentioning because they left their effect not only in the country concerned but also in other countries.

Nearly half the citrus trees in the world had been destroyed by the citrus trisetza virus up to the year 1959 and millions are annually destroyed every year. Panama disease or Fusarium wilt of banana is considered one of the six most destructive plant diseases in recorded history (Agrios,1988). In Jamaica alone it had destroyed 4 million plants within 25 years of its appearance there. The epidemic of southern corn leaf blight in the United States had caused a loss of $1 billion. To eradicate citrus canker in Florida (USA) millions of trees had to be destroyed in 1910s and again in 1980s. The fire blight of apple kills numerous tress annually in the United States.

In India, wheat rusts were considered to cause a loss of over Rs. 40 million annually. In the years of epidemics there were losses amounting to Rs. 500 million or more. During the wheat season of 1956-57, average yield of wheat in the state of Bihar was reduced from 900 kg/ha to 50 kg/ha due to rust epidemic. Although introduction of dwarf high yielding varieties has reduced the losses caused by rusts to a great extent even now the farmers lose 8–10% of the expected yield due to rusts. The loose smut of wheat (*Ustilago segatum tritici*) is estimated to cause an average loss of 3% (about Rs. 50 million) every year. This figure is based on only 3% incidence. Individual fields not adopting proper seed treatment may show 23–30% smutted heads. Different smuts of sorghum are responsible for an annual loss of Rs. 100 million. Five to 75% loss in chickpea (gram) due to Ascochyta blight was reported from Rajasthan during 1982. Wilt of pigeonpea causes 5–10 per cent loss every year in U.P. and Bihar. At a time when there is shortage of pulses in the country control of only wilt and sterility mosaic of pigeonpea could increase the production by 15–16%. The first epidemic of apple scab that occurred in Kashmir Valley in 1973 ruined apples worth Rs. 5.4 million in a single season. In the Shimla Hills, 16–34% fruits are rendered worthless annually due to scab. Losses due to this disease in Himachal Pradesh are estimated at Rs. 15 million every year (cf. Singh, 2000).

Worldwide losses due to nematodes are estimated at $1 million (Perry, 1996). Of about 10 per cent crop losses in India due to nematodes, there is a loss of about Rs. 20 million every year in coffee alone due to the attack of *Pratylenchus coffeae*. The 'Molya' disease (cereal Cyst nematodes) of wheat and barley prevalent in most parts of Rajasthan, causes a loss of Rs. 30 million in barley and Rs. 40 million in wheat every year losses varying from 10–26% in different vegetable crops by root knot nematodes are common every year.

Other plant diseases, such as red rot and wilt of sugarcane, potato viruses, rice blast and blight, Karnal bunt of wheat, root knot of tomato, brinjal and cucurbits, apple scab, mango malformation, bunchy top of banana, and sandal spike are responsible for huge losses. Epidemics of these diseases have been frequent.

In addition to direct loss in yields and monetary returns to the farmer, the plant diseases affect the society in many other ways. When foodgrains are attacked by fungi they may contain toxins (such as aflatoxins) which cause insanity, paralysis, stomach disorders and liver cancer. The money spent on management of plant diseases is also a loss because in absence of diseases this money could be saved. The expenditure on raising the crop before it is attacked by the pathogens is also a waste. When there is less production transport industry may suffer due to lack of goods for transport. Industries that consume raw agricultural materials (cotton, jute, oilseeds, vegetable and fruits for processing) face difficulty in utilizing their installed capacity when there is less production due to plant diseases. In order to make up for the loss of foodgrains and other agricultural products such as oilseeds the governments have to import these commodities which means loss of foreign exchange at the disposal of the government. Plant disease management requires use of toxic chemicals. Excess use of such chemicals may lead to environmental pollution affecting human health.

THE SCIENCE OF PLANT PATHOLOGY

Plant pathology or Phytopathology (*phyton*– plant; *pathos*– ailments; *logus* – knowledge) is that branch of agricultural, botanical or biological science which deals with the cause; *etiology*, resulting losses and management of plant diseases. It is a science similar to that of medicine and veterinary which deal with diseases of man and animals, respectively. The science of plant pathology has four major objectives:

1) to study the living, non-living and environmental causes that induce disease,
2) to study the mechanisms of disease development by pathogens,
3) to study the interactions between the plants and the pathogen,
4) to develop strategies for managing the diseases and reduce the losses caused by them.

A plant pathologist or phytopathologist has to achieve these objectives. For the plant world his place is the same as that of a physician for human beings. He protects the plants from their enemies, enables them to live their full life and express their maximum ability for production. This provides the plant pathologist a special status in the society since the latter has to depend on health of the plants. The plant pathologist is a practical scientist. His aim is not only to destroy the enemies of the food crops or fruits trees but also to increase the productivity of the crop and profits of the growers. Disease management is only a means of achieving these aims.

It appears that in early days plant disease management developed as an art practiced by farmers on the basis of their and their ancestros' experiences. No science was involved. This perhaps continued for centuries until some attempts were made to understand the nature of disease and relationship between the disease and its causal agent. The observations reported by biologists during the latter part of the nineteenth century gave the subject some appearance of science comparable with other branches of biology. The science of plant pathology is related to most of the old and new branches of science. It was related to physics, chemistry, botany and zoology in the early stages of its development but now it is similarly related to molecular biology and biotechnology, bacteriology, virology, nematology, plant anatomy, physiology, genetics, biochemistry and soil science including soil microbiology. In many of these biological sciences plant pathology has contributed in basic research.

TERMINOLOGY

Disease and Disorder: These two terms carry the same meaning, i.e., malfunctioning in the plant body. However, while the term disease includes all types of harmful physiological changes in the plant, often the non-infectious plant diseases due to abiotic causes such as adverse soil conditions are termed disorders. Components of a disease are the pathogen and the host interacting in a given set of environments.

Pathogen: In a literal sense a pathogen is any agent that causes *pathos* (ailment, suffering) or damage. It should, therefore, include all living organisms associated with the disease, viruses as well as the abiotic agents such as adverse atmospheic or soil conditions. However, the term is generally used to denote specific living organisms (fungi, bacteria, MLO, nematodes) and viruses but not such agents as nutritional deficiencies.

Parasite: Organisms which derive the material they need for growth from living plants (the host or the suscept) are called parasites. Most, but not all, pathogens are parasites, similarly most, but not all, parasites are pathogens.

Biotrophs: Organisms which, regardless of ease with which they can be cultivated on artificial media, always obtain their food in nature from living tissues on which they complete their life-cycle are called biotrophs or **obligate parasites**. Fungi causing rust, smut and powdery mildew are typical examples.

Saprophytes: Organisms that derive their nutrition from dead organic matter.

Parasites and saprophytes may have the faculty to change their mode of nutrition. A parasite may be **hemibiotroph**, i.e., it will attack living tissues in the same way as a biotroph but will continue to grow and reproduce after the tissue is dead. Such parasites can also be called **facultative saprophytes**. A parasite is **necrotroph** when it kills host tissue in advance of penetration and then lives saprophytically. *Sclerotium rolfsii* is an example. Similar to necrotrophs are **facultative parasites** which live as saprophytes but under favourable conditions attack living plants and become parasites. The necrotrophs are also known as **perthotrophs** or **perthophytes**.

Pathogenicity is the ability of a pathogen to cause disease while **pathogenesis** is the chain of events (steps) that lead to development of disease in the host. **Virulence** is a measure or degree of pathogenicity of an isolate or race of the pathogen. The term **aggressiveness** is often used to describe the capacity of a pathogen to invade and grow in its host plant and to reproduce on or in it.

This term, like virulence, is also used as a measure of pathogenicity.

Infection is establishment of parasitic relationship between two organisms following entry or penetration. In immune or highly resistant plants penetration may occur without infection.

Systemic infection: The pathogen grows from the point of entry to varying extents without showing adverse effect on tissues through which it passes.

Invasion and Colonisation: After infection the pathogen grows through the tissues of the host to varying extent drawing nutrition during its growth.

Immune: Immunity of a plant against a given disease is an absolute quality. It denotes that the pathogen cannot establish parasitic relationship with the host. The extent to which a plant prevents the entry or subsequent growth of the pathogen within its tissues or the extent to which a plant is physically or physiologically damaged by a pathogen is used to measure the **resistance** or **susceptibility** of the plant. High resistance or law susceptibility approaches immunity. **Hypersensitivity** is extreme degree of susceptibility in which the host cells react violently to the invading pathogen and as a result there is rapid death of cells in the vicinity of the invading pathogen (*infection court* or *site of infection*). This may halt the progress of the pathogen although it may or may not die immediately. Thus, hypersensitivity is a sign of incompatibility or very high resistance approaching immunity.

Resistance can be **horizontal** (uniform) when it is evenly spread against all races (strains) or a pathogen, or **vertical** (differential) when it is effective against one or few races but not against others. Since the number of races of the pathogen is indefinite, new races being evolved often, the resistance may be found lacking even in varieties with uniform resistance. Resistance may be **monogenic** (single gene), **polygenic** (determined by many genes) or oligogenic (several genes). **General resistance** is known under a number of synonyms such as partial resistance, race nonspecific resistance, horizontal resistance, field resistance, dilatory resistance, slow rusting, slow mildewing and slow blighting.

Systemic acquired resistance: Resistance triggered in the plant during its life time is acquired resistance. It can be local (LAR) confined to few cells or tissues or systemic (SAR) having been moved throughout the plant.

Tolerance is a type of defence that minimises crop loss without restricting disease development. **Disease escape** is the ability of an otherwise susceptible plant to avoid damaging disease stress because of the way it grows. This is considered a useful system of protection in natural ecosystem.

Elicitors are signals, presumably chemical signals, generated by an organism to bring about metabolic shifts in another organism. It is usually applied in phytoalexin production. **Inducer** is more general term. Elicitation is the first step in cell-cell communication when the infection propagule comes in contact with the host cell surface.

Recognition system: Cell recognition is defined as the event of cell-cell communication which elicits a defined biochemical, physiological or morphological response. The term is used to explain the earliest events in host-parasite interaction or pathogenesis when molecules of the host cell surface recognise like molecules in pathogen surface and accept it for establishment of parasitic relationship. If the pathogen molecules are unlike, they are rejected and no relationship is established. Similarity at molecular level or genetic matching leads to **compatibility** between host and parasite and dissimilarity leads to **incompatibility**. At the molecular or genetic level the interactions are highly specific giving specificity to the host or the pathogen in the recognition system.

Incubation period: The time lapsing between inoculation and appearance of symptoms (sign of establishment of parasitic relationship) is known as incubation period. It varies with pathogen, host and environmental conditions.

Latent, dormant or **quiescent** parasitic relationship is a condition in which the pathogen spends long periods during host's life in a quiescent stage until, under specific cirumstances, it becomes active. In epidemiology the period from spore landing until the parasitized tissue produces new spores is known as **latent period.**

Queiscence of an organism is the stage when under the influence of some biological, physical or chemical stress the organism or its propagules undergo an inactive state. This is known as dormant, anabiotic or queiscent stage. This is an adaptation for survival under adverse conditions.

Symbiosis: Living together of two unlike organisms is known as symbiosis, the two organisms being called **symbionts**. This includes parasitism also when the attacked plant or organism remains alive. In associations where there is some or high dergee of adjustment between the two organisms whereby each gets some benefit the term **mutualism** is used. In contrast to mutualism is **antagonism** in which one organism is injured by the other either through competition for food or other demands or through secretion of toxic substances (**antibiosis**) or by direct injury (**parasitism** and **predation**).

2

History of Plant Pathology

Man started depending on plants for food as security against starvation at least 12000 years ago, if not earlier. There is evidence suggesting that he was practicing some form of crop (plant) culture as early as 7000 BC. In the beginning he was a hunter and food gatherer. Whatever grew in the wilds, he tested, tried and consumed as food. Later, when he noticed that plants could be grown and multiplied he started domesticating them, first by vegetative propagation and then by planting seed. The man suffered from different kinds of diseases and he was worried about them. His sufferings prompted him to observe similar problems in his cattle that were part of his life. During his association with plants and his domesticated animals he had noticed diseases of plant and considered them harmful to the plant and therefore to his food security. It appears that in the ancient times the learned individuals in the society first started their observations and attempts to mitigate the sufferings of humans and cattle and then tried to apply the same knowledge to the sufferings of the plants. At least in the Indian ancient literature almost all thinkers, composers and writers were physicians becoming experts in agricultural subjects.

Ancient Agriculture

Probably, the development of agriculture started in the orient. There is mention of Chinese practicing crop rotation as early as 3000 BC and in the first century BC they were supposed to keep the field fallow for a year if the field gave poor yield in the second year (cf. Singh, 2001) Seed health was mentioned in China by Fan Sheng-Chih in the first century BC. These records suggest that agriculture was fairly developed in China centuries before the Christian Era.

In the ancient western literature (the Greek and Roman civilization) the central figures of observations on plant life and plant diseases was Theophrastus (384-322 BC) a Greek Philosopher and writer on various subjects. He was disciple of Plato (428-348 BC) and Aristotle. Although most of the work of Theophrastus is lost, two of his books, *Historia Plantarum* and *De Causis Plantarum*, still find a place as reference. In his writings, Theophrastus elaborately mentioned plant diseases (rusts, mildews, blight, etc.) but expressed the opinion that these diseases were due to bad nutrition and bad air. There was no mention of association of living organisms with plant diseases. Others during that time (Varro, 116-27 BC, Maro, 70-19 BC) did mention specific treatments of seed. While writings of Theophrastus are considered more of philosophy than science, Varro gave some of the thoughts of Theophrastus a scientific approach although he also believed in worship of the rust god Robigus. In his *Rerum Rusticarum*, the best Roman treatise on agriculture of that time, Varro included 12 councillor gods including Robigus and Flora. The time of Theophrastus was nearly preceded by the period of the Mauryan Empire of Magadh in India (c. 400 BC) when agriculture was fairly developed and was the main concern of the government. Kautilya, also known as Chanakya and Vishnugupt, had written Arthashastra (Science of Source of Livelihood, 400 BC) which incorporated many observations and recommendations for healthy crop culture.

Much earlier than the time of Theophrastus in the west and Kautilya in India (3700 BC and later) sages had composed the four Vedas (religious verses). These were composed and, in absence of a script, were passed on through word of mouth from generation to generation for centuries. Rigveda (3700 BC) is considered the oldest composition of religious hymns*. It was followed by Yajurveda (hymns and rituals), Samveda (Rigveda recomposed for singing) and Atharvaveda (3000 BC and later) which contained charms and spells for warding off evils and diseases. This composition specifically mentioned blight as a disease and its control. During the Vedic period agriculture in India was fairly developed. Ploughs and other agricultural implements were in use. In addition to such diseases as blight, powdery mildew, rust and tumors on trees, fungi (mushrooms) and algae are also mentioned in the Vedas. In Rigveda (3700 BC) a number of verses are devoted as prayer to the Sun God for purifying and protecting everything and for destroying the tiny, invisible or visible creature that poison the food and cause disease. Similarly, fire (Agni) was worshipped for destroying the poisonous beings and purifying articles used in religious activities. Spoiled foodstuff was recommended not to be eaten. Clean water was recommended for men and cattle and vigorous, clean seed was recommended for planting. Obviously, thousands years before the time of Theophrastus the civilization in India was aware of living beings (Krimi) that caused disease in man, cattle and plants and also spoiled foodstuff. In Yajurveda, serpent-like Krimi are often mentioned. Probably, the intestinal worms (*Ascaris*) had been observed. Two facts that come out from the verses in the Vedas are (i) visible and invisible creatures entered the body and caused disease and (ii) sunheat and fire kill these poisonous creatures.

---

*The citations about the Vedic literature are mainly based on various publications of **Asian Agri-History Foundation**, Hyderabad, India (1999-2000).

More specific references to plant health and disease are found in written literature after the four Vedas when attempts were made to prevent and cure diseases. Susruta, the great Indian pioneer in medicine and surgery who wrote Susruta-Samhita (c. 400 BC) and Charak, also a man of medicine, were aware of diseased conditions in plants and often compared them to diseases in man. Although, Susruta did not deal with plant disease, while advising men for good health, he wrote "just as the proper season, good soil, water and vigorous seed produce a healthy plant". The oldest text on Indian agriculture-Krishi Parashar-was written by Parashar, probably before the Arthashastra of Kautilya (c. 400 BC). Parashar was also a physician and philosopher. His book is considered the first extensive coverage of ancient agriculture of the Aryans. It contains chapters on most aspects of crop culture from sowing to harvest and seed storage. However, plant protection is mentioned directly only in one verse that mentions powdery mildew, rust, insects, and larger animals as enemies of crops and invokes the Wind God to move them away from his field. Parashar had declared that origin of plentiful yield is the seed, implying seed health. Kautilya (321-296 BC), in his Arthashastra, listed treatments of seed in addition to recommendations for punishment against sale of spurious seed. He also recommended treatment of cut ends of sugarcane cuttings before planting. These composers and writers belonged to the North and North-West region of the Indian subcontinent. In the South, the Tamil poet Tholkappier (200 BC) considered plants as living beings, mentioned monocot and dicot plants, and wrote about benefits of rice-legume rotation.

Varahamihir (505–587 AD) in his Brihat-Samhita included a chapter on science of plant life. Apart for writing about fungi (mushrooms) and algae, he also advocated the importance of good seed and seed treatment for good and healthy seedlings. An interesting recommendation of Varahamihir was that when a piece of land is brought under a crop, sesame should be planted, chopped down before seeds are formed and incorporated into the soil. The recommendation suggests value of green manuring and organic amendment which are now known to reduce root diseases. Sesame is also a trap crop for Striga.

More extensive coverage of ailments of plants is found in the text Vriksharyurveda written by Surapal (c. 1000 AD). This text mainly deals with cultivation of fruit and flower trees. It is based on previous compositions and writings as well as author's own observations. It covers such topics as importance of trees, their location, soil types, methods of propagation, tree nutrition, diseases and their treatment. Plant protection was already recognized as an important agricultural activity when Surapal wrote his Vrikshayurveda. Surapal was also a physician and had divided plant diseases, like human diseases, into two categories, internal and external. The internal diseases were caused by inroads of foreign organisms in the plant body while the external diseases were attributed to non-parasitic injuries by heat, frost, high winds, soil acidity, water stress, poor quality seed, etc. To avoid both internal and external diseases Surapal prescribed treatment of pits for planting trees, treatment of seed and treatment of standing trees.

The materials used for treatments of seed, field crops and trees during the period of Parashar to Surapal appear to be exclusively organic sources. The list includes cowdung, clarified butter (ghee), decoction of root of five trees, mustard, hog fat, cattle horn, milk, honey, liquid manure, oil of *Madhuca*, ash, cow urine, human urine, beef, extract of *Embelia ribes* (most commonly recommended) and many other things. These were available in plenty those days and farmers could use them at no cost. They may look strange but modern science is gradually explaining their utility (cf. Singh, 2001). The application of cow dung to seed, pits, and tree trunks in even now practiced in certain areas and is effective. In the Kumaon Hills of U.P. the apple tree owners apply a mixture of cowdung and mud to cut ends of branches after annual pruning which protects the wound from infection until self-healing. The decomposed excreta of animals have been found to reduce incidence of powdery and downy mildews of grapevine in Germany and other countries (Weltzien and Ketterer, 1986; Schlosser, 1994). Plant oils have been used effectively against powdery mildew of apple giving as much as 90% control. Mechanically emulsified rape oil applied as spray is comparable to the use of Karathane against apple powdery mildew (Northover and Schneider, 1993) and grapevine powdery mildew (Azam, *et al*, 1998). Applying a coat of pure mustard oil to ripe mango fruits gives 90% control of Aspergillus rot. Apart from promoting development of biocontrol agents in soil and on plant surfaces, such organic substances are also agents of inducing systemic acquired resistance (cf. Singh, 2001).

Discovery of the Role of Fungi

In 1675 the Dutchman Leeuwenhoek developed the first microscope and in 1683 he described bacteria seen with this microscope. The Italian botanist Micheli was the first scientist who in 1729 studied fungi and saw their spores under the microscope. He also proved that if these spores are placed on a piece of fruit they grow into a new thallus of the fungus. Although this was a successful experiment it was not universally accepted.

In 1775 the French botanist Tillet published a paper on bunt or stinking smut of wheat (now known as *Tilletia tritici*). In this paper he described well planned field experiments and proved that such wheat seeds that contained a block powder on their surface (he did not know that this powder represented spores of the fungus) produced more diseased plants than clean seeds. While emphasizing that the bunt was a contagious disease he observed that its occurrence could be reduced by seed treatment. However, Tillet believed that not the fungus but some toxin produced by the black powder caused the disease.

Although scientists like Persoon (1801) and Fries (1821), who were busy with classification and nomenclature of fungi, believed that microorganisms originated from disease, the French scientist Benedict Prèvost pioneered the germ theory of plant disease and proved in 1807 that diseases are caused by microorganisms. Like Tillet he was also working with bunt of wheat. In addition to other details of the disease he studied the germination of spores. By mixing spores with clean seeds he could reproduce the disease. The credit for discovering the life-cycle of the bunt fungus goes to Prèvost. In his opinion bunt spores did not germinate in copper sulphate solution, hence this could be used as a chemical treatment for control of the disease. He also mentioned the fungicidal and fungistatic properties of chemical treatments. These were major discoveries, later confirmed by many scientists. Tulasne brothers (R.L. Tulasne and C. Tulasne) of France observed pleomorphism in rust fungi and produced illustrated description of rust and smut fungi. Tulasne had also confirmed the findings of Prevost. But because of firm belief in the theory of spontaneous generation among majority of the contemporary scientists, the acknowledgment came only after 40 years.

During 1830–1845, when late blight of potato was fast spreading in England, Ireland, and the continental Europe, there was no one opinion amongst the scientists about the disease-fungus relationship. While acknowledging the fact that the fungus was invariably associated with late blight the majority believed that the fungus developed from the disease rather than it caused the disease. Berkeley (1846), Morren.(1845) and Von Martius (1842) were the few scientists who believed that late blight of potato was caused by the fungus found associated with it. However, they had no experimental evidence to prove it.

The foundation of modern experimental plant pathology was laid by the German scientist Anton de Bary (1831–1888). In 1853 he confirmed the findings of Prèvost. In 1861 he experimentally proved that the fungus *Phytophthora infestans* was the cause of late blight. The credit for a detailed study of the late blight fungus, its nomenclature, and experimental proof of organisms being plant pathogens goes to the work of de Bary. He studied other diseases also which included rusts, smuts, downy mildews, and rots. The discovery of heteroecious nature of rust fungi was made by him. Probably, the first study of physiology of plant diseases was carried out by de Bary when in 1886 he reported the role of enzymes and toxins in tissue degradation caused by *Sclerotinia sclerotiorum*. Brefeld, a colleague of de Bary, developed methods of artificial culture of microorganisms between 1875 and 1912. With these methods the study of infectious microorganisms became easier. After de Bary suggested the role of enzymes and toxins in pathogenesis many workers successfully attempted to explain such effects of infection on plants as rotting and wilting. In 1905 Jones reported the role of cytolytic enzymes in soft rots caused by bacteria and in 1915 Brown recognized the role of pectic enzymes which was followed by discovery of the role of cellulases. Although de Bary had hinted the possibility of toxins in rots and many others had suspected involvement of toxins in leaf spots and wilt diseases the first experimental proof was obtained in Japan by Tanaka in 1933 in black spots of pear caused by *Alternaria*. Subsequently, the role of toxins in vascular wilt disease syndrome and many other blight and similar diseases was established. Now toxins are known to play a major role in pathogenesis of most non-obligate host-specific pathogens.

With the establishment of the role of fungi in plant diseases and observations that there was some degree of variation among different isolates of the same pathogen in their ability to cause disease and among their hosts to suffer different degrees of injuries, attention was diverted to study the genetics of host-parasite interactions and disease resistance. Although Mendel had published his work on genetic of peas in 1866 and by 1898 it was known that resistance to rust in wheat was inherited, the names of Orton (1900–1909) and Biffen (1905–1912) are mentioned as pioneers in this field of resistance breeding. In 1905 Biffen described inheritance of resistance to yellow rust in two varieties of wheat and their progenies on the basis of Mendelian laws of inheritance. In 1909, Orton working with wilt diseases of cotton, watermelon, and cowpea developed varieties resistant to the disease and distinguished disease resistance from disease escape and disease endurance (tolerance). Since then efforts to develop resistant varieties in most crops have been continuing but there are only few crops that have varieties possessing permanent or durable resistance to a disease. One of the causes for this short lived nature of resistance is the variability among the pathogens.

The phenomenon of variability among fungi was first discovered by the Swedish scientist Erickson in 1894 when he reported the existence of physiologic races in the rust fungi. Almost at the same time Ward (1903) and Salmon (1903–1904) also discovered physiologic specialization in fungi causing rust and powdery mildew of cereals. E.C. Stakman of the University of Minnesota, USA took up this aspect of plant pathology for further investigation in the second decade of the last century. After prolonged studies he came to conclusion that due to continuous evolution of races and biotypes in botanical species of the rust fungi their pathogenic capability goes on changing in their favour and as a result the resistance capability of the host also shows changes.

Resistance to disease in plants was earlier considered to the due to presence of some toxic substances in the host. In 1946 Flor, working with linseed (flax) rust advanced the gene for gene concept of disease resistance and susceptibility. According to him susceptibility to a disease depends on compatibility of genes in the host and the pathogen. For every gene controlling resistance or susceptibility in the host there must be matching genes for avirulence or virulence in the pathogen. This gene for gene relationship is now proved in a large number of host-parasite systems. Wherever genetic information is sufficient for both the host and the pathogen it is usual to find a gene-for-gene relationship between the avirulence gene in the pathogen and the resistance gene in the host. In 1963, Vanderplank suggested that there are two kinds of resistance: one, controlled by few 'major' genes is strong but race specific (vertical resistance) and the other determined by many 'minor' genes is weaker but effective against all races of a pathogen species (horizontal resistance). The plant cell structures and substances that impart resistance are controlled by genes.

In 1902, Ward had observed that attempted colonization of resistant *Bromus* spp. by *Puccinia dispersa* was accompanied by necrosis of cells of the host adjacent to the fungus. In 1915, E.C. Stakman, working on the resistance of wheat to *Puccinia graminis*, also observed rapid cell death around the sites of penetration in resistant hosts and used the term "hypersensitive" to describe the response. The phenomenon of resistance through hypersensitivity was later elaborated by Gaumann in 1946. In late 1990s, possible connection between plant disease and programmed cell death (PCD), cell suicide or apoptosis associated with hypersensitive reaction was reported by many (*cf.* Gilchrist, 1998). In 1940, K.O. Mùller and H. Borger first defined phytoalexins as the antimicrobial compounds synthesised and accumulating at sites of infection or stress. This was based on their research into resistance of potato to late blight. Cruickshank in (1963) confirmed accumulation of antimicrobial plant metabolites during pathological processes and their role in resistance.

Resistance breeding by conventional methods (mating of plants of the same species) does not usually allow interspecific crosses and takes years to get a desired resistant variety. During the lapsing period the pathogen may develop new races against which the finally obtained variety may not be resistant. To overcome these shortcomings pathologists and breeders have been trying to utilize the techniques of tissues culture and genetic engineering. In 1970s it was demonstrated that plant cells and protoplasts could be selected in culture for resistance to a pathogen toxin and that plants with an altered response to infection by the pathogen could be regenerated from these cultured cells. Since then the techniques have improved much and a few varieties have also been developed from cell culture. However, the progress has been slow for various reasons, one being that not enough is known about the basic biochemical and genetic events that occur in diseased as well as healthy plants. Of particular interest in these techniques for developing resistant varieties are. protoplast fusion methods, ovule and embryo culture and *in vitro* fertilization, and uptake of organelles, chromosomes and DNA by protoplasts. Techniques of meristem culture have been used to obtain virus-free material in sugarcane and potato. Microtuber production in potato is likely to change the culture of this crop through rapid production of virus-free seed stock.

Discovery of the Role of Bacteria

The discovery that bacteria can act as specific infectious agents of disease was first made in animals through the study of anthrax disease. Rod-shaped bacteria had been seen in the blood stream of diseased animals as early as 1850. Conclusive demonstration and irrefutable proof of the bacterial etiology of anthrax disease was provided by Robert Koch in 1876. He gave the famous Koch's postulates for proving that a particular organism is the cause of a particular disease. He also demonstrated the *biological specificity* of disease agents. Every bacterium does not cause disease and those that cause disease are specific to certain types of organisms.

In 1882, T.J. Burrill of USA for the first time reported that a plant disease (fire blight of pear) was caused by a bacterium (now known as *Erwinia amylovora*). This report was soon followed by reports of bacterial etiology of yellows disease of hyacinth in 1883 by Wakker and olive knot disease in 1886 by Savastano. By 1909, over a score of plant diseases had been shown to be caused by bacteria. Erwin Frink Smith of USA was the main contributor to the discovery of most of these bacterial plant diseases since 1895. He is considered founder of phytobacteriology for his discoveries and the methodologies he introduced for study of bacterial plant diseases during 1905-1920. Alfred Fischer of Germany, who had studied under de Bary, did not agree with Smith and all others who had claimed to have seen bacteria in plant cells. He was of the opinion that since bacteria can not enter plant cells they can not cause disease. The bacteria seen by others entered the plant after the disease as accidental invaders. The heated controversy between Smith and Fischer became one of the best documented cases of scientific disagreement. But methodological studies conducted by Smith won the battle. Smith was also among the first to notice and study the crown gall disease (1893-1894). He considered crown gall similar to cancerous tumors of humans and animals. Later, in 1977, it was demonstrated by Chilton and his team that the crown gall bacterium transforms normal plant cells into tumor cells by introducing into them its plasmid, part of which becomes inserted into the DNA of chromosomes of the plant cell. Subsequent pioneers who made significant contributions to plant bacteriology during the first half of twentieth century include C. Elliott (1930-1951), P.A. Ark (1937), H.W. Burkholder (1930-1948), W.J. Dowson (1949-1957) and C. Stapp (1956-1961). Since the 1960s an explosive world-wide development in the field of phytobacteriology was noticed. The bacteriological studies during latter half of the twentieth century emphasised molecular and genetic aspects and biochemical taxonomy. The application of biochemistry, molecular biology, and genetics to phytobacteriology had actually begun in the late 1940s, slowly at first but accelerating markedly in the 1960s (Starr, 1984). Notable among studies are gene transmission systems, plasmid biology, the pigments, nucleic acid homologies, bacteriophage relationships, enzymology, phytotoxins, serological characterization, molecular basis of host-bacterium interaction.

Role of Nematodes

Existence of nematodes in nature has been traced to millions of years. A stylet bearing nematode was found in 26 million years old fossil of the insect *Drosophila* (Poinar in *J. Parasitol.* 70:306. 1984). The Guinea worm and roundworms as parasites of man were known to Egyptians as early as 1553-1550 BC. A plant disease with which association of a nematode could be noticed was first reported by Needham in England in 1743 AD. He described the wheat gall nematode now known as *Anguina tritici*. However, for about 100 years after Needham no attention was given to the role of nematodes in plant diseases. In 1857, the life cycle of *Anguina tritici* was studied by C. Devaine.

Root knot nematode was the second phytonematode to be discovered and reported. In England, M.J. Berkeley in 1855 noticed galls (knots) on roots of greenhouse cucumbers. On examination he found the white larvae and eggs of the nematode which he called *Vibrio*. However, the first specific mention of root knot nematode was that of Cornu in 1879. He described them under the name *Anguillula marioni*. In 1887, E.A. Goeldi published a description of *Meloidogyne exigua* (coffee root knot). In 1889, Atkinson published a preliminary report on the life cycle of these nematodes under the name *Heterodera radicicola*. This name persisted for decades until the name *Meloidogyne*, earlier proposed by Goeldi, was accepted.

The stem nematode (*Ditylenchus dipsaci*) was reported by Kuhn in 1857. The cyst nematode of beet (*Heterodera schachtii*) was first noticed in Germany by H. Schacht in 1859. The name was given in 1871 by Schmidt. The cyst nematode of beet was the first nematode of economic importance because beet was a highly valued crop at that time. From 1913 to 1932 N.A. Cobb studied the structure of many nematodes and classified them. He coined the word 'nematology' and developed many techniques for the study of nematodes.

The economic loss due to infestations of nematodes was realized when soil fumigants were developed in the 1940s and with their application high increases in yields were obtained. The association of nematodes with diseases caused by other agents had been noticed as early as 1892 when Atkinson reported that Fusarium wilt of cotton was more severe in the presence of root knot nematodes. In 1901 Hunger showed that bacterial wilt of tomato was facilitated by root knot nematodes. An outstanding discovery on nematodes in relation to plant diseases was the finding of Hewitt and associates in 1958 when they discovered that the grapevine fan leaf virus was transmitted by a nematode. This started studies on nematodes as vectors of viruses. Now atleast 18 plant viruses are known to be transmitted by species in four genera of nematodes. Developments in phytonematology have been more rapid than in other branches of plant pathology and now the science occupies an independent position in many research centres.

DISCOVERY OF THE ROLE OF VIRUSES

The origin of viruses is unknown. In absence of a cellular body their fossils are not found. However, the viruses we encounter today are not of relatively recent origin. Many diseases of man and abnormalities in plants, which have now been proved to be caused by viruses, have actually been known for centuries. Much before the causal relationship of microorganisms (such as fungi) with plant diseases was established in the latter half of the nineteenth century, pictorial records and some written references did exist of plant abnormalities that eventually proved to be of virus origin. Yellow leaves of *Euparatum chinensis* were praised in Japanese poems as early as 752 AD. The disease in now known as yellow vein mosaic. The oldest pictorial record of an abnormality in plants induced by a virus is of "broken tulips" or the famous Rembrandt tulips. No virus disease has been more beautifully illustrated than these flowers with variegated colours, painted during the period from the year 1600–1660 by Dutch artists.

The year 1882 may be considered as the beginning of the era of plant virology when scientific studies were initiated by Adolf Eduard Mayer, a German scientists working in Netherlands. Between 1882 and 1886 he reported that the tobacco mosaic disease was neither due to a microorganism nor due to nutritional imbalance. He demonstrated the contagious nature of the causal agent by artificial inoculation and also showed that boiling of the sap of infected leaves destroyed infectivity of the causal agent. In 1892 Dimitrii Ivanowski, a Russian botanist working in Crimea, reported that he had confirmed the findings of Mayer regarding transmission of tobacco mosaic agent and, in addition, had found that the causal agent could pass through filters with pores small enough to retain bacteria. The filtered sap remained infective for months. Martinus Wilhelm Beijerinck, a Dutch scientist, further confirmed the findings of Mayer and Ivanowski in 1898 and concluded that the causal agent of tobacco mosaic was something other than a microbe. The agent could pass through porcelain filters and could diffuse through agar gel. He was convinced that the agent was not a bacterium but a *'contagium vivum fluidum'*, a contagious living fluid. This was a revolutionary idea at a time when all substances could be classified only into two groups, corpuscular (bacteria and blood cells) and dissolved such as salts and other molecules in solution. The idea that the tobacco mosaic agent was fluid, and therefore dissolved, but at the same time living and capable of reproduction and, therefore, infectious in plants, seemed extraordinary. Beijerinck is considered founder of virolgy for it was he who firmly established the novel characteristics of tobacco mosaic mosaic agent distinct from known agents of infectious diseases. Soon after, many other plant diseases were found to be caused by similar agents.

In 1926, about 40 years after the work of Mayer, the biochemist Maurice Mulvania suggested that the tobacco mosaic virus (TMV) might be a protein of very simple nature having characters of an enzyme. In 1935, W.M. Stanley of USA was able to obtain a crystalline protein by treating juice of TMV infected leaves with ammonium sulphate. This crystalline substance remained infective. He concluded that the virus was an autocatalytic protein that could multiply within living cells. Studies carried out by F.C. Bawden and N.W. Pirie in Britain in 1936 showed that TMV was a nucleoprotein and contained phosphorus. The specific nucleic acid in TMV was identified as ribonucleic acid (RNA). In 1929 H.O. Holmes had provided a method by which the quantity of the virus in tissues could be estimated. He showed that the amount of virus present in a plant sap preparation is proportionate to the number of local lesions produced on an appropriate host plant leaf rubbed with that sap. The leading role of the nucleic acid of plant viruses in the infection process was discovered through the study of bacterial viruses (bacteriophages). In 1956, Gierer and Schramm from Germany and Fraenkel-Conrat and his coworkers from USA showed that TMV nucleic acid free from its protein coat, could alone cause infection provided it was protected from inactivation.

Although the TMV was considered a rod-shaped particle on the basis of various biochemical and biophysical tests its visual observation was possible only when the electron microscope was developed in 1939 and Kausche and colleagues for the first time saw the virus particles with the help of this microscope. The crude pictures obtained confirmed the rod shape of TMV particles. The shadow casting technique developed in early 1940s using heavy metals enabled the scientists to obtain a clear picture and determine the overall size and shape of particles. Widespread use of ultracentrifuge, the electron microscope, eletrophoresis, and serological techniques during 1940-1950 further helped in the understanding of plant virus structure, chemistry, replication, and genetics. Quick and accurate detection and identification of viruses became easier with the development of agar double diffusion serological tests in 1962 and enzyme-linked immunosorbent assay (ELISA) in 1977 as well as production of monoclonal antibodies in 1975. These techniques are now being used for detection and identification of mycoplasmas and bacteria (*Xanthomonas*) and some fungi also. Subsequently, more sophisticated techniques like immunosorbent electron microscopy (IEM) were also developed. Prior to 1960 the viruses studied were shown to consist of single stranded RNA. With the advances in technique for study some were found to have double stranded RNA (1963), some double stranded DNA (1968) and some single stranded DNA (1977). The cauliflower mosaic virus (ds DNA) was the first virus for which the exact sequence of all its 8000 base pairs was determined in 1980. In 1982 the complete sequence of bases in tobacco mosaic virus (ss RNA) and some viroids was also determined.

The discovery of viroids and virusoids were additions to virology after 1970. In 1971, T.O. Diener reported that potato spindle tuber disease was caused by a small, naked, single stranded circular molecule of infectious RNA which he called viroid. There are now more than a dozen plant diseases known to be caused by viroids. These include potato spindle tuber viroid (PSTVd), citrus exocortis viroid and coconut cadang-cadang viroid. In 1982, a circular, single stranded viroid- like RNA was found encapsidated together with a single stranded linear RNA causing velvet tobacco mottle disease. This RNA molecule was called virusoid which seems to form an obligatory association with the viral RNA in many plant viruses. Viroids are the smallest nucleic acid molecules to infect plants. So far they have not been found in animals although similar proteinaceous infectious particles known as 'prions', were reported in 1982 in scrapie disease of sheep and goats. The prions were later mentioned in connection with the "mad cow disease" in UK.

MYCOPLASMA-LIKE ORGANISMS (MLOs) AND RICKETTSIA-LIKE BACTERIA (RLBs)

Fungi, bacteria, nematodes, and viruses were considered the main incitants of plant diseases upto 1967. The discovery of virus, so small that it could not be seen under the light microscope and could pass through bacterial filters, induced a feeling that nothing could be smaller than these agents. This, together with the observation that only few types of bacteria were plant pathogens and whatever passed through a bacterial filter must be a virus were responsible for delay in uncovering of a variety of new types of phytopathogenic bacteria (prokaryotes) that were later called the fastidious or hidden vascular pathogens. Fastidious because they were very exacting in their food requirement and did not grow on routine bacteriological media and hidden because they remained confined to plant vessels and could not be seen. Mycoplasmas were known to medical science during the closing years of the nineteenth century. The contagious bovine pleuropneumonia was believed by Luis Pasteur to be caused by a specific microorganism which could neither be seen nor grown in culture. In 1898, E. Nocard and E.R. Roux had succeeded in growing the organism in artificial medium and the organism is now known as *Mycoplasma mycoides*. The organism was placed in the order Mycoplasmatales among bacteria. The organisms in this order differed from true bacteria in having no true cell wall and not responding to antibacterial antibiotics like penicillin which act on cell wall but responding to tetracyclines. In 1967 Doi and his colleagues in Japan observed that mycoplasma-like bodies were constantly present in the phloem of plants suffering from leafhopper transmitted yellows diseases till then considered as virus although virus particles had not been seen. The same year Ishiie and colleagues reported that the mycoplasma-like bodies temporarily disappeared when the plants were treated with tetracycline antibiotics. Since then a large number of plant diseases were conclusively proved to be caused by such organisms. The true mycoplasmas associated with animal diseases were culturable but none of those associated with plants could be grown in cell free media. Thus, they could not be characterized and Koch's postulates could not be proved in most of these diseases. Therefore, they are called mycoplasma-like organisms (MLO) or phytoplasma. In 1972, Davis and his colleagues observed a motile, helical, wall-less microorganism associated with corn stunt disease. They called it *Spiroplasma* and this organism could be cultured and characterized. Therefore, it has been assigned to a separate family. MLO and *Spiroplasma* are phloem-inhabiting bacteria. During closer examinations of phloem for presence of other such organisms another group of fastidious prokaryotes was discovered in 1973 in citrus plants attacked by greening disease. The organism has definite cell wall and is susceptible to both penicillin and tetracycline antibiotics. This suggested that the organism is a true bacterium, not MLO. Gram-negative character of the organism was confirmed in 1974. In late 1990s the organism was identified as *Liberobacter asiaticum* for Asian citrus greening and *Liberobacter africanum* for African citrus greening. Search for similar organisms in xylem enabled the discovery of another Gram-negative bacterium in Pierce's disease of grapevine. The bacterium was identified as *Xylella fastidiosa* in 1987. Strains of this xylem-limited fastidious bacterium cause many diseases of peach, plum, almond etc. The causal agent of ratoon stunting disease of sugarcane, long thought to be a virus disease, was identified as a Gram-positive xylem inhabiting bacterium (*Clavibacter xyli*) in 1980 although association of a coryneform bacterium with the disease had been known since 1973. The bacterium associated with Sumatra disease of cloves was identified as *Pseudomonas syzygii* in 1990. These fastidious bacteria are often referred as Rickettsia-like (RLB) only because of some superficial resemblance to *Rickettsia* but they are not *Rickettsia*.

PROTOZOAN DISEASES OF PLANTS

In 1909 flagellate protozoa had been found in latex bearing cells of plants in family Euphorbiaceae but were thought to be living in the latex without causing a disease. In 1931 Stahel found flagellates infecting the phloem of coffee trees and causing abnormal phloem formation and wilting of trees. It was further confirmed in 1963 by Vermeulen. In 1976, flagellates were reported to be associated with several diseases of coconut and oil palm trees in South America and Africa.

DEVELOPMENTS IN CHEMICAL CONTROL OF PLANT DISEASES

By trial and error men had found that chemicals could be useful in disease management much before the discovery of Bordeaux Mixture. Sulphur was known as a pest-averting material in 1000 BC and was probably in general use by 1800. It was recommended for the powdery mildew of peach by Robertson in 1824. Admixture of lime to reduce phytotoxicity of sulphur was first recommended by Weighten in 1814. Boiled lime sulphur was recommended by Kenrick in 1834 and Grisson in 1852. In 1902, Lowe and Parrott noticed that apple scab was controlled by lime sulphur. By 1906 it was recommended as a general fungicide. Commercial formulations of lime sulphur continue to be common protectant fungicides even to-day.

Copper sulphate was recommended for wheat seed treatment against bunt by Prèvost in 1807 who for the first time established the fungitoxic value of the compound. It had been used as wood preservative in 1767. Copper sulphate was suggested as a possible control of late blight of potato during the 1844 epidemic and was recommended for roses in 1862. The use of copper sulphate admixed with lime was introduced by Dreisch in 1873 as a treatment for wheat seed infested by bunt. While basic information on copper sulphate with or without lime as a fungitoxicant was known, the modern era of chemical control of plant diseases started with the discovery of Bordeaux mixture during the epidemic of grape downy mildew in France during 1879–1882. The mixture of lime and copper sulphate had been sprinkled on the vines to deter pilferage. Prof. Millardet of Bordeaux University noticed that vines thus sprinked had little downy mildew. He developed the mixture which dominated chemical plant disease control for more than half century since 1885. Burgundy mixture, using sodium carbonate in place of lime, was introduced in 1887. Later, soluble copper fungicides (copper oxychloride) were discovered and gradually replaced Bordeaux mixture.

In late 1940s and during the 1950s a large number of synthetic compounds were introduced in the market as fungicides. These included chlorinated hydrocarbons, organophosphates, carbamates, phenoxys and acetamines. By 1990s, there were 113 active ingredients registered as fungicides worldwide (cf Knight et al, 1997). However, all are not equally effective and safe. Salts of toxic metals and organic acids, organic compounds of mercury and sulphur, quinones and heterocyclic nitrogen compounds have been major protectant fungicides in the latter half of the twentieth century.

The introduction of systemic fungicides, that could penetrate tissue and work from within the plant, in 1966, was a major landmark in the history of fungicidal management of plant diseases. It started with the discovery of oxathiins in 1966 by von Schmeling and Kulka. It was soon followed by confirmation of systemic activity of pyrimidines (1968) and benzimidazoles (1968, 1969). These fungicides are not effective against Oomycetes. Metalaxyl, effective against Oomycetes (Peronosporales), was developed by Ciba-Geigy in 1973 and came in use as a fungicide in 1977. During the same period, an organic phosphate fungicide, fosetyl-Al, was also developed and used against Oomycetes. The efficacy of these systemic fungicides made them highly acceptable to the farmers and for some time greater attention was given to them rather than to protectant fungicides. However, since they are narrow spectrum and site-specific in action against the pathogen, the latter soon developed resistance to them. Nearly 73 species of fungal pathogens have been reported resistant to 62 fungicidal compounds all over the world (cf. Singh, 2001). Thus, their dominance over protectant fungicides was soon over.

Antibiotics also have been used in plant disease control ever since the discovery of streptomycin. The antibiotics used in plant disease control belong to groups known as streptomycin, tetracyclines, polyenes, cycloheximide and griseofulvin. However, because of rigid dosage and danger of phytotoxicity none could become popular except streptomycin. This antibiotic was first used against fire blight of apple and pear in early 1950s. As in the case of systemic fungicides, resistance to streptomycin in the fire blight bacterium was also reported in 1971. Although this happened after the antibiotic had been in used for almost 20 years.

Information about the toxicity of chloropicrin (tear gas) to nematodes was known in 1919 when Matthews successfully controlled root knot nematode population in pots. Field application was demonstrated in pineapples in 1932. Nematicidal properties of dichloropropene-dichloropropane mixture (DD) were for the first time reported by Carter in 1943. It was later found by Newhall and Lear in 1948 that dichloropropene was the actual toxic fraction in the mixture. Almost at the same time, in 1945, Christie reported excellent response of ethylene dibromide (EDB) against root knot nematode. Rapid commercial manufacture of these two fumigants began in 1946. Soon after the success of DD and EDB a large number of other halogenated hydrocarbons were introduced which included chlorobromide, chlorobutene and chloropropane. In 1954, McBeth and in 1955, Raski separately reported successful use of dibromo chloropropane (DBCP). This nematicide could be used in standing crops and was generally accepted but had to be withdrawn because of health hazards. Attempts were also made to bring out combination products to broad base the effective control spectrum.

The fumigants were volatile chemicals and were sold as liquid. Their method of application was quite cumbersome. Most of them had to be applied days or weeks ahead of planting. This prompted the concept of chemicals that could be formulated as granules and which could quickly dissolve in soil water and act against the nematode. These were the non-fumigant contact or systemic nematicides such as carbofuran, phorate and aldicarb. Aldicarb was introduced in 1965 by Union Carbide. The advantage with these granular nematicides-insecticides was that they could be applied more conveniently and when systemic they could be absorbed by plant roots and remain effective for several weeks.

PLANT PATHOLOGY IN INDIA

The development of science of plant pathology in India, as in other countries, followed the development of mycology. Upto 1930, there was more attention to study of fungi than to diseases caused by them. The study of fungi in India was initiated by Europeans in the nineteenth century. They used to collect fungi in India and send the specimen for identification to laboratories in Europe. During 1850-1875, D.D. Cunningham and A. Barclay started identification of fungi in this country. Cunningham made a special study of rusts and smuts. K.R. Kirtikar was the first Indian scientist who collected and identified the fungi in this country.

Organized researches on fungi and plant diseases, based on long term planning, were started in this country only in the first decade of the twentieth century when the then British government established the Imperial Agricultural Research Institute at Pusa (Bihar). This institute, now known as Indian Agricultural Research Institute (IARI), was shifted to Delhi in 1934. It was in this institute at Pusa in 1901 that E.J. Butler initiated an exhaustive study of fungi and diseases caused by them. During the 20 years Butler stayed and worked in this country, he made a scientific study of most of the fungal plant diseases known in India at that time. In addition, he trained a team of plant pathologists who took over the work from him. The diseases, a detailed account of which was given for the first time by Butler, included wilts of cotton and pigeonpea, different diseases of rice, toddy palm, sugarcane, potato and rusts of cereals. He studied and wrote a monograph on Pythiaceous and allied fungi. His very important contribution to plant pathology in India still exists in the form of the classic *Fungi and Disease in Plants*, written by him and published from Kolkata in 1918.

J.F. Dastur (1886-1971), a colleague of Butler, was the first Indian plant pathologist who is credited with a detailed study of fungi and plant diseases. His special field of study was the genus *Phytophthora* and diseases caused by it in castor and potato. He is internationally known for the establishment of the species *Phytophthora parasitica* from castor. During the period when butler was here, plant pathologists trained by him, or his associates had by 1920, made a detailed study of a number of other diseases. G.S. Kulkarni published exhaustive information on downy mildew and smuts of sugarcane and pearl millet and S.L. Ajrekar studied wilt of cotton, smut of sugarcane, and ergot of sorghum. At that time the plant pathologist in India were more inclined towards a descriptive phase of the disease including the pathogen involved and had not paid much attention to the control aspect.

E.J. Butler left India in 1920 to take over as the first Director of the Imperial Mycological Institute (later Commonwealth Mycological Institute, (CMI) in England. He had trained a good number of mycologists and plant pathologists who took over the work on plant diseases with more emphasis on control aspects. B.B. Mundkur started work on control of cotton wilt through varietal resistance which ultimately resulted in reduction of losses from this disease in Maharashtra to a great extent. He was also responsible for identification and classification of a large number of Indian smut fungi. The most significant contribution of Mundkur to plant pathology in India will be remembered through the Indian Phytopathological Society which he started almost single handed in 1948 with its journal *Indian Phytopathology*. He also authored a textbook *Fungi and Plant Diseases* which was the second book of its type after the classic work of Butler. Dr. K.C. Mehta of Agra College made outstanding contributions to the knowledge of disease cycle of cereal rusts in India during the first half of the twentieth century. J.C. Luthra and associates developed the solar heat treatment of wheat seed for the control of loose smut. Before the advent of systemic fungicides this was the only sure method of control of this disease. Dr. R. Prasada, trained by Dr. K.C. Mehta, continued the work on rusts and added to the knowledge of linseed rust. Strong schools of fundamental plant pathology, especially biochemistry of host-parasite relationship, were started at Lucknow and Madras Universities under the leadership of S.N. Dasgupta and T.S. Sadasivan, respectively. Dasgupta carried out extensive studies on the controversial mango black tip disease. Sadsivan's school developed the concept of vivotoxins and worked out the mechanism of wilting in cotton due to *Fusarium oxysporum* f. sp. *vasinfectum*. The production of fusaric acid by the fungus was demonstrated by them. M.K. Patel, V.P. Bhide and G. Rangaswami pioneered the work on bacterial plant pathogen in India M.J. Thirumalachar conducted exhaustive studies of smuts and rusts. His association with the Hindustan Antibiotics Limited at Pimpri (Pune) resulted in discovery several antifungal antibiotics. In more recent period, a notable development was the establishment of the International Crop Research Institute for semi Arid Tropics (ICRISAT) at Hyderabad. Although confined to work on specific crops, this Institute is responsible for introducing new trends in plant pathological research in the country.

Teaching of plant pathology as a major subject in Indian Universities started rather late. The first India Universities that were established in 1857 at Kolkata, Bombay and Madras emphasized fungal taxonomy. Probably, the University of Madras was the first to take up plant pathology as a university science. University of Allahabad (est. 1887) and University of Lucknow (est. 1921) also took up plant disease aspects of mycology. Organised teaching in mycology and plant pathology as part of agricultural science was being conducted by the Indian Agricultural Research Institute at Delhi which later grew up and started giving post-graduate degree in the subject. However, before this the Agra University had introduced post-graduate degree programme in plant pathology at the Government Agricultural College at Kanpur in 1945. This college was the first agricultural college in the country established in 1906. After the establishment of agricultural universities in the country in 1960 and thereafter, teaching in plant pathology with its supporting courses in mycology, bacteriology, virolgy and nematology has become an important part of graduate and post-graduate programmes in agriculture. In the field of research publications the country showed phenomenal increase in numbers. After Indian Phytopathology, started in 1948, Indian Society of Mycology and Plant Pathology was started in 1971 with Indian Journal of Mycology and Plant Pathology (now Journal of Mycology and Plant Pathology) as its publication. Same year Nematological Society of India was also established and started Indian Journal of Nematology as its publication. Subsequently many other societies and journals were started in the country.

3

Causes of Plant Diseases

Injury to plants, affecting their life and productivity, can result from many causes. These include damage by natural calamities such as hail, storm, snow, lightning, mechanical injuries during cultural operations and damage by members of animal kingdom. All these causes cannot be brought under one subject and, therefore, plant pathology includes only the causes that have a definite type of relationship with the plant. The conditions defined as plant disease in the introductory chapter are separate from other harmful conditions of the plants. Disease is manifestation of reaction between the plant and the challenging disease inciting agent. If a plant part is damaged by hail or by any other mechanical means the injury to the plant is not disease because there is no reaction between the cause of the injury and the plant. Similarly, when most insects and other animals damage the plant, the resulting condition is not disease because the insects eat the plant parts without any reaction between the two. However, certain members of the animal kingdon, such as nematodes, incite disease because of their particular method of feeding and because their invasion initiates a chain of reactions in the plant. There are many commercial practices, such as drawing of latex for natural rubber or extraction of resin, in which the tree is injured or wounded but the condition is not a disease because the productivity of the tree is not affected. A diseased condition of the plant often results in enhanced beauty of the plant and is no more called a disease. In many virus infections, the variegation of the foliage or flowers enhances the beauty of the plant.

Typical plant diseases are caused by inanimate (non-living) or animate (living) causes. In addition, there are viruses and viroids that are neither living nor non-living. They cause some destructive diseases of plants. Genetic abnormalities are also cause of plant diseases but are rarely mentioned as disease. They are constantly eliminated through artificial and natural selection and hence have no place of significance in plant pathology. The commonly accepted causes of plant diseases can be grouped under following categories.

INANIMATE CAUSES

The inanimate, abiotic, nonparasitic or non-infections causes of plant diseases include (i) adverse climatic conditions such as very high or very low temperatures, unfavourable intensity of light, excess of humidity or rains, (ii) chemical injuries caused by (a) faulty application of fungicides, insecticides, weedicides and plant nutrients, (b) atmospheric impurities or pollutants such as phytotoxic components of smog, ozone, sulphur dioxide, ethylene etc. and (iii) adverse soil conditions including low, high or unbalanced soil moisture, poor soil structure affecting root growth, aeration and water holding capacity, poor oxygen supply and, most important, deficiency, excess or imbalance of nutrients, injurious salts and soil reaction.

Adverse Climatic Conditions: Unfavourable temperature is a major cause of disorders in field and orchard crops. Plants usually grow at a temperature range of 1°–40°C, the normal range for growth of most plants being 15°–30°C. Sudden and sharp variations from these limits causes injury to organs or the entire plant body. Damage to plants is much more by low temperature than by high temperature. Warm weather plants are usually more susceptible to low temperature. Freezing injury in potato tubers and frost injury in winter crops are common examples. In potatoes, the subfreezing temperatures cause ring-like necrosis of the vascular elements which may expand to the flesh if there is prolonged exposure to freezing. The frost injury to winter field crops is characterized by death of meristematic tissues, apical portion and leaves showing the maximum damage. Low temperature usually kills the tissues by ice formation in and between the cells which rupture. Young succulent parts with high water content are especially susceptible to frost. The damaged tissues are later invaded by secondary saprophytic organisms inducing rot.

High temperatures fatal for plants rarely occur in nature except in the tropics when occasional high temperatures for short durations may cause death of aerial plant organs. Generally, the adverse effects of high temperature are seen in conjunction with other abnormal environmental conditions such as excess of sunlight, low oxygen supply, drought, high winds, etc. Sun scald of fruits is a common example. The surface of fruits exposed to sun shows necrosis of the skin which may go deeper in the flesh. High temperature usually injures the tissues by inactivating certain enzyme systems while accelerating others. This leads to abnormal biochemical reactions and cell death. It may also coagulate proteins, cause disruption of cytoplasmic membrane and release of toxic products into the cell.

Excess light is rare phenomenon in nature and rarely injures plants except when it is combined with high temperature as during the summer months. Sun scald is common at high altitudes. Low light conditions retard chlorophyll synthesis and promote slender growth. The leaves are pale, almost white if light is completely excluded. Generally, dense plant canopy or dense stand of plants excludes light from around the lower parts of the plant. While low light intensity or absence of light reduces chlorophyll synthesis, excessively bright light destroys chlorophyll.

Unfavourable Soil Conditions: Among adverse soil conditions the moisture content is highly variable and, in addition to affecting parasitic diseases, directly causes injury to plant roots. Low moisture or acute water deficit causes physiological wilting while excess soil moisture results in reduced oxygen supply, accumulation of carbon dioxide and toxic metabolites produced by anaerobic bacteria, and accumulation of soluble salts around the roots and stem base. Under conditions of poor aeration many nutrients are not available to the plant. Low moisture allows temperature to rise and this may result in damage to underground storage organs of the plant.

Soil structure not only affects its moisture holding capacity and aeration, it can retard growth of the roots if sufficiently hard and compact. This results in stunted growth of the plant and their predisposition to parasitic diseases. In orchards, presence or formation of hard pan at certain depth in the soil causes hindrance in root growth and when the trees have grown symptoms of die-back, withering, etc. resulting in drying of trees appear.

Chemical Injuries: Faulty application of agricultural chemicals leads to chemical injury which has become common after introduction of modern methods of production and protection. Overdoses of fungicides, antibiotics or insecticides or application of unsuitable pesticides often results in plant injury. Many crops are highly susceptible to herbicides. Application of weedicides in the fields in the neighbourhood may cause damage to the crop through drift and persistence of residues. The air in industrial areas or near brick kilns in likely to contain such atmospheric impurities as sulphur dioxide, hydrogen sulphide, coal gas and chlorine. These substances are known to produce many diseases such as black tip of mango fruits. Smog (smoke and fog) damage is becoming increasingly serious in some highly populated areas. The main phytotoxic component of smog is peroxyacetyl nitrate (PAN) which is reaction product of ozone and waste hydrocarbons released by automobiles, tractors and diesel engines of water pumps. Ozone itself is produced by photochemical action of ultraviolet light on exhaust fumes. Other atmospheric impurities are ethylene, hydrogen fluoride, nitrogen dioxides and aldehydes.

Nurtitional Imbalance: Mineral deficiency is the most common cause of non-infectious diseases. Deficiency of essential minerals in the soil or their non-availability to the plants, even if not deficient in soil, develops hunger signs in the crop. This is especially true for some crops. Nitrogen, phosphorus and potash are major nutrients while copper, zinc, iron, magnesium, manganese, boron, calcium, molybdenum and sulphur are minor but essential elements for balanced nutrition of plants. Their deficiency results in poor growth, poor fruiting and poor yield. Excess of some of these elements is also toxic to plants. Faulty mineral nutrition often produces symptoms which are close to symptoms of diseases caused by virus and many fungi.

VIRUSES AND VIROIDS

Plant viruses are a group of submicroscopic entities showing obligate relationship with living cells of the host and ability to cause specific disease. These pathogens are not parasites in the usual sense of the term since they do not possess the necessary enzyme system for biological activities performed by living organisms. Their parasitism is at the genetic level within the host cell. Although they show increase in their quantity within the host cell they do so not by division or formation of reproductive structures but by a system of replication of their genome and synthesis of the full particle. The particles are made up of nucleic acid and protein, the latter forming a coat around the former. There is only one kind of nucleic acid (RNA or DNA) in a specific virus and in most viruses only one kind of protein. Virus particles are of various shapes such as rigid rods, long flexuous rods and many sided quasihedral spheres.

Viruses enter the host through damaged cells including leaf hairs or are introduced into the plant cells through some vector (insects, nematodes and fungi). During infection the nucleic acid of the virus particle must come in contact with the living protoplasm of the host. Some injury to host surface or rupture of the cell wall is essential for their entry. Viruses have no system capable of dissolving cell walls or causing direct penetration. However, gross mechanical injury resulting in death of protoplasm may prevent infection. After entry into the host cell the virus particle uncoats (loses the protein coat) and the nucleic acid is released. Replication of the nucleic acid (genome) then occurs through malfunctioning of the host nucleus from the material present in the cell. Movement of the virus particles in the host occurs mostly through the phloem by cytoplasmic stream. Cell to cell movement is through movement of particles via plasmodesmata between cells. The virus particles move along with carbohydrates to energy requiring organs of the plant. In tobacco mosaic virus and potato virus X the speed of movement is about 0.1–18 cm per hour while in beet curly top it may be 35–152 cm per hour.

Viral infection of plants causes distinct morphological, histological, cytological and metabolic changes in the host. Mosaic and mottle are most common symptoms. Distortion of organs and tissues is also seen in some diseases.

Transmission of plant viruses occurs through the agency of insect vectors (plant to plant), seed, vegetative propagation of infected plants, sap inoculation and grafting. There are many viruses which are transmitted by nematodes, mites and fungi.

Viroids are small, low molecular weight, naked strands of RNA that can infect plant cells, replicate and cause disease. Smaller than viruses, they have no protein coat but are highly resistant to adverse environments including high temperature.

ANIMATE CAUSES

The animate causes of plant diseases include such groups of organisms as bacteria (including mycoplasmas and spiroplasmas), fungi, algae, protozoa and nematodes. These organisms are characterized by their cellular nature as distinct from viruses and viroids. Among animate causes fungal diseases of plants are most common followed, in order, by bacteria, nematodes, algae and flowering plant parasites.

Mollicutes (Mycoplasma-like Organisms or Phytoplasmas): Among the cellular organisms that cause plant disease smallest are the prokaryotes (bacteria). In this group the most primitive appear to be the mycoplasma-like organisms or phytoplasmas. Their discovery as plant pathogens dates back to 1967. Before that many leafhopper transmitted diseases of yellows and witches, broom type were considered to be caused by viruses. The conception was based on the fact that they were transmitted by leafhoppers like the viruses, they were ultramicroscopic and filterable. But no virus particles had been seen under electron microscope. Although presence of fairly large infective particles in aster yellows disease was reported in 1943, it was in 1967 that Doi, et al., (1967) and Ishie, et al., (1967) in Japan proposed that the causal agent of the yellows type of diseases may be mycoplasma-like or chlamydia-like cellular organisms. Their observation was based on discovery of such bodies in the phloem of plants infected with several leafhopper transmitted diseases, absence of true virus particles in the phloem, and therapeutic effect of tetracycline antibiotics which caused remission of symptoms and disappearance of the bodies from the host cells temporarily. Since then more than 60 diseases, previously considered viral in nature, have been described as mycoplasmal diseases. These include aster yellows, sandal spike, brinjal little leaf, chilli little leaf, sugarcane grassy shoot and white leaf, rice yellow dwarf, potato witches, broom purple top roll and marginal flavescence and mulberry dwarf.

Mycoplasma was earlier known as one of the agents causing animal diseases. True mycoplasmas have the following principal characteristics: They are very small, filterable, unicellular, usually non-motile, and highly pleomorphic showing small coccoid bodies, ring forms and fine filaments. The filaments could be branched. Rigid cell wall is lacking and instead there is a triple layer unit membrane. Multiplication is by binary fission. Due to absence of cell wall they are not affected by penicillin but are inhibited by tetracycline. The true mycoplasmas could be grown in cell-free media of complex composition. The mycoplasmas seen in plants could not be grown in artificial media and in absence of proper characterization they have been grouped as mycoplasma-like organisms (MLO) or phytoplasmas. The pathogenic mycoplasmas that were found to have self motility and could be grown in culture were later identified as spiroplasmas and put under a separate group. Examples of plant diseases caused by *Spiroplasma* are citrus stubborn and corn stunt diseases

True Bacteria: Many more diseases are caused by walled bacteria or eubacteria than by the Mollicutes. Some of the bacterial diseases are exceedingly destructive. Examples are fire blight of apple, citrus canker, leaf blight of rice, black arm of cotton and wilt and brown rot of potato and other solanceous vegetables. In general most bacteria are thermophilic in nature and hence more destructive in warm regions. They grow better in alkaline media. The bacteria can not directly penetrate into the host and need some type of natural opening or wound for effective penetration. These organisms are all culturable on cell-free media.

From among the taxonomically established genera of bacteria many had remained unseen and hidden in the vascular elements of plants for long. Diseases caused by them also were considered of viral etiology. When MLOs were established as causal agents of plant diseases some of these diseases were suspected to be of mycoplasmal etiology. Search for MLOs in the phloem and exploration of the xylem subsequently revealed the presence of bacteria with definite cell wall. These bacteria did not grow on routine bacteriological media. Due to their extreme selectiveness for food, which was provided only by the vascular elements of the host plant or the hoemolymph of their insect vectors, these bacteria were called xylem-restricted or phloem-restricted fastidious prokaryotes. Most of them have been grown in highly specific media and characterized. They do not constitute a separate taxonomic group and, as known so far, belong to four genera, viz. *Xylella, Pseudomonas, Liberobacter* and *Clavibacter.*

The true bacteria are characterized as typically unicellular organisms, the cells being usually very small and lacking the definitely organized nucleus found in the cells of fungi and higher plants or animals. Plant pathogenic true bacteria are mostly rod-shaped, non-motile or motile by means of one or more flagella on their body. Multiplication is by binary fission. Stain reaction is Gram-positive in *Clavibacter, Curtobacterium, Bacillus, Arthrobacter, Streptomyces, Rhodococcus* and *Nocardia.* Most plant pathogenic bacteria, such as *Xanthomonas, Pseudomonas, Agrobacterium* and *Erwinia* are Gram-negative.

Bacterial infections of plants cause specific diseases such as blight, soft rot, leaf spot, tumors and galls, canker and vascular wilt. These diseases are mainly spread through the agency of seed, soil, insects, air and water. Insects sometimes play a major role as carriers of bacterial cells. Bacterial cells are usually covered in slime, extracellular polysaccharide, which protects them from drying, helps in dissemination by water drops and air, and also determines their virulence as a pathogen.

Fungi: Fungi are microorganisms having chlorophyll-less, nucleated, unicellular or multicellular filamentuous bodies (thallus) which reproduce by division of vegetative cells or by well defined asexual or sexual spores. Evolutionary trends are better marked in fungi than in prokaryotes. In the primitive forms of the fungi the thallus may be plasmodium-like, unicellular and holocarpic

(*Plasmodiophora, Spongospora,* etc.). In better evolved forms (oomycetes and higher fungi) reproduction is carried out by specialized organs or gametes, the thallus may be multicellular or at least well developed and branched. In asexual reproduction the propagules may be simple vegetative cells or well defined spores cut off from the vegetative filaments (conidia) or in specialized fruit bodies (sporangia, pycnidia, etc.). Sexual reproduction is brought about by a variety of methods in which morphologically different or similar gametes fuse resulting in the formation of sexual spores. In lower fungi relatively simple structure (oospores) are formed while in higher fungi (Ascomycetes) the sexual spores are naked or enclosed in different types of fruit bodies (perithecia). The asexual spores are mostly the repeating spores that spread the pathogen in plant populations while the sexually produced spores are mostly resting structures. Chlamydospores are asexual spores formed by hyphal or conidial cells and serve as resting structures. In may fungi aggregations of filaments form hard structures (sclerotia) that are resting structures or they may form rope-like strands (rhizomorphs).

The plant pathogenic fungi survive through the agency of soil, seed, alternate or weed hosts and are dispersed by these and other agencies such as insects, wind, water and animals including man. Unlike bacteria, many fungi are biotrophs, i.e., they are active only in the living host. However, a very large number of them are hemibiotrophs or perthotrophs. Entry into the host is through natural openings (stomata, lenticels, hydathodes), through wounds due to mechanical or insect injury as well as through direct penetration.

Protozoa: Flagellate protozoa were seen in the latex-bearing cells of some plants of the family Euphorbiaceae as early as 1909. However, they were thought to be confined to the latex without causing a disease. Abnormal phloem formation and wilt of coffee trees due to infection of a flagellate protozoan was reported in 1931. Subsequently, many more diseases of coconut and oil palm in South America and Africa were ascribed to flagellate protozoan agents.

Algae: Plant diseases are also caused by algae, the best known being *Cephaleuros mycoides* which attacks mango, guava, papaya, tea, citrus, cashew and coffee causing fruit and leaf spots. Often the diseases caused by algae are known as "red rust". Parasitic algae spread through their sporangia which are air-borne and produce zoospores. These spores enter the host through stomata and other natural openings and grow as chains of algal cells. In India algal diseases of mango and papaya are common in warm humid areas.

Nematodes: The nematodes, organisms resembling roundworms in human intestine, are natural fauna of soil and water. They constitute the largest group of the animal kingdom in soil. Majority of them are free-living, subsisting on microscopic organisms present in soil. However, many of them are destructive parasites of plants causing such diseases as root knot of numerous crops, root rot, tree decline etc. Although nematodes as animals have mouth parts, their study is included among plant diseases because they produce abnormalities resembling common diseases caused by bacteria and fungi. During their feeding in or on the plant they secrete enzymes and other metabolites that react with the host system. Such abnormalities as stunted growth, proliferation of roots, chlorosis, hypertrophy and hyperplasia of tissues result from these interactions.

The life-cycle of a nematode starts from oval eggs. Four molts occur during the life cycle resulting in five stages. One molt occurs within the egg. When the egg hatches second stage larvae are released. These cause infection of the host. Body of an adult nematode is always cylindrical in case of males and often sac-like, oval, much swollen in case of adult females. They have well organized mouth parts bearing a stylet (spear). Sensory organs are also present around the mouth. The stylet is used to pierce the host tissue, inject saliva by a pumping action and ingest the host cytoplasm. Reproduction is through copulation between male and female, fertilization of ova and production of eggs. Parthenogenetic reproduction is also common. The infective larvae feed on the host from outside (ectophytic) or inside (endophytic). The feeding nematodes may be migratory or sedentary. Full development of the nematode occurs only in the host because these pathogens are obligate parasites and cannot obtain their nutrition from any other source except the host cells.

Flowering Parasitic Plants: Parasitic species of angiosperms are destructive pests of several economically important fruit trees such as mango and of several field crops such as mustard, legumes, tobacco, berseem, lucerne, etc. Among important families containing parasitic species are Loranthaceae, Orobanchaceae, Convolvulaceae, Scrophulariaceae, Lauraceae, Santalaceae and Balanophoraceae. These parasites damage the plants through exhaustion of nutrients and sometimes through restriction of growth of the plant. Some of them produce toxins also.

The parasitic angiosperms produce seed which are dispersed by wind, birds and other animals and also through soil. In *Cuscuta* (dodder) the chlorophyll-less stem pieces can also be carried to new hosts and produce a new parasitic plant. In *Orobanche* and *Striga* prolonged survival in soil can occur through seeds and underground stem. The parasitic flowering plants can be total or partial parasites of stem and roots. *Cuscuta* is a total parasite of stem while *Orobanche* is total parasite of roots. *Striga* is a semi-parasite of roots while *Dendrophthoe* (Loranthaceae) is a semi-parasite of stems. The semi-parasites possess leaves and synthesize the carbohydrate portion of their food. These parasites establish relationship with the host vascular elements to draw nutrients.

CLASSIFICATION OF PLANT DISEASES

Plant diseases are grouped under two categories on the basis of the cause:

1. Non-infectious or non-parasitic diseases: These are diseases with which no animate or virus pathogen is associated. Therefore, they remain non-infectious and cannot be transmitted from a diseased plant to a healthy plant. It is wrong to call them physiological diseases because physiological disorders are present in all types of diseases, parasitic or non-parasitic. These non-infectious diseases are due to disturbances in the plant body caused by lack of proper environmental conditions of soil and air. Low and very high temperatures, unfavourable oxygen relations, unfavourable soil moisture, and pH, presence of toxic gases in the atmosphere, mineral excesses and deficiencies in the soil and absence or excess of light are major causes of these non-infectious diseases. Common examples of such plant diseases in India are tip rot or necrosis of mango fruits due to boron deficiency caused by atmospheric impurities, black heart of potato due to unfavourable oxygen relations in the stores and in the field, and Khaira disease of rice due to nonavailability of zinc to the plant.

2. Infectious diseases: Diseases incited by attack of parasitic organisms or virus and viroids under a set of suitable environment are infectious. Association of a specific pathogen is essential with such diseases. In other words these diseases are due to inroads of parasitic organisms or virus pathogens into the body of the plant. These diseases are always infectious, sometimes contagious, and are transmitted from diseased to healthy plants in the field and from one place to another through various agencies. Under this group the diseases are further categorized as those caused by (a) parasitic organisms like bacteria, MLOs, fungi, nematodes, etc. and (b) viruses. There are certain infectious diseases in the first category which are caused by parasites but it is not necessary that the parasite should be bodily present in the plant. It can produce toxins around the roots in quantities enough to produce pathological effects.

Infectious diseases are often classified according to their occurrence in following groups:

1. Endemic diseases: The word "endemic" means " prevalent in, and confined to, a particular area, country or district" and is applied to disease. These diseases are natural to one country or part of the earth. When a diseases is more or less constantly present from year to year in moderate to severe form in a particular area it is called endemic to that area. The causal agent is well established in the fields or in the locality by virtue of its ability to survive through soil or other means for long durations and the environmental conditions are generally favourable for its survival and multiplication. All those diseases which become persistent through their survival on alternate or wild hosts from one crop season to the next are also included in this group.

2. Epidemic or epiphytotic diseases: The term "epidemic" is derived from a Greek word meaning "among the people" and in true sense applies to those diseases of humans which appear very virulently among a large section of the population. The term is also used for such diseases among cattle and poultry. To carry the same sense in the case of plant diseases the term "epiphytotic" was coined. An epiphytotic disease is one which occurs widely but periodically. It may be present constantly in the locality but assumes severe proportions only on occasions. This is because the environments or conditions favourable for rapid development of the diseases occur only periodically. It is also possible that the environments may be favourable but the pathogen causing the disease may be irregular in its occurrence or the inoculum of the pathogen has not reached the suitable concentration to cause the disease.

3. Sporadic diseases: Sporadic diseases are those that occur at very irregular intervals and locations and in relatively few instances.

A given disease may be endemic in one region and epidemic in another. When a disease is prevalent throughout the country, continent or the world in a severe form it is known as **pandemic disease**.

For practical purposes the diseases are also classified according to their symptoms, plant organs they affect or the mode of survival and dispersal of their propagules. A disease may be *localized* affecting only the special organs or parts of the plant, or it may be *systemic* affecting the entire plant, i.e., the pathogen moves through the entire plant. According to symptoms the diseases may be *rusts, smuts, wilt, blight, canker, mildew, root rot* and *fruit rot*. When the disease symptoms appear most conspicuously on special organs they may be *leaf spot, stem lesions, root rot, fruit rot,* etc. According to host plants the diseases may be grouped as *cereal diseases, forage crop diseases, flax diseases, root crop diseases, plantation crop diseases,* etc. When a disease-causing agent survives and spread through soil it is known as soil-borne pathogen. Diseases or pathogens perpetuated through seed are known as seed-borne disease (loose smut of wheat, barley) or seed-borne pathogens (covered smut of barley, smut of sorghum, etc.). When dispersal is through the agency of air, the pathogen is known as air-borne. The same pathogen may be soil-borne, seed-borne and air-borne. Since pathogen and disease are two different things and disease occurs only on the living plant it is itself not soil-, air, or seed-borne. Only infectious diseases are classified in this manner. Soil, seed and air are the media for dispersal of the pathogens not the disease. Seed and soil also serve as media for survival. Some seed-borne diseases, such as loose smut of wheat, result from intraseminal or embryonic infection and in such cases the disease itself is seed-borne because the seed is diseased.

4

Symptoms and Indentification of Plant Diseases

During invasion, the pathogens induce reactions in the body of the host. As a result of these reactions certain abnormalities appear on the plant. In addition, the pathogen itself may become visible on the host surface giving it an abnormal appearance. The abnormalities, signs or evidence of the disorder, are known as **symptoms** of the disease. Since the origin of these symptoms is mainly from internal disorders and many kinds of pathogens can produce same type of disorder the symptoms resulting from host-pathogen interactions are not very reliable basis for identification of plant diseases although they do help to some extent. On the other hand, when the pathogen itself becomes visible on the host surface, it gives more reliable information about its own identify and identification of the disease is easy.

Following are the symptoms of plant diseases due to the characters and appearance of the visible pathogen, its structures and organs:

Mildew: In these diseases the pathogen is seen as white, grey, brownish or purplish growth on the host surface. In **downy mildew** the superficial growth is a tangled cottony or downy growth while in **powdery mildew** enormous numbers of spores are formed on the superficial growth of the fungus giving the host surface a dusty or powdery appearance. Black, minute fruiting bodies may also develop in the powdery mass.

Rusts: These diseases produce rusty symptoms. The rusts appear as relatively small pustules of spores, usually breaking through the host epidermis. The pustules may be either dusty or compact and red, brown, yellow or black in colour. The colour is due to mass of the spores of the fungus.

Smut: The word smut means a sooty or charcoal-like powder. In plant diseases known as smut, the affected parts of the plant show a black or purplish-black dusty mass of spores. In some smuts this mass of spores may be compact and held under the epidermis appearing as black streaks or blotches on the host surface. The symptoms of smut usually appear on floral organs, particularly the ovulary and the pustules are usually considerably larger than those of rusts. Smut symptoms may also appear on stems, leaves as well as roots.

White blisters: On leaves of crucifer and many other plants numerous white, blister-like pustules are found. They break open the epidermis and expose white powdery masses of spores. Often, such symptoms have been called white rust. Since there is nothing common between them and the true rusts they are more appropriately called white blisters.

Blotch: This symptoms consists of a superficial growth giving the fruit a blotched appearance as in the sooty blotch of apple fruits.

Sclerotia: A sclerotium is a compact, often hard, mass of dormant fungus mycelium. In some diseases, as in ergot of cereals, the sclerotium assumes a characteristic horn-like shape but in others the shape may be more variable. The sclerotia are most often dark coloured (dark brown to black). Presence of these sclerotia on the host surface or within the tissues helps in identification of the pathogen and the disease.

Exudations: In some fungal diseases the invasion of the pathogen causes exudation of a gum-like material on the stem and the collar. This is known as **gummosis**. In most bacterial diseases (bacterial blight of rice, bacterial leaf streak of rice, fire blight of pear and apple, bacterial cankers of stone fruits) masses of bacterial cells come out as ooze on the affected host surface where they may be seen as drop or smear. On drying they form a crust.

Symptoms resulting from internal disorders in the host plant may appear in one or more of the following forms.

Colour Changes: Change of colour from the normal, or discolouration, is one of the most common symptoms of plant diseases. The green pigment may disappear entirely and its place taken by yellow pigment. When this yellowing is due to lack of light it is called **etiolation**. A similar conditions results from the influence of low temperature, lack of iron, excess of lime or alkali in the soil and infection of viruses, fungi and bacteria. These conditions interfere with synthesis of chloroplasts. Such a yellowing is known as **chlorosis**. Change of colour to red, purple or orange is **chromosis**. Sometimes the leaves are devoid of any pigment and look bleached or white. This conditions is known as **albinism**.

Overgrowth and Hypertrophy: Many pathogens, through their biochemical activity induce hormonal imbalance. This results in excessive growth of host tissues and causes abnormal increase in size of affected organs. It is brought about by one or both of the two processes known as **hyperplasia** and **hypertrophy**. Hyperplasia is the abnormal increase in the size of plant organs due to increase in number of cells comprising the tissues of the organ. The cell division is increased and so the number of cells at a given location is much higher than normal. In hypertrophy the increased size of the organ is due to increase in the size of cells of a particular tissue. The pathogen may dissolve the intervening walls between the cells and its products may cause the cells to increase in size. Often the cells increase in size to accommodate the fungal structures. The overgrowth and its effect are seen in galls, curl, pockets or bladder, hairy root, witches' broom, intumescence, etc.

Atrophy or Hypotrophy or Dwarfing: In many diseases one of the results is the inhibition of growth resulting in stunting or dwarfing. The internodes fail to elongate. The whole plant may be dwarfed or only certain organs may be so affected. Sometimes hypertrophy and atrophy occur together in the same organ such as flowers.

Necrotic Symptoms: Necrosis indicates the condition in which death of cells, tissues and organs has occurred as a result of the parasitic activity. Necrosis is most commonly defined as cell death that results from exposure to highly toxic compounds, severe cold or heat stress, or traumatic injury that lead to immediate damage to membranes or cellular organelles (Gilchrist, 1998). It does not require the active participation of the cell in its own death. The characteristic appearance of the dead area differs with different hosts and host organs and with different parasites. Thus, the necrotic symptoms include the following:

Die-back and wither tip: As the name indicates such diseases are characterized by drying of plant organs, especially twigs and branches, from the tip backwards.

Spots: The cells are killed in definitely limited areas and dead tissues usually become some shade of brown. Yellowing may precede the death of cells. Most common diseases in this group are the leaf spots. Yellow halo is often present around the necrotic spots. It indicates presence of pathogen toxin(s).

Streaks or stripes: These are elongated but relatively narrow lesions containing dead cells. They are also initially yellowish in colour before cell death occurs.

Canker: A canker is a dead area in the bark or cortex of the stem of woody and sometimes on leaves also. Usually they are large with a definite margin. Some cankers are only superficial or they may involve all the tissues except the fibres. The surface of the cankers may be smooth but generally in later stages of development they become rough with a sunken centre. On woody parts cankers cause splitting of the bark which may peel away. In some diseases the fruiting bodies of the fungus are seen after the bark is destroyed.

Blight: The term refers to sudden death of leaves, blossoms or twigs. The dead organs usually turn brown or black and may soon disintegrate.

Damping off is a condition in which the tender stem is attacked near the soil line. The affected portion becomes constricted due to necrosis of the cortical tissues and the stem is unable to bear the load of upper portion. The seedlings collapse.

Burn, scald or scorch: Areas in the succulent organs of the plant die and turn brown due to effect of high temperature.

Rot: The affected tissues die, decompose to a great extent and turn brown. The condition is brought about by fungi and bacteria which dissolve the middle lamella between cell walls by means of enzymes.

Anthracnose: A particular group of fungi causes ulcer-like lesions on fruit, pods and stems.

Wilt: In many diseases the characteristic symptom is gradual or sudden drying or wilting of the entire plant. The leaves lose their turgidity, become flaccid and droop. Later the young growing tip or the entire plant wilts. Wilting may be the result of injury to the root system, to partial plugging of the water conducting vessels or to toxins produced by the pathogen and carried to delicate tissues.

Miscellaneous symptoms: (*i*) *Alteration in habit and symmetry:* Some plants which under normal conditions are prostrate or creeping become ascending or even erect under the influence of some parasites. Leaves may become lobed from being simple. The inflorescence is changed from a head to a spike. (*ii*) *Transformation of organs* or replacement of organs by new structures: Sometimes the normal plant parts are replaced by fungus structure or by some abnormal plant structure. In ergot of cereals and millet and in false smut of rice the grains are replaced by sclerotial masses of the fungus structures. In green ear disease (downy mildew) of pearl millet the flowers are converted into green leafy structures (*phyllody*). The term *virecence* is also used for reversion of floral organs to vegetative organs, i.e., greening of petals and conversion of petals to leafy structures.

Disease Diagnosis

Diagnosis of plant disease is a field science and practice is the most sound method of identifying a disease in the field. Tentative identification in the field can be made with the help of illustrated guides and charts. It is easy to identify many diseases in which the pathogen produces specific growth such as in downy mildew, rusts, smuts, powdery mildews and many sclerotial diseases. However, in many other diseases the symptoms are not so useful and may cut across the area of many diseases. For example, chlorosis or yellowing of leaves is a symptom which can be caused by nutritional disorder, virus, fungus or a bacterium. For effective and economical management of any disease, accuracy in diagnosis is important for choosing the proper control strategy.

In those cases where symptoms are such that may be caused by a variety of living and non-living disease incitants, the first step is to determine whether the disease (pathogen) is infectious or non-infectious. Field observations on the pattern of development of the disease in the plant population and possible spread of symptoms on other plants helps in distinguishing the two. If the disease is spreading in the plant population it is infectious. An infectious disease may be caused by fungi, bacteria, viruses and nematodes. These can be determined first by visual observations of the affected parts for presence of fungal structures, bacterial ooze, or nematode cysts or females and then by laboratory studies. If present on the host and examined under the microscope the fungal structures may reveal presence of a particular fungus. If the fungus is not a biotroph it can be cultured on artificial media and then tested for Koch's postulates. Bacteria can also be detected in a similar manner. Apart from the ooze often seen on the host surface, examination of cut pieces of the affected parts in water under microscope will reveal streaming of bacterial cell masses in water.

Following steps constitute the Koch's postulates:

1) The pathogen must be invariably found in affected plant or it must be associated with it in some form.
2) The pathogen should be separated from the host and grown in artificial culture.
3) The pathogen from the artificial culture should be able to reproduce the disease when inoculated on a healthy plant of the same kind from which · it was isolated. Symptoms produced should be identical with those seen on the plant from which the isolation was made.
4) The artificially produced disease should yield the same pathogen on re-isolation.

Nematodes, if present, can be seen on the host or in macerated tissues examined under microscope. Nematodes cannot be grown on artificial media to obtain material for inoculation. By separating the eggs and larvae from the host and multiplying them on a susceptible host under sterile conditions sufficient number of larvae can be obtained for proving the Koch's postulates.

Viruses can not be seen or grown in culture. They generally produce symptoms similar to those of nutritional deficiencies. However, the latter condition is non-infectious and does not spread from plant to plant while viruses are infectious and spread in the population. If in artificial culture no pathogen is obtained and in tissue examination, including the vessels, no fungal structures of a biotroph or fastidious organism or structures of nematodes are.seen but the disease incitant is infectious, it can be expected that the disease is caused by a virus or MLO. The plant pathogenic MLO are not easy to culture. Transmission tests by grafting, sap inoculation or use of insect vectors can help in final diagnosis. The differentiation between viruses and MLO can be made by electron microscopy and by spraying streptocycline antibiotics which mask the symptoms of MLO diseases but not of virus diseases.

5

Pathogenesis

The infectious diseases of plants are caused by viruses and viroids, bacteria (including mycoplasma-like organisms, MLO), fungi, and nematodes. In addition, plants are also attacked by other parasitic flowering plants. The number of pathogens and diseases caused by them is very high. Only on tomato 80 fungal species, 10 types of viruses, 16 types of bacteria and many forms of phytophagous nematodes are found. They are pathogenic but diseases caused by them are not equally important or harmful. Potatoes and apple trees and fruits may be attacked by 200 different diseases. The total number of plant diseases may be more than 80,000. Nearly 8000 different fungi are associated with the development of these diseases. There are about 180 types of bacteria and 500 types of viruses that cause plant diseases. More than 500 species of nematodes are known to feed on plants.

Infectious pathogens have the characteristics of reproducing or multiplying rapidly on or in the host and spread to other plants in the population by different means. In this and subsequent chapters of the book the conditions that influence parasitism, pathogenicity and pathegensis of pathogens are discussed. To obtain effective management of a plant disease knowledge of these conditions related to disease development is essential.

PARASITISM AND PATHOGENICITY

When an organism (microorganism, plant or animal) obtains food for its life activities by living in or on another organisms it is known as a **parasite**. It is not essential that this host-parasite relationship should invariably result in diseased condition of the plant. Parasitic pathogens are mainly bacteria (including mycoplasma-like organisms), fungi, algae, phanerogams and nematodes. Viruses and viroids are pathogens but not parasites in the same sense as the fungi, bacteria, etc. They lack the Lipman enzymatic system for conversion of high energy into potential energy required for biological activities. Their parasitism is at the genetic level. Thus, while inducing malfunction of the host nucleic acid they do not directly draw any food from the host. A parasite is not necessarily a pathogen and a pathogen is not necessarily a parasite.

A parasite of plants establishes close relationship with its host and draws nutrients from host cells for its own development. The drain on nutrients in the host body affects growth of the plant and its reproductive capabilities. Thus, there is reduction of yield. This relationship between the parasite and the host is known as **parasitism**. When this parasitism results in visible disease symptom, the capacity of the parasite or the pathogen to cause disease is known as **pathogenicity**. Since parasitism mostly results in disease development there is not much difference between parasitism and pathogenicity. However, there are many exceptions. The association of some fungi with plant roots to form mycorrhizae is initially an act of parasitism but the relationship later turns into a mutually beneficial relationship. The plant roots support the fungi and the fungi help in absorption of mineral nutrition from the soil. The formation of roots nodules in legumes due to nitrogen fixing bacteria (*Rhizobium* spp.) is actually due to parasitic relationship but the bacterium lives in symbiotic relationship with the plant, giving more benefit than causing harm to the plant. Therefore, there is no expression of diseased condition. In this case the symbiotic relationship is not pathogenicity although the bacterium enters the roots as a parasite. The symbiotic relationship is an example of very high degree of genetic and phsyiological synchrony between a parasite and the host. There are many rhizobacteria that are deleterious to plants without being actually a parasite. They produce toxic compounds around the roots or interfere with absorption of nutrients by roots from the soil.

The total loss of the plant or the amount of injury in a diseased condition in the plant is mostly much higher than the loss expected from drain on nutrients by the pathogen. The drain on nutrients could weaken the plant, check its growth and prevent formation of new shoots, buds, flowers etc. But due to biochemical interactions between the pathogen and the host there are other abnormalities such as tissue necrosis, dissolution of cell walls, overgrowth etc. The substances involved in these abnormalities are pathogen produced toxins, enzymes and hormonal imbalances resulting in change in growth habit. The host plant reacts to invasion of the pathogen and its products and often forms defensive structures that can also affect growth of the plant. Following abnormalities in the plant are observed:

1) Increase in respiration of the host tissues.
2) Disintegration and death of cells.
3) Wilting of the entire plant or its parts.
4) Abnormal cell differentiation and overgrowth.
5) Disintegration of specific cell organelles such as chloroplasts.

These abnormal conditions do not necessarily help the pathogens but they invariably harm the host. Therefore, pathogenicity is the capacity of the pathogen to induce malfunctions or interfere with the physiological activities of the plant. Parasitism mostly, but not always, plays a major role in pathogenicity. The activity of infectious parasites depends on how successfully they can infect the active host, draw nutrition from it and multiply in or on it. Some pathogens and infectious parasites such as viruses, downy and powdery mildew and rust fungi can remain active only on the living host. They become inactive or are killed as soon as the living host ceases to remain active. These disease incitants are known as **obligate parasites, obligate pathogens** or **biotrophs**. The biotrophic fungi generally avoid causing too much tissue disintegration, not more than their requirement for nutrition. The relationship is mainly nutritional. The biotrophs normally cannot be grown on artificial media but some have been cultivated on specially prepared media suggesting that if nutritional conditions similar to those in the host cell can be created they can be cultivated.

Some parasites spend a major part of their life as active organisms on the living active host (ascending parasitic phase) but when the host is dead or removed from the field or under the conditions of adverse environment they can live as **saprophytes** on dead organic matter. These are **facultative saprophytes**. If the parasite is mainly a saprophyte and becomes a parasite only in favourable conditions it is known as **facultative parasite**. These parasites, by virtue of their ability to produce hydrolytic enzymes and toxins, have a tendency to cause extensive damage to host tissues which is probably more than their nutritional requirement.

The degree of parasitism of a pathogen is not always correlated with the severity of the disease it causes. Disease severity is determined by many other factors in addition to parasitic ability of the causal agent. There are many diseases that are caused by weak parasites (mostly facultative parasites) which incite high disease incidence and severity if other factors are favourable for the parasite and unfavourable for the host. The wound parasites are example. There are certain fungi (such as sooty molds) that are saprophytes but cause disease on the plant without becoming parasite. *Fusarium moniliforme*, although a parasite, sometimes produces disease without entering into a host-parasite relationship with the plant such as maize. While growing as a saprophyte in the vicinity of roots it liberates toxins that damage the roots and suppress their development. As a result the plant suffers from malnutrition, remains stunted and yellow and may finally die.

Obligate parasites and facultative saprophytes or parasites differ in the mode of infection and absorption of nutrients from the host. Facultative parasites are basically saprophytes in their feeding mechanism which involves high degree of tissue disintegration. They first liberate enzymes that disintegrate the tissues (**killing in advance**). Due to this effect on cell walls and, sometimes, due to effect of toxins produced by the pathogen the cells die. Then, the parasite (fungus or bacterium) draws its nutrition from the dead cells. The wound parasites first colonize the cells killed by the wounds, multiply, increase their mass and thus produce increased amount of enzymes. The enzymes dissolve middle lamella of more cells and then the parasite moves in the killed area.

The fungus body actually never comes in contact with living cells and lives a true saprophytic life. All the obligate parasites and facultative saprophytes do not kill the host cells before or immediately after infection. If cells are killed, it is not because of enzymes or toxins of the parasite but due to acute reaction of the host cells (**hypersensitive response**). This may involve the programmed cell death in which when the cells find themselves under stress or become useless a self destruction mechanism is triggered leading to cell suicide. The obligate parasites never kills the host otherwise the parasite will also be killed or inactivated. These parasites enter the host, grow intra- or intercellularly, establishing intimate relationship or continuity with the host cells and absorb nutrients. The intercellular hyphae of fungi send haustoria into the host cells and absorption of nutrients occurs through membrane of the haustorium. The exhaustion of nutrients may ultimately cause death of the host but by that time the parasite may have formed its survival structures.

The differences among plant pathogens are also determined by host species, the susceptible organs of the host and the kind of tissues and their age. Some pathogens attack only one plant species while others infect plants of a particular genus. Other plant pathogens have no special affinity for a particular type of host and can infect plants of different species, genera and families. Plant organ specialization is also common. Some pathogens are only root inhabiting while others may specially be present in stems, leaves or fruits. Similarly, growth of pathogens in the host may vary according to specific tissues. Some pathogens are exclusively confined to xylem (**fastidious xylem limited bacteria**), some exclusively to phloem (viruses, MLOs and some bacteria), some mainly confined to xylem (vascular wilt pathogens) or they may be restricted only to the meristem.

Usually, the obligate parasites are host specific. This specialization or selectivity for the host is attributed to the coevolution of the parasite with the host. Nutritional and genetic compatibility was established during this period. These pathogens have specialized for utilizing only the nutrients present in the particular host species and variety they attack. The particular specialization is mediated through genetic compatibility. Although the facultative sparophytes or parasites have no host specificity to the same extent as rust and powdery mildew fungi there are a number of facultative saprophytes which attack only a very limited range of plant species or genera. Most of the vascular wilt causing species of *Fusarium* are host specific to the extent that they infect only one or few very closely related host species. The parasitism or pathogenesis of a pathogen may undergo change due to genetic changes and thus their liking for hosts may also be altered.

PATHOGENESIS

Development of a disease in the plant is not a sudden effect. A chain of events are responsible for causation of any disease. The events start taking place before the contact between the plant and its pathogen has occurred. The symptoms and manifestation of injury to the plant is the last link in this chain of events. Before symptoms appear, the pathogen independently or in association with the host has passed through several stages. These stages, reactions and interactions arranged in a sequence lead to disease development and the entire chain of events leading to disease development is known as **(pathogenesis)**. It is also called **disease-cycle**. The disease cycle tells a practical plant pathologist about the source of perennation of the pathogen, mode of dispersal from the source of survival and during spread of the disease and structures involved in survival and in spread in the plant population. These informations are essential in the formulation of effective management strategy.

The events constituting pathogenesis occur in seven well defined stages, one after another, to complete the disease cycle. The pathogen survives or perennates (over-summering or over-wintering) at some location during absence of the cultivated host (Ch. 6). It is then transported to the cultivated host in the field through different agencies (Ch. 7). Attack and infection follows this stage and the pathogen breaks down host barriers to establish infection (Ch. 8). This is followed by effects on the host physiology, damage to the plant and expression of symptoms (Ch. 10). The pathogen finds exit from the infected plants and spreads the disease in the plant population (epidemiology, Ch. 9). When the living or dead substrates are exhausted the pathogen undergoes the dormant stage for survival. At every stage in the disease cycle or pathogenesis, the environment including climatic factors, cultural operations, and physical, chemical and biotic conditions of the soil play a decisive role (Ch. 15).

THE INFECTION CHAIN

Gaumann (1950) had used the term "infection chain" for the chain of events leading to completion of pathogenesis. The infection chain can be of two types. In **continuous infection chain** the pathogen continuously lives in active form by moving from one host to another. Most viral diseases and some diseases caused by fungi have this type of infection chain. The **intermittent infection chain** is found in diseases caused by bacteria, fungi and nematodes. After harvest of the crop these pathogens survive in dormant state or as active saprophytes to maintain continuity of the chain. When the pathogen survives on only one genus or species of plants it is known as **homogenous infection chain**. If many plants species are involved in the disease cycle the chain is **heterogeneous**. The heterogeneous infection chain can be facultative or obligate in nature according to the nature of the fungus or the bacterium.

Time taken by a pathogen to pass through the different stages mentioned above is the length of a single disease cycle. It is not essential that disease cycle and nuclear life cycle of the pathogen should be same. In several disease, such as in rust of cereals, pea and lentil, disease cycle is repeated only by the diplophase of the fungus. All the events during completion of a disease cycle are influenced by environment which may determine the length of each cycle. Shorter the cycle greater are chances for spread of the disease and build up of an epidemic. Two types of disease cycle are seen in diseases with intermittent infection chain. In **polycyclic pathogens** the cycle is repeated many times in the same season on the host through spore-infection-spore chain. These pathogens are responsible for epidemics if weather favours. Examples are late blight of potato and apple scab. In **monocylic pathogens** there is a single cycle after infection and there is no repetition of the cycle during the season. Examples are loose smut of wheat and covered smut of barley.

The knowledge of disease cycle is a basic requirement for success of management strategies. In a disease cycle the means of survival of the pathogen during absence of the crop and method of its dispersal in the standing crop are two most critical stages. Control of these two stages ensures success of control measures for the disease. Disease cycle of some common diseases are explained below:

Damping off of seedlings: Species of *Pythium* are the major cause of damping off of vegetable seedlings in the nurseries. These fungi survive in soil through their sexually produced oospores. The initial inoculum may be high or low

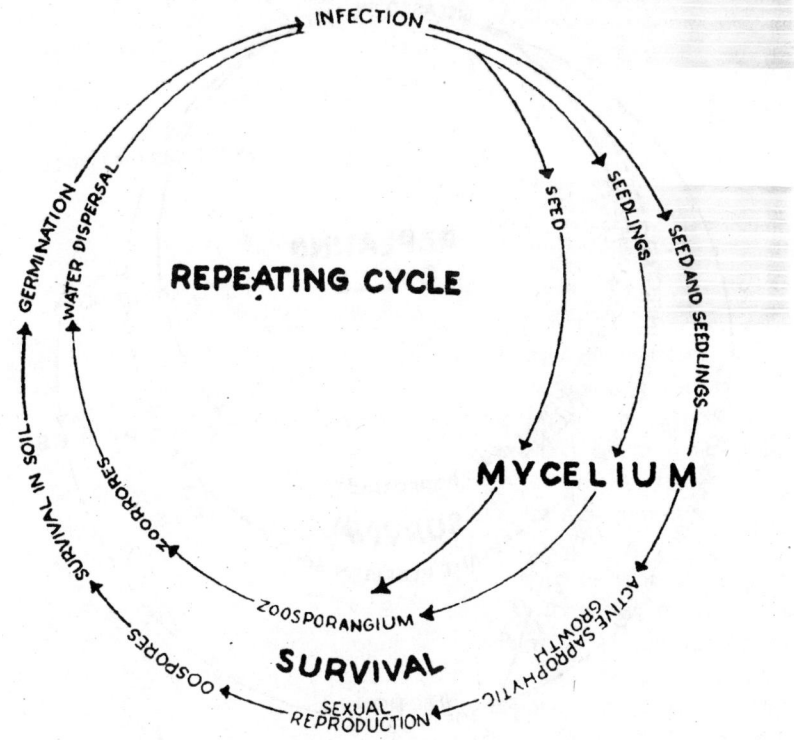

Figure 1. Disease Cycle of Damping off of Seedling (*Pythium* spp.)

depending on treatment of the nursery soil. When seeds are planted in wet soil at relatively high temperatures the oospores germinate and produce zoospores. In presence of moisture these zoospores move and infect the seeds and germinating seedlings causing pre- and post-emergence damping off. On the rotting seeds and seedlings the pathogen produces its a sexual stage with hyphae and sporangia and again zoospores. The cycles continue until either there is no more host tissue available for attack or the host develops structural resistance. The fungus hyphae soon get replaced by oospores. Since it is known that oospores in nursery soil initiate the infection the necessary treatments are taken at this stage.

Late blight of potato: *Phytophthora infestans* mainly survives through infected tubers. Sexual reproduction is known but it requires the presence of mating types in the area. Since all tubers harvested from a field are not infected and badly infected tubers generally rot before they can be used for seed, the initial inoculum of the pathogen is low. When infected tubers germinate, the mycelium moves into the stem and produces the first lesions. These bear the sporangia and zoospores. These spores are disseminated by wind, water and leaf

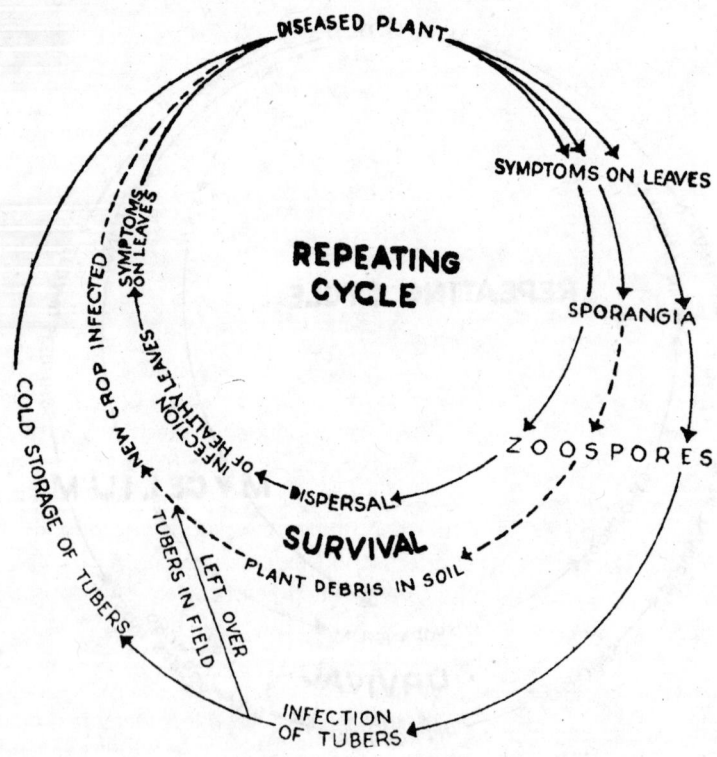

Figure 2. Disease Cycle of Late Blight of Potato (*Phytophthora Infestans*)

contact. In wet weather with suitable temperature new infections are caused and new crop of spores is produced in 7–8 days. If favourable weather prevails this 7–8 day cycle may be repeated several times during the season. Spores falling on the soil surface move through soil pores and reach and infect the tubers.

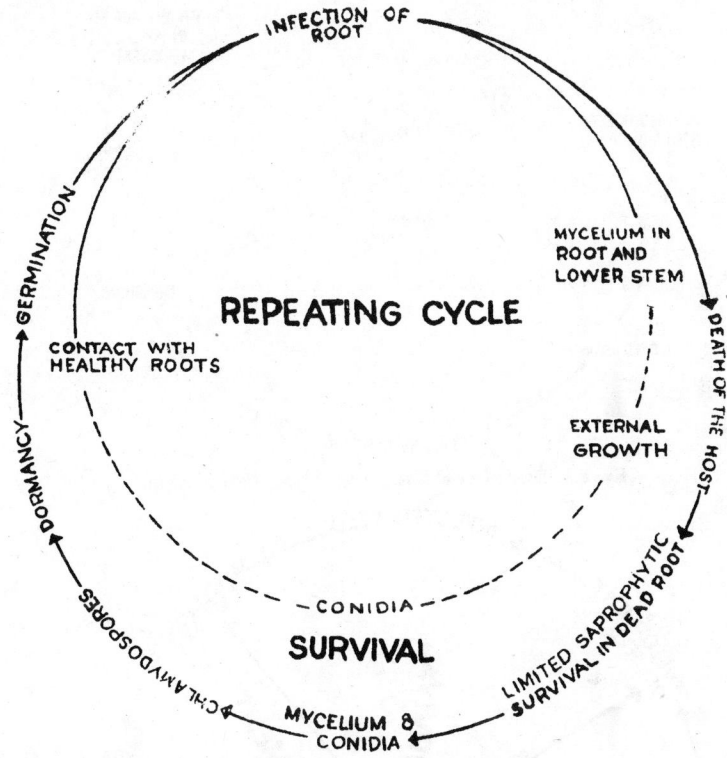

Figure 3. Disease of Cycle of Pigeonpea Wilt (*Fusarium Udum*)

Apple scab: The apple scab fungus *Venturia inaequalis* has very large amount of initial inoculum. It survives through its pseudothecia as saprophyte on fallen leaves during winter. These pseudothecia produce very high amount of inoculum as ascospores which cause infection of the emerging leaves and the disease is initiated in early spring. Infected leaves produce crops of conidia. The generation time is 8–10 days. Meanwhile fresh showers of ascospores from pseudothecia also continue. The pathogen has at its disposal 3–4 months to repeat the cycles.

Root-knot of tomato: The root knot nematodes, *Meloidogyne* spp., survive through their egg masses in debris of galled roots. In vegetable plots, where suitable host is generally present, the initial inoculum is high. Second stage larvae hatched from eggs infect tomato roots. The females develop and induce

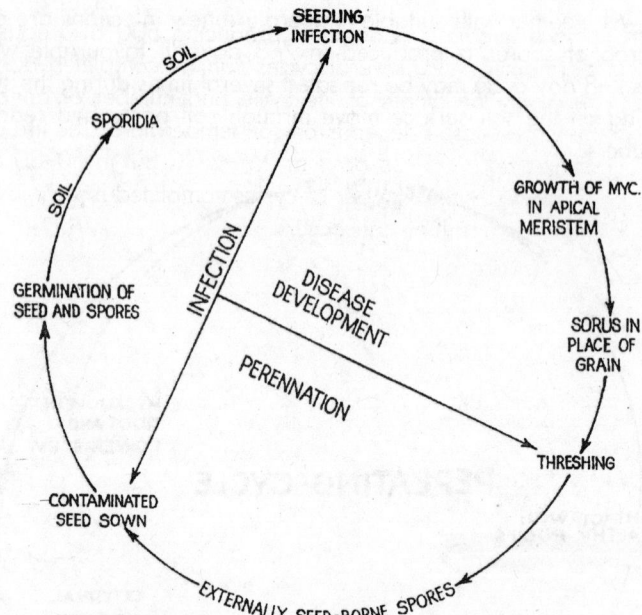

Figure 4. Disease Cycle of Covered smut of Barely (*Ustilago Hordei*)

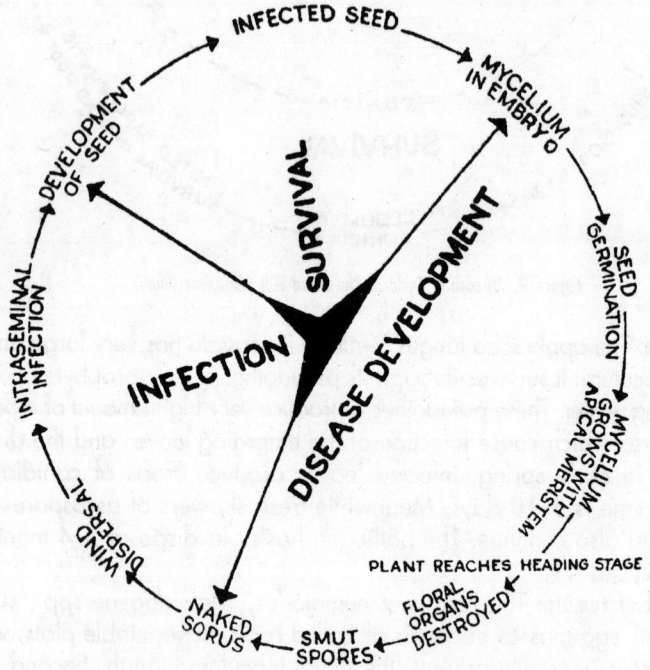

Figure 5. Disease Cycle of Loose Smut of Wheat (*Ustilago segatum tritici*)

gall formation. Each female is capable of producing 500–600 eggs which hatch without dormancy. Thus, several cycles can be completed during the crop season. However, the length of life cycle and number of generations completed in one crop season depends on soil temperature. The life cycle of *M. javaica* is completed in 21–25 days at 26°–27°C but in 50–80 days at 14°–16°C. Thus, in winter the number of cycles completed is very low while at 25°–28°C many generations can occur.

6

Survival of Plant Pathogens

In absence of their cultivated host, animate pathogens must find some alternate source of their survival to maintain the infection chain. The same holds true for viruses. These sources of survival of the pathogens or the sources for renewal of the infection chain can be grouped as below:

(i) Infected host as reservoir of inoculum
(ii) Saprophytic survival outside the host
(iii) Dormant spores and other structures in or on the host or outside the host.

The means of survival are the first link in the infection chain or the disease cycle. The initial infection that occurs from these sources in the crop is **primary infection** and the propagules that cause this infection are called **primary inoculum**. It is also possible that the source of survival (diseased material) may itself be used as seed or propagating material. Since this material is already infected, primary infection has already occurred on the propagating material. Examples are viruses of potato and sugarcane and ratoon stunting disease of sugarcane. Potato tubers from a virus infected plant are already infected. When they germinate a diseased plant is produced. After initiation of the disease in the crop, the spores or cells or other structures of the pathogen are sources of **secondary inoculum** and cause **secondary infection**, thereby spreading the disease in the field. In pathogens not producing such structures (as in viruses, MLOs residing in the phloem, or fastidious vascular bacterial pathogens) the secondary inoculum is spread by insect vectors. As an example, the fungus causing late blight of potato (*Phytophthora infestans*) survives in seed tubers. Infected tubers bring the primary infection in the field. Then the fungus produces spores on leaves. These spores are dispersed by wind and water and reach healthy plant surfaces to cause new infections. This is secondary infection. Primary inoculum of many pathogens is present in soil and causes primary infection of the crop raised from healthy seeds. The primary inoculum may also be brought by wind or irrigation water from neighbouring fields or from long distances. Then, the pathogen produces propagules for secondary infection. The primary infection initiates the diseases and the secondary infection spreads the disease.

The methods of survival or perennation differ with the type of pathogens. Therefore, the survival mechanisms are described separately for each group.

SURVIVAL OF FUNGAL PATHOGENS

The fungal pathogens have more elaborate mechanisms of survival than other pathogens like viruses, bacteria, nematodes and phanerogamic parasites. Many of these mechanisms directly or indirectly apply to other groups also.

Infected Host as Reservoir of Primary Inoculum: In plant pathology generally the host on which the fungal pathogen is overwintering or oversummering is not only infected but diseased also, i.e., it is not only a substrate for survival of the pathogen but exhibits symptoms also. If the pathogen is present in the host without producing symptoms the host is known as a **carrier** of the pathogen.

The infected hosts serving as reservoir of active inoculum can be considered in three groups, viz:

(i)   the cultivated host
(ii) wild hosts of same family as the host (collateral host)
(iii) wild hosts of other family (alternate hosts)

Cultivated hosts are seasonal or perennial. The perennial cultivated plants are good examples of the first group. Most diseases of fruit trees maintain the active or dormant primary inoculum throughout the year. Brown rot fungi of pome and stone fruits (*Monilinia* spp.) survive through cankers and dead twigs, leaves and dried (mummified) fruits left hanging on the tree. In powdery mildews of apple (*Podosphaera leucotricha*) and grapevines (*Uncinula necator*) and in downy mildew of grapevines (*Plasmopara viticola*) the pathogens survive in dormand buds and under the scales. *Botryosphaeria ribis* and *B. obtusa* that cause root rot and cankers survive in cankers on the trees. Sugarcane is an annual crop but has no dead season. The crop is found throughout the year. In red rot disease of this crop (*Colletotrichum falcatum*), although the pathogen can survive for some time in soil through resting structures (appressoria), the main source of survival and transmission is the cane cuttings taken from a diseased crop and planted immediately as a new crop.

The fungi causing rice leaf spot (*Drechslera oryzae*) and rice blast (*Pyricularia grisea*) can survive in soil and seed but under the Indian conditions these sources do not play any significant role. The primary inoculum comes as windborne spores from external sources such as early sown rice crop and some grass hosts of the pathogen. In powdery mildew of cucurbits the pathogens *Erysiphe cichoracearum* and *Sphaerotheca fuliginea* can produce perithecia as resting structures but under Indian conditions they are not common in the plains. The pathogens survive in their conidial stage in off-season cucurbits growing in shaded places or in cooler hills in the north where perithecia are also found. The conidia are wind blown to the main crop in the plains. Same is true for the powdery mildew of pea (*Erysiphe polygoni*). Viruses of cucurbits also perennate in a similar manner.

The resting structures of the cereal rust fungi (teliospores) do not survive the heat of the northern plains of India. The pathogens survive in active uredial stage on wheat and collateral hosts in the hills in the northwest and in the hills of south India. The spores are carried to main wheat crop by wind currents.

Alternate hosts are not as important as the collateral hosts. But when a pathogen has very wide host-range, cutting across many plant families, and is tolerant to wide range of weather conditions the alternate hosts become very important source of the pathogen. There are many fungi such as *Fusarium moniliforme*, *Sclerotium rolfsii* and *Rhizoctonia solani* that have rather a wide host range and thus alternate hosts may come into play in their survival.

Saprophytic Survival of Fungal Pathogens: In absence of the living host, the facultative parasites are capable of surviving as saprophytes. Soil and plant debris serve as media for this saprophytic survival. Species of *Pythium*, *Sclerotium* etc. survive in soil in absence of their host for considerable length of time. Although in this survival, resting structures like oospores and sclerotia play the major role, the ability of such fungi to attack and colonize the dead plant material enables them to remain active as saprophytes for some time and develop their mycelium and more resting spores. Saprophytic survival of such fungi is dependent on the precolonization of the substrate before the latter is approached by other microbes. Antagonism by other soil microflora reduces their ability to continue saprophytic activity unless they have strong competitive saprophytic ability (aggressive growth, production of antibiotics) and can colonize organic substrates more successfully than other soil saprophytes. Usually these fungi sooner or later from dormant structures which remain so due to existence of widespread natural fungistasis in soil. In *Pythium aphanidermatum*, causing foot rot of papaya, the fungus colonizes the host tissues which fall down to the soil where the fungus continues feeding on the dead tissues and rapidly forming oospores. These spores are the major survival structures. In *Sclerotium rolfsii*, the fungus colonizes the host, goes to the soil where it continues growing on the colonized residues. Sclerotia are formed and these can germinate under the influence of volatiles from the decaying residue. But the germlings that develop are challenged by antagonists and again form sclerotia without growing farther. But in the meantime they form more resting structures.

Another category of saprophytic survival is of those fungi that are slightly more tolerant to soil antagonism and usually predominate in the rhizosphere. The root region of plants is occupied by a large number of qualitatively and quantitatively different microorganisms some of which may be pathogens but do not necessarily infect the plant. In this region these pathogens are restricted to pioneer colonization of dead substrates and are tolerant to competition by other soil microflora. In habitats where recently dead materials are fairly continuously provided, such pathogens may survive actively and multiply in absence of the host. However, if organic matter already colonized by other microorganisms is added to the soil near the root region the survival ability of these pathogen is reduced.

The third category of saprophytic survival in soil is of those pathogens that have low competitive saprophytic survival ability and survive saprophytically for only a short time. These are described as root inhabiting fungi. They are defined as having a declining saprophytic phase which limits their survival in soil. These pathogens remain in active saphrophytic phase only so long as the host tissue in which they were living as parasites is not completely decomposed. Many vascular wilt causing species of *Fusarium* (*Fusarium oxysporum* f.sp. *vasinfectum, Fusarium udum*), *Verticillium* spp., the root rot pathogen *Phymatotrichum omnivorum* fall in this category. After the host roots have completely decomposed these pathogens are displaced by more aggressive soil saprophytes and then they live only as dormant resting structures.

The active saprophytic survival of facultative saprophytes and facultative parasites in soil is affected by soil structure, moisture, organic matter, pH, antagonism, etc. These are discussed later.

Dormant Organs of Fungal Pathogens as Source of Survival and Primary Inoculum: Among plant pathogens, fungi are the only organisms that produce spores and other resting structures for their inactive survival. In most fungal pathogens these dormant stages are the major source of survival. Since majority of plant diseases are caused by fungi, the dormant survival as a source of primary inoculum is of much importance.

The dormant structures of survival can be grouped in the following categories:

1) *Soil-borne fungi:*
   (a) Dormant spores (conidia, chlamydospores, oospores, perithecia, etc.)
   (b) Other dormant structures such as thickened hyphae and sclerotia.
2) *Seed-borne fungi*
   (a) Dormant spores on seed coat
   (b) Dormant mycelium and spores under the seed coat or in the embryo
3) *Dormant fungal structures on dormant or active host including dead plant parts remaining on the trees.*

in the plasmodial fungus, *Plasmodiophora brassicae,* (clubroot of crucifers) the pathogen survives through its resting spores lying free or in crop debris in the soil. Although no dormancy period is required for their germination they can remain viable for upto to 10 years. *Synchytrium endobioticum* (wart disease of potato) is a similar persistently soil-borne fungus in which resting spore play the important role of survival and dissemination. The fungus is one of the most persistent soil invaders. Similarly, in the powdery scab of potato (*Spongospora subterranea*) the fungus may survive through its spore balls although subsequent infection and disease development is strongly influenced by specific environmental conditions. Oospores in species of *Pythium*, *Phytophthora*, downy mildew fungi and *Albugo*, cleistothecia of powdery mildew fungi, chlamydospores of *Phytophthora* and *Fusarium* and sclerotia of *Sclerotinia*, *Rhizoctonia*, *Claviceps* and *Sclerotium* are dormant structures that can survive in soil in free state or with the crop debris. Such structures are generally formed in response to external conditions. When the substrate is exhausted and is unable to support active vegetative growth the fungi respond by forming resting structures. Thus, in downy mildew when the infected tissues start getting exhausting, oospores are formed. In ergot diseases (*Claviceps purpurea* on wheat and *Claviceps fusiformis* on pearl millet) the parasites remain active in their honey dew stage producing conidia. When the ovaries of the host are totally destroyed the mass of hyphae turn in to hard sclerotia. In vascular wilt fusaria, the fungus lives as chlamydospores formed by cells of conidia and hyphae. Metabolites of bacteria in soil stimulate their formation. However, in some cases the formation of resting structures like oopsores is the intrinsic character of the fungus. *Pythium* species saprophytically growing on a substrate take in more food than is required for their normal growth. They start producing oospores while continuing feeding on the substrate. Thus, the substrate yields little of vegetative mycelium and abundance of oospores. Often the stromatic mass of fungal hyphae, such as in Cercospora leaf spot, also serves as dormant structure of survival. Telia of rust fungi and smut spores on seed and in leaf tissues are other dormant structures of survival. Conidia of some fungi-imperfecti serve as resting structures. In *Alternaria solani* (early blight of tomato and potato) the conidia remain viable in crop debris for months. Under laboratory conditions at room temperature they have been found to remain viable for 17 months. Formation of resting structure ensures soil survival under conditions of intense antagonism because these structure are not affected by competition, and antibiosis.

The spores of smuts and many other fungi are present in dormant state on the seed surface. Examples are covered smut of barley (*Ustilago hordei*), grain smut of sorghum (*Sphacelotheca sorghi*) bunt of wheat (*Tilletia* spp.). In some diseases the fungus is present in dormant state within the seed, often in the embryonic tissues. In loose smut of wheat (*Ustilago segatum tritici*) the pathogen is carried by seeds as dormant mycelium in the embryo. In Ascochyta blight of chickpea the pathogen (*Ascochyta rabiei*) is carried as dormant mycelium and pycnidia subepidermally and also in cotyledons (cf. Singh, 1998). In Phomopsis blight of eggplant, the pathogen (*Phomopsis vexan*, perfect stage *Diaporthe vexans*) is externally as well as internally seed-borne. Pycnidia and mycelium are present in the seed coat and embryo.

In addition to soil and seed, the dormant organs of fungi can also be present on the host. Thus, in powdery mildew of apple (*Podosphaera leucotricha*), the fungus survives as dormant mycelium and encapsulated haustoria in the dormant terminal and lateral shoot buds and in blossom buds produced and infected in the previous growing season. In many pathogens of fruit trees, the fungus is found in dormant state in mummified fruits and dead twigs. The fruit mummies are formed when the fungus, such as *Monilinia* spp. in brown rot of pome and stone fruits and *Phytophthora* in guava fruits, permeates through out the flesh and exhausts the moisture. The fruit dries and shrinks. The fungus mycelium survives in dormant form.

Factors Affecting Survival in Soil: From the above description of means of survival of fungal plant pathogens it is obvious that soil is one of the very important and common medium for their active or inactive carry over from one season to the next. Seeds are the next in importance and serve as substrates for only dormant survival. While the survival in or on the seed and on the plant is affected by only temperature and moisture conditions during the survival, the survival in complex environment of the soil is invariably subjected to a wide variety of physical, chemical and biotic factors that are interrelated with each other. These environmental factors can be inimical to survival in soil or may favour survival.

Extreme cold in winter (-12°C) kills the mycelium and encapsulated haustoria of *Podospheaera leucotricha* and apple buds become free of infection of powdery mildew. Under suitable temperature and moisture conditions, in fruit tree diseases, the dead plant parts harbouring dormant fungal structures, may disintegrate and decompose allowing destruction of the pathogen.

In discussing the operation of factors inimical to microorganisms in soil it is important to recognise the distinction between activity in soil and survival in soil. Some adverse agencies cause loss of viability in the active phase of growth while others affect the inactive phase of the pathogen. Often the features inimical to active survival indirectly prolong survival of the fungal pathogens by inducing them to form dormant spores or sclerotia. The active phase is mainly restricted by high temperature, irradiation, desiccation, anaerobiosis caused by excess moisture or intense competition for oxygen under conditions of high microbial activity, toxic chemicals of biotic origin (antibiosis), lack of nutrients due to competition with other soil microorganisms for the same substrate, and finally, parasitism and predation by microflora and fauna (fungi, bacteria, protozoa, nematodes, amoeba, etc.). When these agencies affect active survival it is not necessary that the fungus is completely eliminated from the soil. Usually under the influence of these agencies (except parasitism and predation) the active vegetative mycelium forms spores (conidia, chlamydospores, sclerotia) which carry on the survival in dormant stage. These structures are tolerant to the physical and chemical soil factors but not to biotic factors. There are numerous reports of amoebae perforating fungal conidia and amoebae or fungal hyperparasites destroying sclerotia in soil. Senescence of the inactive stages may be expected to occur in soil and may account for a progressive decline in the number of viable propagules and thereby the population of the pathogens incapable of multiplying in absence of their host. A more important cause of decline in the number of dormant survival structures in cultivated fields is inducement of their germination, formation of vegetative hyphae and then their destruction by forces mentioned above as inimical to active stages. Lysis of hyphae by saprophytic bacteria is most common. The factors that induce germination are those that help in breaking down the natural fungistasis which prevents the germination or activation of dormant structures. Thus, application of nutrients to soil, exudates of roots of the host and, sometimes, nonhost crops, physical treatment of the soil to reduce microbial antagonism are factors that stimulate germination of spores thereby subjecting them to antagonistic elimination. The factors that favour survival are continuous cultivation of plants that help in multiplication and reproduction of the pathogen in the soil, thus continuously adding fresh spores, soil fungistasis that prevents germination of spores thus preventing them from untimely germination and continued physical environments favourable for the pathogen.

SURVIVAL OF PHYTOPATHOGENIC BACTERIA

Phytopathogenic bacteria do not form resting spores or similar structures except in such opportunistic genera as *Bacillus* and *Clostridium* which form resting endospores or in *Streptomyces scabies* which forms conidia. Most bacteria have very limited soil survival even in debris of the infected host if it is buried deep. They mostly survive in mild or vigorous form on the host.

Survival with Seed: Association of phytobacteria with crop seed is known since 1909. Bacterial cells are not subject to dormancy and their survival corresponds to seed viability. However, some pathogens die before the seed loses its viability (cotton blight bacterium) while many survive beyond the longevity of seed such as in legumes. Most species of *Xanthomonas* and *Clavibacter* are seed-borne having only limited soil phase, that too only in association with crop debris. Depending on seed storage conditions, *X. axonopodis* pv. *phaseoli* (common blight of bean) can survive in bean seed for 13–15 years. *X. vesicatoria* can survive in pepper seed for 10 years. Other xanthomonads that have prolonged survival in seed are *Xanthomonas oryzae* pv. *oryzae* (leaf blight of rice), *X. oryzae* pv. *oryzicola* (leaf streak of rice) and *X. campestris* pv. *campestris* (black rot of crucifers). Cotton blight bacterium (*X. axonopodis* pv. *malvacearum* survives in cotton seed for only one year. *Clavibacter michiganensis* subsp *michiganensis* (canker of tomato) and *C. michiganensis* subsp, *insidiosum* in alfalfa seeds survive for 3 years. Most of the leaf blighting pseudomonads are well known seed-borne pathogens. *Pseudomonas syringae* pv. *tomato* (bacterial speck of tomato) can survive in tomato seed for 20 years and remain infective. Seed and weed hosts are only means of its long term survival. Often weed seeds are an important source of survival and transmission of phytobacteria.

The level of seed infection or contamination varies with crops and environmental conditions. However, even a low level of seed-borne inoculum, if successful in initiating the disease in the crop, can lead to serious epidemics under favourable weather conditions because of the high rate of multiplication of bacteria and their easy dissemination. Even 12 infected seeds of bean per acre, carrying *Pseudomonas savastonoi* pv. *phaseolicola* (halo blight of bean) can lead to an epidemic in the crop (*cf.* Singh, 1989).

Success of seed-borne bacteria is dependent on their location in the seed. Access to vascular elements promotes survival and subsequent pathogenesis of many bacteria. The testa of legume seed contains vascular elements that may be extensive in larger seeds. Once the bacteria gain entry in the testa, through hilum, micropyle or threshing injuries, they have easy access to vascular elements in such seeds. In bean, cotton, cucurbits and tomato the vascular tissue (raphe), a continuation of the funiculus, provide favourable sites for internal spread of the bacteria. They may also invade the embryo.

Survival in Plant Residue: The survival of phytobacteria through diseased crop debris is mostly dependent on the speed with which the debris decomposes. Thus, when the debris is buried deep survival is less. In wet areas the debris decomposes early limiting the survival of bacteria. Plant pathogenic bacteria are weak competitors and have high susceptibility to the activities of microorganisms associated with decomposition of debris. Under experimental conditions survival is better in sterilized than in unsterilized soil.

Coryneform bacteria are not considered soil-borne pathogens but some apparently survive a year or more with plant debris. *Clavibacter michiganensis* subsp. *insidiosum* survives in dry alfalfa stem left on the soil for 10 years. *Curtobacterium flaccumfaciens* over-winters in temperate regions in host debris and also on straw of nonhost plants if left on the soil surface. Burying of the debris hastens its mortality. The capacity of *Clavibacter michiganensis* subsp. *michiganensis* to survive in plant debris is well known. In infected stem pieces in soil it survives for 3 months in a granite sandy loam and for 7 months in a clay loam soil. At constant temperature of 22°C and low soil moisture it can survive for 11 months (Mofett and Wood, 1985).

Species of *Xanthomonas* have the poorest ability to survive outside the host. In crop debris in soil they show a rapidly declining phase depending on the rate of decomposition of debris. In arid regions where decomposition of debris is slow, *X. axonopodis* pv. *malvacerum* can survive for long in cotton residue but in wet regions the survival may not be even up to the next sowing season. In India, the bacterium is reported to survive for 6 months on soil surface but for only 3 months if the debris is buried 15 cm deep. A similar situation exists in *Xanthomonas axonopodis* pv. *citri* (citrus canker). Lesion bearing fallen leaves and twigs on soil surface under wet conditions or if buried 3–6 cme deep in soil support viable cells only for 2–3 weeks but if the soil is kept dry the survival may be for 2–3 months. *X. campestris* pv. *campestris* survives in infected cabbage stem tissue for 244 days (*cf.* Singh, 1989). The rice leaf blight bacterium can survive on rice straw in dry but not in moist conditions for some months. The species of *Pseudomonas* behave much in the same manner as the species of *Xanthomonas* in their survival with plant debris. The persistence of *Ralstonia solanacearum* (potato and tobacco bacterial wilt) in infested debris depends on the environment since either desiccation or the antagonistic effect of secondary invaders decreases their population rapidly.

Survival in Soil: The bacterial plant pathogens inhabiting the aerial parts of the host reach the soil through diseased plant debris, raindrop or sprinkled irrigation washings and dry flakes of the bacterial ooze displaced by wind and movement of the plant parts. The bacteria that mainly colonize the underground plant parts such as tubers, bulbs and roots are released in the soil with disintegration of these parts. Most bacteria do not survive well in the soil when separated from the plant tissue. Different categories of soil-bacterium relationship have been described (*cf.* Singh, 1989). **Transient visitors** consists of bacterial species whose populations are developed almost exclusively on the host plant where maximum number of generations are produced. When such bacteria reach the soil with rain water or with debris the populations decline rapidly and do not remain important source of primary inoculum for the next season. Being poor competitors most phytopathogenic bacteria fall under this category. Examples are most species of *Xanthomonas*, non-soft rot species of *Erwinia*, and *Clavibacter michiganensis*.

The bacterial species in the category of **resident visitors** also have their maximum generations in the host but their numbers only gradually decline in soil. If the populations enter the soil at a sufficiently high rate, as usually happens in case of underground bacterial diseases, the slow decline would permit net increase of such bacteria from season to season. The long term persistence of such bacteria in soil is host dependent and their increase, decrease or total extinction depends on the cropping practices. Pathogens with extended soil phase include *Agrobacterium tumefaciens* (crown gall) and *Ralstonia solanacearum* race 1 (bacterial wilt of solanaceae), *Streptomyces scabies* and *Streptomyces ipomoea*. These can be considered true soil-borne pathogens or soil inhabitants. *R. solanacearum* can survive in soil in free state under bare fallow for 4–6 years and upto 10 years in soil cropped to non-host or non-susceptible crops. Susceptible weed hosts prolong its soil survival.

There are very few plant pathogenic bacteria which are truly saprophytes with permanent soil phase. The group of bacteria classified as soil saprophyte is typified by those whose populations are largely produced in soil, including the rhizosphere and whose relation to plant disease is only ephemeral. The examples are rhizosphere bacteria, the green fluorescent *Pseudomonas* causing soft rots, the species of *Bacillus* pathogenic on plants and pectolytic soft rotting

Clostridium which are opportunistic pathogens.

Survival in Perennial Hosts: Different organs of the same plant and different plant species have characteristic microflora, including bacteria, on their surfaces and these live on the plant exudates. The surface flora are known as **epiphytes** and are present on roots **(rhizoplane)**, buds **(gemmiplane)** and leaves **(phylloplane)** as well as on seed surface. The characteristic flora generally exists in a balance. There is evidence that some of the epiphytic bacteria are antagonistic to pathogenic bacteria (*cf.* Singh, 2001) but often, the latter constitute a dominant component of the surface microflora and could be a source of primary inoculum.

In citrus canker (*Xanthomonas axonopodis* pv. *citri*) the bacterium survives in cankers on the infected trees. It has only limited survival with fallen leaves in the soil. However, there are reports that the bacterium colonizes the exposed as well as underground roots where it can survive for long (*cf.* Singh, 2000). In the bacterial canker of stone fruits (*Pseudomonas syringae* pv. *syringae*) the bacteria survive in cankers, buds, and systematically inside other host tissues with no symptoms. The bacterium is a good epiphyte and survives on foliage of weeds growing in the orchard. Similarly the bacterium of fire blight of apple and pear (*Erwinia amylovora*) survives in dead twigs and as epiphyte in the buds of apple during the dormancy of the tree.

Epiphytic populations of *Pseudomonas savastonoi* pv. *phaseolicola* on bean and of *Xanthomonas vesicatoria* on tomato have been reported. *X. axonopodis* pv. *phaseoli* var. *fuscans* (fuscus blight of bean) survives on phylloplane of primary leaves but not on unifoliate leaves of bean. *X. axonopodis* pv. *malvacearum* is transmitted from germinating infected seed to the cotyledons and then a resident population of the bacterium is found only on the first and second leaves. This resident population serves as primary inoculum of susceptible leaves under favourable weather conditions (Moffett and Wood, 1985). The epiphytic survival is influenced by host resistance also. *Curtobacterium flaccumfaciens* pv. *flaccumfaciens* survives epiphytically better on susceptible leaves of *Phaseolus vulgaris* than on resistant leaves *X. axonopodis* pv. *malvacearum* is reported to survive in the rhizoplane of more than a dozen weeds that grow in cotton fields. *X. vesicatoria* can also survive in rhizoplane of host (pepper) and non-host plants. The cucurbit angular spot bacterium (*Pseudomonas syringae* pv. *lachrymans*) survives in rhizosphere for four growth cycles in the same soil and maintains pathogenicity to cucurbit seedlings (*cf.* Singh, 1989). *Erwinia carotovora* subsp. *carotovora* and subsp. *atroseptica* have been isolated from rhizosphere of numerous cultivated and non-cultivated plants from agricultural and non-agricultural areas in diverse geographic regions.

Association with Insects: Some bacterial plant pathogens use insects as their host and source of survival and dispersal. The potato black leg bacterium, *Erwinia carotovora* subsp. *atroseptica* can live in all stages of the seed corn maggot (*Hylemya platura*) and persist in the intestinal tract inspite of its ability to survive through tubers and soil. Since the pathogen survives pupation the emerged adults may contaminate eggs as they are laid. A similar situation exists in soft rot of crucifers and cabbage maggot (*Hylemya brassicae*). Fruit flies (*Drosophila melanogaster*) also provide short term survival to erwinias. The corn wilt bacterium (*Pantoea stewartii* subsp. *stewartii*) lives through winter in the body of cucumber beetle (*Diabrotica undecimpunctata*) and corn flea beetle (*Chaetocnema pulicaria*). It not only survives in these beetles but is also distributed over long distances. The cucurbit vascular wilt bacterium *Erwinia tracheiphila* is totally dependent on cucumber beetles (*Acalymna vittatum* and *Diabrotica undecimpunctata*) for its survival between seasons. This information has been used to forecast the incidence of the disease. During acute winters the beetles are killed. Their prevalence in the next season can foretell the extent of incidence of the disease.

SURVIVAL OF PLANT PATHOGENIC NEMATODES

Some nematodes maintain continuity of infection chain through active parasitic living on variety of hosts while others, majority of the phytophagous nematodes, survive through their dormant structures (eggs, cysts, galls, cockles formed from host tissue). These structures are formed in the normal course of the life cycle. Eggs, cysts and galls are present in soil and sometimes in the seed lot.

Quiescence is an adaptation for survival of nematodes under adverse conditions. The nematodes become dormant or anabiotic. In this condition they can live for many years. The best example of survival through quiescence is that of *Anguina tritici*, the wheat gall nematode. The nematode has only one cycle in a season. The eggs hatch in the seed galls and with onset of summer (in India) when the crop is ready for harvest the larvae become quiescent. They are activated only when the galls are soaked in water. Thus, in a cultivated soil where irrigation is periodically given nematode cannot survive for long in soil. But in quiescent state the second stage larvae of *A. tritici* are known to remain viable for up to 28 years in cockles. Quiescence due to dessication has also been reported in *Ditylenchus dipsaci* (fourth stage larvae), *D. phyllobius*, *Tylenchus balsamophilus*, *Aphelenchoides ritzemabosi* (adults) and *Anguina agrostis*.

In root knot (*Meloidogyne* spp.) the nematode has a very wide host-range in Solanaceae, Cucurbitaceae, Cruciferae and Malvaceae. One or more of the vegetables from these plant families are present throughout the year and hence the nematode can find suitable hosts for survival if susceptible vegetables are grown in rotation in the same field. The larvae that hatch from eggs early in the season find the host easily and cause infection. Those that are hatched late in the season live for some months in the soil provided the temperature does not rise very high. These larvae can cause infection of the next crop. Eggs are generally protected in the galls. This also gives some chance of survival.

The cyst nematodes are host specific and mainly survive through their cysts for many years. Since some hatching factor from host roots is required for release of larvae in some species, the cysts may retain viable larvae in the absence of a host for considerable time. The cysts of *Globodera rostochiensis* (potato cyst nematode) continue to contain viable larvae for 7–8 years in absence of the host. Even in presence of the host, the cysts may continue releasing viable larvae for several years because all the larvae in the cyst are not released in the first year.

SURVIVAL OF PHANEROGAMIC PLANT PARASITES

Phanerogamic parasites produce seeds, just like any other flowering plant, and through these seeds they can live in dormant stage, sometimes, for years. The parasites attacking perennial plants survive in the active state. In *Dendrophthoae falcata* (giant mistletoes), the parasite survives and is transmitted through seeds and vestiges of its haustorial system on the trees. In *Orobanche ramosa* (broomrape), seeds of the parasite are present mostly in the top 5 cm of soil and remain viable if the soil does not dry. The seeds of dodder (*Cuscuta* spp.) fall on the ground and remain viable in dormant state until a favourable season returns. The witchweed (*Striga* spp.) survives through its seeds which require a minimum dormancy of 15–18 months. Each plant of witchweed produces 50,000 to 500,000 seeds and all the seeds do not germinate in the same season even in presence of a host.

PERENNATION OF PLANT VIRUSES

Plant viruses have no dormant stage and maintain a continuous infection chain. They are actively present in the crop host or, in its absence, in some collateral or alternate host. Thus, all viruses of potato are carried by tubers. Viruses of all fruit trees are maintained on the tree. The only exceptions are the viruses vectored by fungi and nematodes. The viruses transmitted by some fungi are intimately associated with their vector and remain viable so long as the fungus vector is living. Same is true for nematode vectors although in this case period of retention in the vector is not long.

7

Dispersal of Plant Pathogens

The dispersal, transport or transmission of the pathogen or the disease is important not only for spread of the disease in the population but also for continuity of the life-cycle and evolution of the pathogen. The living host or other organic substrates cannot indefinitely provide space and nutrition for the growing population of the animate pathogens. This compels them to move out to new, hitherto uninhabited, sites otherwise the pathogen would die due to starvation resulting from exhaustion of the host and due to overcrowding. Nature has provided the pathogens the necessary mechanisms for exit from the host and move to new locations. In fungi, production of asexual and sexual spores follows the active vegetative growth in or on the invaded host tissues. These are dispersed mechanically in time and space by various means. In bacterial diseases, the bacterial cells come out on the host surface as ooze or the tissue may be disintegrated to such an extent that the bacterial mass is exposed and then dispersed by various physical and biological agencies. Viruses have no such organs. They are transmitted by insects and man. Nematodes are themselves motile and try to move out of the exhausted and overcrowded sites and then may be moved with soil, water and other means.

Pathogens of all infectious diseases are transmissible, i.e., they are carried by various means from the source of survival or from a diseased individual in the population to the healthy individual. Some diseases may even be **contagious** i.e., they can be transmitted by contact between diseased and healthy plants. Many mechanisms described for survival of pathogens apply to dispersal of pathogens also. In general the seed-borne pathogens or diseases are dispersed by the movement of seed, soil-borne pathogens through movement or displacement of soil and through root contacts, and those surviving on wild hosts (collateral or alternate hosts) in active stage through wind-borne spores, other structures and insects. However, in majority of diseases a combination of mechanisms, aided by external physical and biological forces, operates in dispersal (as well as survival). The same pathogen may be carried by soil as well as by seed and its transport may be aided by water, wind, insects, etc.

The two links in the infection chain of an animate pathogen, viz., survival through dormant structures and dispersal of the pathogen are very closely bound with each other. Actually, the dormant structures of fungi provide means of **dispersal in time**, i.e., the pathogen is retained in a viable condition over a period of time enabling it to be transported through physical agencies without being harmed. In lower fungi, many pathogens are transported only through their resting structures in soil. The resting structures of fungi on seed or in plant debris are other examples. Among soil fungi, fungistasis provides a large, well distributed reserve of inactive spores able to colonize new substrates quickly when available.

The dispersal of pathogens is accomplished in the following manner:

1. Direct (active or autonomous) dispersal, such as dispersal in and by soil and by seed and planting materials.
2. Indirect (passive) dispersal involving the role of man, insects, nematodes and other animals, water and air.

The knowledge of these methods of dispersal is essential for effective management of plant diseases because possibilities of preventing dispersal and thereby breaking the infection chain exist.

AUTONOMOUS DISPERSAL

Autonomous dispersal of bacteria, fungi, viruses and nematodes is accomplished through the agency of soil, seed and plant organs during normal agronomic operations. There is no primary role of external agencies like insects, wind, water etc. in this type of dispersal.

Soil as a Means of Autonomous Dispersal: Pathogens may survive through soil. This is specially true for soil-borne facultative parasites or facultative saprophytes. The dispersal may be by movement of the pathogen in the soil or by its growth in the soil (**dispersal in soil**). Both are rare possibilities because plant pathogens can rarely compete with other soil microbiota and move far. They generally form resting structures (oospores, chlamydospores, sclerotia) not only on the host but also in the soil when under stress. Movement of the soil containing the pathogen (**dispersal by soil**) is more common method.

Among plant pathogens only nematodes are capable of locomotion and can bodily move in soil. The nematode larvae are capable of wriggling through pores spaces having some water and reach their preferred host to cause infection. They have some ability to migrate towards higher moisture content, optimal temperature and chemicals released by the host roots (chemotactic response). There is no evidence to suggest that nematodes show any response in their active movement to soil texture, light and gravity. Limitations are the optimum moisture in the pores and diameter of the pore. It must be enough to allow the body of the larvae pass through. However, this active movement of nematodes is not of much epidemiological importance because when measured in cms it is few cms to less than a meter in several months. But the chances are that within this space they may encounter their suitable host and then the speed of movement is accelerated under the influence of root exudates. The ability of locomotion in nematodes mainly helps in finding a suitable site on the host surface for penetration.

Many fungi such as species of *Pythium* and *Phytophthora* form zoospores provided with flagella for locomotion and with this facility they can swim in films of water present in the soil. During rains, the zoospores of *Pythium aphanidermatum* or *Phytophthora parasitica* in papaya plants affected with root rot, collar rot or stem rot are capable of swimming in water and reaching healthy plants to cause fresh infections. This results in rapid spread of the disease in papaya plantations. In collar and crown rot and gummosis of citrus caused by *Phytophthora parasitica*, *P. palmivora* and *P. citrophthora* the zoospores move in soil with water in irrigation channels to long distances. If the soil does not dry between irrigations their movement is continued and new roots are infected. Long intervals between irrigations cause the soil to dry sufficiently to prevent movement of zoospores.

Only fungi and, perhaps, actinomycetes can grow to new sites from the site of survival or activity in the soil. However, in absence of the host and precolonized substrates, soil environments are highly inimical to active survival and growth. Thus, the growth of the pathogenic fungi in soil is non-existent or extremely slow. Most of the serious soil-borne pathogens such as *Fusarium solani* f.sp. *phaseoli* (root rot of bean), *Fusarium oxysporum*, species of *Pythium*, *Sclerotium rolfsii*, etc. grow in soil to only a limited extent and are, therefore, immobile. The root inhabiting fungi (wilt causing species of *Fusarium* and the root rot pathogen, *Phymatotrichum omnivorum*, with low saprophytic competitive ability cannot grow far from the roots of their host. For practical purposes, it can be assumed that their range of spread is limited to roots that are adjacent or in contact with the infected roots. Some pathogens such as *Sclerotium rolfsii* grow to some distance depending on the size of the food base from which they have originated and also on the extent to which the soil antagonism permits them to grow. Under the influence of antagonism lysis of the hypha sets in and the fungus produces secondary sclerotia. All such fungi use soil mainly as sites for survival. Another example of dispersal in soil through growth is that of the rhizomorphic fungus (*Armillaria mellea*) which produces strands of interwoven hyphae (rhizomorphs) protected by a covering of modified hyphae that enable them to grow in soil unharmed and reach uninfected roots at a distance. In Armillaria root rot of citrus, the rhizomorphs must remain connected to the infect roots (food base) for their growth through soil. However, they can obtain nutrients from soil also rich in organic matter (cf. Singh, 2000). These rhizomorphs can be fragmented and distributed in the field during tillage operations. The pathogen produces antibiotics that prevent any harm to it by antagonistic soil microflora such as *Trichoderma*. It is only when the pathogen is weakened by some treatment that this antagonist effectively checks it.

Soil particles laden with bacterial cells or plant residue containing bacterial cells are splashed by rain drops. They reach the foliage, dry and protect the bacteria from desiccation. The soil particles can also be blown by wind and disseminate the pathogen over various distance. The movement of bacteria through the soil (dispersal within soil) can occur through any agency than can move the soil particles. This includes water, different cultural operations, movement of nematodes and earthworms and even fungus hyphae growing in the soil.

Wind blown soil particles as dust are known to carry such dormant structures as cysts of nematodes and resting structures of some fungal pathogens. In Rajasthan, the cereal cyst nematodes (*Anguina tritici*) leave cysts in the soil after crop harvest. There are dust storms during summer. This distributes the cysts with dust not only locally but also to distances for away. In *Synchytrium endobioticum*, the resting spores, whether in crop debris or free in soil, are main survival structures. In USA and Canada, the disease is generally confined to home gardens with small plots which optimize conditions for the pathogen through monoculture, infected seed tubers, and crop debris left on soil surface after crop harvest as well as potato peels on compost heaps. Resting spores of the pathogen have been detected in dust on windscreen of automobiles exiting from such areas (Hampson, 1996).

From the above description of dispersal in or by soil the following three stages of dispersal become evident:

(i) *Contamination of soil:* The soil receives the pathogen by gradual spread of the pathogen from an infected site to new sites, by introduction of contaminated soil into new pathogen-free soil and by transfer of infected plant debris or seed.

(ii) *Growth and spread of the pathogen in the soil:* Once the pathogen has reached soil it can grow and spread in the soil depending on several factors. Except actinomycetes, other plant pathogenic bacteria rarely grow or multiply in soil. Phytonematodes also are obligate parasites and do not multiply outside or away from their host. Only fungi have some possibility of growing in soil. The factors that determine the possible multiplication and spread are specific characters of the pathogen, presence of susceptible hosts, and the modern technology of intensive cropping. Among characters of the pathogen its adaptability to soil environment including saprophytic survival ability are most important. Survival ability, in turn, is governed by such factors as high growth rate and rapid spore germination, good enymatic complement and potential, capacity to produce antibiotics and tolerance to antibiotics produced by other soil microorganisms. On the basis of this competitive survival ability, the plant pathogens in soil can be specialized facultative parasites, unspecialized facultative parasites and obligate parasites. The non-specialized facultative parasites (*viz. Pythium* and *Rhizoctonia*) can pass their entire life in soil. Specialized facultative parasites (or saprophytes) can pass their life in soil in absence of the host but depend more on the residue of their host plant. Examples of such pathogens are *Armillaria mellea, Ophiobolus graminis, Phymatotrichum omnivorum* and most other root inhabiting fungi. The soil-borne obligate parasites in this category are few such as *Plasmodiophora brassicae* and *Synchytrium endobioticum*. Nematodes and viruses also come under the same category. For the activity of these obligate parasites the presence of active host (or soil borne vector in case of viruses) is essential.

(iii) *Persistence of the pathogen:* This aspect of dispersal has already been discussed under survival.

The above three stages help a pathogen to enter the soil and it may grow and **spread in the soil.** However, since the soil is likely to be disturbed and displaced they are mostly **spread by the soil.** During cultural operations in a field soil is moved from one point to another within the field by agricultural implements, workers' feet, erosion, etc. Propagules are thus spread throughout the field. There is evidence that when in attempts to pulverise the soil excessive tilling is done, sclerotia of fungi and root bits containing roots pathogens are distributed all over the field. Movement of farm implements and animals from one field to another or from one farm to another may likewise transport the contaminated soil and contaminate the pathogen-free soil. However, this type of dispersal is highly erratic.

The most important method of dispersal of pathogens by the soil is transfer of soil from one place to another along with plant parts and propagating materials. Thus, transfer of papaya seedlings from a nursery infested with *Pythium aphanidermatum* or *Phytophthora parasitica* (stem and foot or root rot) can introduce the pathogens in new pits for transplanting of seedlings. Grafts and seedlings of fruit trees transported with soil around their roots can transport pathogens present in the nursery to the orchards. Root rot and gummosis of citrus (*Phytophthora parasitica, P. palmivora* and *P. citrophthora*) and collar rot of apple (*P. cactorum*) are examples. Most commercial nurseries are infested by these pathogens. Dispersal or spread of root knot to fields where vegetables had not been grown or of clubroot of cabbage (*Plasmodiophora brassicae*) by use of seedlings raised in infested nurseries is a common phenomenon. Soil adhering to tomato seedlings is one of the most important means of introduction of root knot in new fields. By this method, pathogens are spread not only locally but also to far distant places.

Seed as Source of Autonomous Dispersal: The seeds serve as medium of autonomous dispersal of pathogens in the same manner as the soil with the difference that seeds are dormant structures of the plant and rarely a fungus or any other pathogen survives or exists in active form on them. Since most of the cultivated field crops are raised from seed the transmission of disease and transport of pathogens by seeds has significant importance for the farmer and also for the plant pathologist.

The dormant structures of the pathogen (*viz.* seeds of *Cuscuta*, sclerotia of ergot fungi, cereal grain cockles and cysts containing nematode larvae, smut balls, etc.) can get mixed with seed lot and get dispersed as seed contaminants. The bacterial cells and spores of fungi can be present on the seed coat, such as in smuts of barley, sorghum, etc., and be transported to long distances. Similarly, the dormant mycelium and spores may be present in the interior of the seed and go undetected and transported.

As in the soil-dispersal, the first stage in dispersal by seed is the contamination of the seed. In the dispersal of the weed *Cuscuta* the seeds of the parasite and the host get mixed during harvest of the crop. In many crops the identity of the seeds of the two entities is difficult to separate. In ear cockles of wheat (*Rathayibacter tritici*), smut of pearl millet (*Tolyposporium penicillariae*), ergot of cereals (*Claviceps purpurea*) and ergot of pearl millet (*Claviceps fusiformis*) it is easy for the cockles, smut sori and ergots to mix with the seed lot during harvest and threshing. In many smuts such as Karnal bunt of wheat (*Neovossia indica*), bunt of rice (*Neovossia horrida*), leaf smut of rice (*Entyloma oryzae*) the infected kernels and pieces of leaves bearing smut sori are mixed with the seed. In many downy mildew diseases (*Peronosclerospora, Sclerospora*) and whire rust (*Albugo candida*) the pathogens produce oospores in tissues of pods and these may go with the seed if the latter are not cleaned. In all these examples the pathogens move with the seed lot as separate contaminants without being in intimate contact with the viable seed.

Close contact between structures of the pathogen and seeds is seen in diseases like loose smut of wheat and barley, covered smut of barley, and sorghum, stinking smut of wheat, Phoma blight of eggplant (*Phomopsis vexans*), Ascochyta blight of chickpea (*Ascochyta rabiei*), bacterial blight of cotton (*Xanthomonas axonopodis* pv. *malvacearum*), bacterial leaf blight of rice (*Xanthomonas oryzae*) and many bacterial diseases of beans where the pathogen gets lodged on the seed coat during growth of the crop or at the time of harvest or processing of seed through dust. It may penetrate the seed coat. In loose smut of wheat (*Ustilago segatum tritici*) there is blossom infection and the fungus penetrates the embryonic tissue where it becomes dormant. In Phoma blight of eggplant the pathogen enters the seed in infected fruits to different depths, even upto the embryo. In Ascochyta blight of chickpea also a similar situation is seen. In black arm and blight of cotton (*Xanthomonas axonopodis* pv. *malvacearum*) the bacteria in infected bolls move along the fuzz and reach the seed. During delinting process the bacteria from the fuzz contaminate the healthy seeds also (secondary contamination). In bacterial leaf blight of rice (*Xanthomonas oryzae* pv. *oryzae*) the bacteria are present in the seed which may produce diseased plants in the nursery (the kresek phase) and spread the disease in the field. The infected seeds also release the bacteria into the soil for soil dispersal. Bacterial pathogens of bean may be transferred to healthy seeds with the dust during threshing.

More than 100 plant viruses are transmitted by true seed. Seed transmission is not necessarily a quality of the particular virus. The same virus may be seed-borne in one host and not in others. Most nematode-transmitted polyhedral viruses show low to high percentage of seed transmission. There is no phloem connection between embryo and the mother plant. Thus, phloem limited viruses and also the MLOs that are transmitted only by phloem feeding leafhoppers cannot reach the embryo and are not seed transmitted. Even if they persist in the seed coat (having vascular connection with the mother plant) they do not infect the seedlings because they are not mechanically transmitted. The viruses that are carried by embryo persist as long as the seed remains viable. Seed transmission of viruses also occurs on seed tissues outside the embryo. Immature seed coat, consisting of the integuments and nucellar remnants and also the perisperm are part of the mother plant and subject to systemic invasion by viruses. However, viruses susceptible to desiccation are not transmitted in this manner.

TABLE 1: Some Seed and Pollen Transmitted Viruses.

Virus	Host	Seed transmission (%)	Pollen = transmission
Arabis mosaic	Capsella bursapastoris	6–33	No
Cacao necrosis	Phaseolus vulgaris	1–24	No
Cherry rasp leaf	Chenopodium quinoa	10-20	Yes
Mulberry ringspot	Soybean	10	No
Raspberry ringspot	Strawberry	50	No
Strawberry latent ringspot	Mentha arvensis	70	No
Tomato ringspot	Soybean	76	No
Cowpea mosaic	Cowpea	1–5	No
Cowpea mottle	Cowpea	10	No
Squash mosaic	Cucurbits	1–94	No
Alfalfa mosaic	Lucerne	10–50	Yes
Bean yellow mosaic	Pea	10–30	No
Bean common mosaic	Beans	83	Yes
Cowpea aphid-borne mosaic	Cowpea	0–20	Yes
Cucumber mosaic	Cowpea	4–28	No
Pea seed-borne mosaic	Pea	30–90	Yes
Soybean mosaic	Soybean	0–68	No
Barley stripe mosaic	Barley	15–100	Yes
Cucumber green mottle mosaic	Cucumber	5–8	Yes
Tomato mosaic	Tomato	upto 94	Yes

Apart from some viruses, the potato spindle tuber viroid is also transmitted by pollen although it is not epidemiologically of any importance.

The growth and spread of the pathogen in seed-dispersal depends on weather conditions for development on the seed coat or for embryonic infection and the frequency of use of infected or infested seed in the same field. If contaminated seeds are repeatedly used build up of inoculum continues and the pathogen gets well established in the field, well distributed in the locality and in the subsequent seed lots produced in the locality.

Persistence of the pathogen in the seed was discussed under survival. Although the seed may not remain viable for so many years, pathogens being carried by them are known to remain viable for surprisingly long period protected by the seed coat. In many pathogens even the externally seed-borne structures such as smut spores can persist for many years due to their inherent capacity for long survival. The spores of *Tilletia tritici* (stinking smut of wheat) remain viable even after 18 years and those of *Ustilago avenae* in oats for 13 years.

Seeds of cultivated crops and some other plants are distributed mainly by man and sometimes by other animals. Thus, these are the main agencies of dispersal of the pathogens through seed. By getting mixed with the seed or establishing itself in or on the seed the pathogen manages its own transport. The distinction between such terms as seed-infection, seed-infestation and transport or transmission by seed must be recognized in this context. As the term "infection" implies that the seed is infected only when the pathogen has grown in or on it and has established its relationship with the seed tissue. On the other hand, in all those examples where the fungus or other pathogen is present on the seed coat and in the seed lot it is only transport of the pathogen and the seed is infested.

The Plant and Plant Organs as Means of Autonomous Dispersal: The third method of autonomous dispersal consists of plants, plant parts other than true seed that are used for vegetative propagation, raw garden produce and plant debris that accumulates during the course of cropping. The classic examples are almost all the virus and viroid diseases of vegetatively propagated crops (potato, banana, citrus, etc.), late blight of potato and citrus canker. All fastidious prokaryotes in vegetatively propagated plants are also transmitted by propagating material.

The late blight of potato (*Phytophthora infestans*) originated in Central Mexico where potato was first domesticated. The host and the pathogen co-existed in that area. The migration of the pathogen outside Mexico first took place in 1842 to the USA and probably simultaneously to Europe. The pathogen is seed-tuber borne. The tuber-borne inoculum is destroyed by constant exposure to high temperature for several days. During the period when quick means of ocean transport were not available the tubers carried from Mexico to the cold climate countries were automatically disinfected by equatorial heat. With the introduction of quicker ocean transport and use of ice for storage, the pathogen was facilitated to reach the northern hemisphere without losing viability.

Citrus canker (*Xanthomonas axonopodis* pv. *citri*) originated in south-east

Asian region. Citrus also originated in these areas. The movement of citrus seedlings and grafts from this region to South Africa and Florida (USA) in the first decade of twentieth century introduced the disease in these countries where it became destructive. Similar introduction of the disease with planting material into Australia is also reported. Potato viruses are transmitted by seed tubers. Potato was introduced into India between 1870 and 1880. Possibly the viruses were also introduced with the seed tubers from outside. The late blight fungus, (*Phytophthora* infestans) and potato wart fungus (*Synchytrium endobioticum*) were also similarly introduced in India.

The phloem-limited bacterium of citrus greening disease (*Liberobacter asiaticum*) and other fastidious xylem-restricted bacteria (*Xylella fastidiosa*) are transmitted by vegetative propagation of the host. In sugarcane, the bacterium of ratoon stunting disease (*Clavibacter xyli* subsp. *xyli*) has no other method of transmission except the cane cuttings used for seed.

The contamination of plants and plant parts occurs in-the fields, orchards, plantations, or nurseries through infection of the plants by the inoculum already present there or introduced from outside. Sometimes infection or contamination of the propagating materials, such as tubers and bulbs, occurs during storage where proper precautions to remove infected materials have not been taken.

PASSIVE DISPERSAL

The passive dispersal of plant pathogens is accomplished through the agency of members of animal kingdom (man, insects, nematodes, farm and wild animals, birds, etc.) and air and water (irrigation and rains).

DISPERSAL BY MEMBERS OF ANIMAL KINGDOM

(i) Man: The most highly evolved member of the animal kingdom, man, can be considered as one of the most important single factor affecting dispersal of plant pathogens in a limited area or over long areas throughout the world. The day-to-day activities of man in normal farming practices and his trade or commerce activities that are helping dispersal of pathogens and introduction of diseases in new areas are:

(1) Vegetative propagation of ornamentals, fruit trees, tuber crops like potato and sweet potato, and plantation crops like banana and sugarcane is a highly efficient means of introducing a disease in new areas. In depending on vegetative propagation for such crops man invariably helps dispersal by moving the diseased propagating material of plants from field to field, orchard to orchard and from one geographic region to another. Late blight (*Phytophthora infestans*), bacterial wilt (*Ralstonia solanacearum*) and viruses of potato, ratoon stunting disease (*Clavibacter xyli* subsp. *xyli*) and mosaic of sugarcane, banana bunchy top virus and Shigatoka (*Cercospora musae*), citrus canker

(*Xanthomonas axonopodis* pv. *citri*) and many diseases of grapevines are some of the examples. Citrus canker originated in South-east Asia and was carried by propagation material to USA and South Africa. Its recurrence in USA and Australia after eradication is also attributed to import of infected propagating stock. The downy mildew of grapevines (*Plasmopara viticola*) originated in North America and was introduced into Europe with the introduction of the planting material. Same is true for the powdery mildew of grapevines (*Uncinula necator*). Role of commercial nurseries is very important. The nurserymen who do not care to follow proper hygienic methods distribute infected or contaminated seedlings, grafts, etc. to growers and thus supply the inoculum of pathogens also.

(2) The seed trade is another means of dispersal of pathogens in which man plays the most crucial role. There is a close relationship between pattern of seed trade and that of dispersal of pathogens (autonomous as well as passive). The import and export of contaminated seed without proper precautions (certification and quarantine) has caused entry of pathogens from one country to another. Entry of soybean and sugarbeet seed from abroad had introduced certain fungal pathogens that were not present in India. Karnal bunt (*Neovossia indica*) was endemic to India. Movement of wheat seed for experimental purposes caused its detection in Mexico and some European countries. Within India also this disease has spread to different states all over the country through seed. The sale of seed from crops badly affected by a seed-borne pathogen or disease is a common method of dispersal of such other destructive pathogens as *Ustilago segatum tritici* (loose smut of wheat), *Sphacelotheca sorghi* (grain smut of sorghum), *Claviceps fusiformis* (ergot of pearl millet) and *Anguina tritici* (ear cockle of wheat) and yellow ear rot (*Rathayibacter tritici*).

(3) Spores and other structures of fungi, bacterial cells in ooze, and even viruses can be carried by workers' clothings, shoes, hands, etc. from plant to plant and from field to field.

(4) The use of contaminated implements, cutting knives, etc. help in transmission of many soil-borne and other pathogens and also the pathogens present in tubers, sugarcane cuttings, bulbs, etc. Transmission of bacterial wilt (*R. solanacearum*), ring rot (*Clavibacter michiganensis* subsp. *sepedonicus*), bacterial wilt of banana (*R. solanacearum*) by cutting knive or matchets during planting and in banana wilt during harvest of fruit bunches is well documented. The ratoon stunting bacterium has no other means of transmission except during harvesting by cutting tools.

(5) Man is responsible for spread of almost all the graft transmissible diseases. Grafting and budding between healthy and diseased plant is the most effective method of distribution of virus pathogens of fruit trees. Careless selection of rootstock and scion of citrus is known to transmit such diseases as tristeza and other viruses.

(6) Cultural operations such as hoeing, weeding, pruning, harvesting and

packing of fruits in the orchards, all carried out by man, not only transmit diseases by contact between the plant and the contaminated tools and by providing wounds on the plant for easy entry of pathogens but also in transporting the pathogen to long distances as contaminants on the packing cases or crates.

(ii) **Insects:** Next to man, insects are the most important agents of dissemination of plant pathogens. They can disseminate bacteria, fungi as well as viruses but are most closely related to the transmission of plant viruses.

Some bacterial pathogens exclusively depend on insects for their survival and transmission. The example is *Erwinia trachiephila* (cucurbit wilt). The striped cucumber beetle (*Acalymna vittatum*) and the 12-spotted cucumber beetle (*Diabrotica undecimpunctata*) carry the bacterium in their body during the winter. The primary infection of the crop during the spring and summer is initiated by feeding of these insects on the new crop. Later, these insects help in secondary infection also by feeding on diseased and healthy plants and inoculating the bacterium in the vascular bundles. The corn wilt bacterium (*Pantoea stewartii* subsp. *stewartii*) is present in the intestinal tract of its vectors *Diabrotica undecimpunctata* and *Chaetocnema pulicaria* (the corn flea beetle). These insects disperse the bacterium over long distances. The bacterium survives through winter in these insect vectors. Black leg of potato (*Erwinia carotovora* subsp. *atroseptica* and subs. *carotovora*) is seed and soil-borne but insects also help in its transmission. A major source of inoculum of these bacteria is the potato cull piles left around the potato fields and discarded rotting tubers in kitchen waste. These sites are common habitat of maggots and fruit flies. The bacteria can live in the body of all stages of seed corn maggot (*Hylemya platura*). The fruit fly (*Drosophila melanogaster*) is another host and vector of these bacteria. The fire blight of pear and apple (*Erwinia amylovora*) is transmitted by bees feeding on diseased blossoms and then visiting healthy blossoms. Spread of citrus canker is facilitated by citrus leaf miners (*Phyllocnistis citrella* and *Thosconyrsa citri*). The fastidious xylem-limited bacterium, *Xylella fastidiosa*, is transmitted by its vectors in a permanent manner. The adults acquire and can immediately transmit the bacteria and continue transmitting them throughout their life. Nymphs also acquire the bacterium but lose it after molting. This is because the location of the bacteria in the insect body is the foregut where they are attached to the lining of the esophagus. This lining is shed during molting. The MLO causing purple top roll of potato is very efficiently transmitted by *Alebroides nigroscutellatus* which itself gets infected by the MLO.

The bacterium-vector biological relationship is often of mutualistic symbiotic type. The apple maggot (*Rhagoletis pomonella*) carries a *Pseudomonas*, that also attacks apple fruits, as an external and internal contaminant. The maggot cannot develop normally in absence of the apple tissue rotted by the bacterium. The bacteria are introduced as contaminants on the surface of the eggs through oviposition wounds at the time of egg laying by the insect. *Erwinia carotovora* causing soft rot of vegetables has symbiotic relationship with several insects such as black onion fly (*Tritoxa flexa*), the onion maggot (*Hylemya antiqua*) and the iris bore (*Macronoctus onusta*). The larvae of black onion fly and onion maggot do not survive on sterile onion tissue and require *Erwinia carotovora* to breakdown the tissue and provide nutrients which are otherwise not available to the insect. These insects deposit externally contaminated eggs and larval feeding provides wounds for entry of the bacterium.

Insects are important media for dispersal of several fungal pathogens of plants. Although the dispersal is accidental it has the chief role in spreading those diseases in which the pathogen produces sugary and sticky substances. Thus, in ergot of pearl millet the honey dew stage attracts insects and conidia get smeared on their body parts. The same insects visiting healthy ears infect the healthy flowers thus spreading the disease. In the Dutch elm disease (*Ceratostomella ulmi*) the spores are dispersed exclusively by insects. Insects also play important role in dissemination of fungal and bacterial pathogens infecting underground parts of the plant in the soil. In addition to their role in dispersal of inoculum, the insects are an important agency of bringing about diploidization of the haploid stage of rusts and many other heterothallic fungi.

The most important role of insects in dissemination of pathogens is in the virus diseases of plants. A vast majority of virus diseases are exclusively transmitted by insect vectors. Largest number of vectors have sucking mouth parts (order Homoptera) such as aphids (Aphididae), leafhoppers (Cicadellidae), plant hoppers (Delphicidae) and white flies (Aleyrodidae). In addition, thrips (having rasping and sucking mouth parts) in order Thysanoptera and mites (having puncturing and sucking parts) in the order Acarina are also vectors of some viruses. The beetles (Coleoptera) and caterpillars which have biting and chewing mouth parts mostly transmit some viruses only accidentally and mechanically. These insects while sucking the plant sap ingest the virus particles also. The acquisition feed or acquisition feeding period is the time for which a virus-free vector actually feeds on a virus infected plant to acquire the virus. While feeding on healthy plants the virus is transferred to the healthy sap. Inoculation access period is the time for which the vector, after acquiring the virus, is allowed to feed on a healthy plant and transmit the virus. The vectors may become viruliferous immediately after feeding on a diseased plant or the virus may require an incubation or latent period in the vector body where it multiplies to achieve infective concentration. The insect vector after acquiring the virus may serially transfer the virus in many plants on which it subsequently feeds or it may lose infectivity after feeding only once on a healthy plant. In non-persistent viruses, the acquisition and inoculation feed period are usually short and there is no latent period. In persistent viruses the relationship is highly specific and there is an intimate biological relationship between the virus and the vector. After the virus is acquired by the vector it usually passes through the alimentary canal, gut wall and circulates in the body fluid before reaching the salivary glands when it can be transmitted. Thus, such viruses have a latent period in the vector.

Most insect transmitted viruses are not transmitted by nematodes. The same insect species can transmit several viruses or only one kind of virus. Similarly, a particular virus can be transmitted by only one or several species of insects. Often different strains of the same virus have specific vectors. The specificity of transmission is due to (i) location of the virus in the host tissues and the depth upto which the vector can probe and feed and (ii) presence of some transmission factor in the plant which enables the vector to pick up and transmit a specific virus. The viral coat protein may have some role as it has been reported in fungal transmission of viruses. Viruses transmitted by leafhoppers (phloem feeders) are generally not transmitted by aphids.

(iii) Nematodes: Nematodes are soil-borne organisms. Although they are closely associated with many fungal and bacterial diseases of roots, the association is mostly for pre-disposition of the host and aggravation of the disease. Because of their limited mobility after they start parasitic action on a plant they are incapable of carrying the fungus or bacterium to other plants. Among bacterial diseases the best examples is that of yellow ear rot of wheat caused by *Rathayibacter tritici*. The nematode causes the ear cockles or seed galls. The bacterial yellow ear rot cannot develop unless the nematode is also present. The nematode acts as a vector of the bacterium as a surface contaminant in the galls. There is involvement of an attractive force of considerable magnitude between the cuticle surface of the nematode and the bacterial cell wall.

Many soil-borne viruses are known to be transmitted by nematodes. Although nematodes had long been suspected as possible vectors of plant viruses, the first experimental evidence of this relationship was reported in 1958 for grapevine fan leaf virus transmitted by *Xiphinema*. Since then more than 18 plant viruses are reported to be transmitted by about 21 species of nematodes. These nematodes belong to four genera of the same group, Dorylaimoidea. The nematode genera are *Xiphinema, Longidorus, Trichodorus* and *Paratrichodorus*. The Longidorids are relatively large nematodes and transmit the isometric (polyhedral) NEPO viruses. The Trichodorids transmit the tubular viruses of the Tobraviruses group also known as NETU viruses. The nematode transmission of viruses resemble the transmission by insects. There is acquisition feed time and inoculation feed time and the viruses are retained in the body of the vectors for different durations. Transmission is equally efficient by the adults and the juvenile stages of the nematode. While in insect vectors feeding time on the diseased plant is important for acquisition of the virus by the vector, in nematode vectors the time required for access to the healthy plant from the diseased plant is important. The site of virus retention varies in different nematodes. In *Longidorus* the virus particles are found in the cuticular lining of the guide sheath of odontostylet. In *Xiphinema* the particles accumulate in lumen of the odontostylet and in the esophagus while in *Trichodorus* the virus particles are seen in the entire esophagus.

TABLE 2: Nematode Species Transmitting Plant Viruses.

Nematode species	Plant viruses
Xiphinema	
americanum	Tobacco ringspot, Tomato ringspot
coxi	Tobacco ringspot, Arabis mosaic (type strain), Cherry leaf roll
diversicaudatum	Arabis mosaic (type and hop strains), Cherry leaf roll, Strawberry latent ringspot, Brome mosaic, Carnation mosaic
index	Grapevine fanleaf
italiae	Grapevine fanleaf
vuiltenezi	Cherry leaf roll
Longidorus	
attenuatus	Tomato black ring (English and German strains)
elongatus	Tomato black ring (Scottish strain), Raspberry ringspot (English and Scottish strains), Carnation mosaic
macrosoma	Raspberry ringspot (English strain), Brome mosaic, Prunus necrotic ringspot
Trichodorus	
cylindricus	Tobacco rattle (European strain)
primitivus	Pea early browning (English isolate) Tobacco rattle (European strain)
minor	Tobacco rattle (European isolate)
similis	Tobacco rattle (European isolate)
viriliferus	Pea early browning (European isolate) Tobacco rattle (European isolate)
Paratrichodorus	
anemone	Pea early browning (European isolate)
allius	Tobacco rattle (American isolate)
christiei	Tobacco rattle (American isolate)
nanus	Tobacco rattle (European isolate)
pachydermus	Pea early browning (Dutch isolate) Tobacco rattle (European isolate)
porosus	Tobacco rattle (American isolate)
teres	Pea early browning (Dutch isolate) Tobacco rattle (European isolate)

Specificity of transmission also occurs. Different strains of the same virus are often transmitted by different species of the same nematode genus. Site of retention or accumulation of the virus in the vector is responsible for some degree of specificity but more important is specific adsorption of virus particles by linings of the nematode body.

(iv) **Dispersal by Other Animals:** In addition to man, insects and nematode other larger animals can also accidentally transmit plant pathogens. Although experimental evidence is lacking, the possibility does exist that birds feeding on insect vectors can transport pathogens to distances not ordinarily covered by the vector itself. Dispersal of spores by birds carrying them on their feather is also possible. The dispersal of phanerogamic parasites by birds is an established fact. Birds feeding on berries of *Dendrophthoe* deposit the seed with their excreta on other trees. Stem fragments of dodder (*Cuscuta*) are carried by birds for preparing their nest and thus these get transported to new locations. Rodents and fur-bearing animals can also be source of transmission of pathogens. Among mammal cattle feeding on contaminated fodder often pass out viable fungal propagules in the dung which can act as source of inoculum when used as manure. Thus, conidia of *Colletotrichum falcatum* (red rot of sugarcane) and sclerotia of many fungi have been detected in cattle dung.

Dispersal by Fungi and Phanerogamic Plant Parasites

Fungi growing in soil are suspected to carry bacterial cells externally and put them in roots of susceptible plants. However, the most interesting interaction is between certain fungi and soil-borne viruses. Many soil-borne viruses are transmitted by members of Chytridiales and Plasmodiophorales of fungi. The established species number five. There are also reports of virus transmission by Oomycetes, Erysiphaceae and Pucciniaceae. The chydrid fungi are characterized by the size of posteriorly uniflagellate zoospores which have a characteristic "jerky" swimming pattern, and by the morphology of their single-celled resting spores. The plasmodiophorales have biflagellate, heterokont zoospores.

The fungus-transmitted viruses generally comprise of heterogeneous groups with respect to their usual properties. Two types of virus-fungus relationships are recognized on the basis of method of virus acquisition and by the location of the virus particles relative to the fungus resting spore. The *in vitro* transmission method (*Olpidium brassicae* x tobacco necrosis virus) involves *in vitro* acquisition, in which virions are not located within the resting spores. The *in vivo* transmission method (*Olpidium brassicae* x lettuce big vein virus) involves *in vivo* acquisition, in which virus enters the thallus as it grows in a virus-infected host, and the virus is located within the resting spores.

In vitro transmission is found with two species of *Olpidium* and with the polyhedral viruses. In the transmission of tobacco necrosis virus by *Olpidium brassicae* acquisition begins when virus-free zoospores released from resting spores or from vegetative sporangia encounter virions in soil water. The virus particles are tightly and specifically adsorbed to the zoospore membrane. This adsorption probably involves receptors in the zoospore membrane and the particular coat protein of the virions. In the transmission of lettuce big vein virus by *Olpidium brassicae* the virus is taken in by the fungus thallus while it is growing in an infected host. The infectivity of air dried soil is retained for 8 years. Similarly, soil-borne wheat mosaic virus and barley mosaic virus are internally zoospore borne in *Polymyxa graminis* and beet necrotic yellow vein mosaic virus in *Polymyxa betae*. Unconfirmed transmission of virus like particles from *Chenopodium quinoa* by powdery mildew and rust fungi and the transmission of virus to *Phaseolus vulgaris* by *Uromyces phaseoli* and to barley by *Erysiphe graminis* are also reported.

Host specialization or host specificity of the fungus pathogen is an important characteristic of fungal vectors of viruses. Host specialization has been known in *Olpidium brassicae*. The lettuce, tomato and red clover isolates of the fungus are plurivorus and transmit the tobacco necrosis virus to tobacco, tulip, bean and potato. More specific fungus isolates from oats, melon and crucifers are less efficient or non-vectors. In *O. bornovanus*, there are at least three host specific strains (cucumber, melon and squash) in the cucurbit group and they vary in their ability to transmit different viruses. *Spongospora subterranea* f.sp. *nasturtii* is specific for watercress. Some isolates of *Polymyxa betae* infect most chenopodiaceous hosts while others are highly specific for *Amaranthus* spp. or *Portulaca* spp.

TABLE 3: Viruses Transmitted by Fungi

Fungal vector	Virus
Olpidium brassicae	Lettuce big vein, Tobacco necrosis, Tobacco stunt, Chenopodium necrosis, Lisianthus necrosis
Olpidium bornovanus	Cucumber necrosis, Melon necrotic spot, Cucumber leaf spot, Cucumber soil-borne virus, Squash necrosis, Red clover necrotic mosaic
Polymyxa graminis	Soil-borne wheat mosaic, Wheat spindle streak mosaic, Barley yellow dwarf mosaic, Oat mosaic, Oat golden stripe, Rice necrosis mosaic, Peanut clump, Indian peanut clump, Rice stripe necrosis
Polymyxa betae	Beet necrotic mosaic, Beet soil-borne virus
Spongospora subterranea	
f.sp. *subterranea*	Potato mop top virus
f.sp. *nasturtii*	Watercress yellow spot, Watercress chlorotic leaf spot

The phanerogamic plant parasite dodder (*Cuscuta* spp.) is a total stem parasite of crops and fruit and roadside trees. It establishes parasitic relationship with plants through haustoria sent into the xylem and phloem of the host. Many viruses and the phloem-limited citrus greening bacterium (*Liberobacter asiaticum*) have been shown to be transmitted by dodder (*cf.* Singh, 1998). Through the haustoria the dodder stems pick up the virus from the plant and when these stem pieces are dispersed and lodged on new host plants the virus or other pathogens are transmitted to the new plant during infection by dodder. In plant species where vascular union by grafting has failed to transmit the virus, dodder has been successfully used to act as a bridge between diseased and healthy plant vascular bundles. Many viruses can multiply in dodder. If this multiplication occurs and if the receiver (healthy) plant is kept in dark transmission by dodder is more efficient.

Tobacco rattle virus (TRV) is transmitted by at least 6 species of *Cuscuta* and there is also infection of the dodder. Cucumber mosaic virus (CMV) is transmitted by at least 10 species. The virus infects the vector and multiplies in it. White clover mosaic virus can infect dodder and transmission can also occur. The alfalfa mosaic virus occurs in at least 5 species of *Cuscuta*. Beet yellows and beet curly top viruses are also transmitted by dodder. *Cuscuta* species transmit the tomato mosaic virus in winter but not in summer. Aphids can pick up a virus from dodder established on a host.

TABLE 4: Some Other Viruses Transmitted by Dodder (*Cuscuta* spp.)

Virus	*Cuscuta* spp.
Arabis mosaic	*C. subinclusa, C. californica, C. campestris*
Barley yellow dwarf	*C. campestris*
Citrus tristeza	*C. americana* (unconfirmed)
Citrus exocortis viroid	*C. subinclusa*
Cucumber green mottle mosaic	*C. subinclusa, C. lupuliformis, C. campestris*
Tobacco etch virus	*C. californica, C. lupuliformis*
Potato leaf roll	*C. subinclusa*

Dispersal by Water

Water provides for short and long distance dispersal of fungus, bacterial and nematode pathogens. Direct dispersal by running water is restricted to the distance covered by the flow. But indirectly, water helps in long distance dispersal of fungi and bacteria with the help of air. Dissemination of the red rot fungus (*Colletotrichum falcatum*), wilt causing fungi (*Fusarium* spp.), bacterial blight of rice (*Xanthomonas oryzae*), root and collar rot pathogens (*Pythium* and *Phytophthora* spp.) that survive through dormant or active structures in soil or plant debris may occur through the agency of water flow in the field. Passive movement of nematode larvae is also facilitated by flowing water.

One of the very efficient methods by which rains spread plant pathogens is splash dispersal. Rain drops or water drops falling with force from sprinkler irrigation on sori, pustules, cankers, bacterial ooze, or even soil surface may splash propagules in small droplets and with the help of air currents enable them to land on neighbouring healthy susceptible surface. Water drops may be carried to long distances by air. Often, the water drops falling on bacterial masses on the host surface form drops full of bacterial cells. When the water evaporates the mass of cells protected by slime forms aerosols that can disseminate the bacteria to long distances.

In many fungi, the spores are embedded in a matrix and can not take off unless wetted by water. The rain drops help liberation of such spores into air currents which carry them to other locations. When spores are present in the atmosphere, rains wet them and bring them down to host surfaces by trapping them.

Dispersal by Air

Unlike soil, insects and, perhaps, water air is not a habitat for any kind of plant pathogen. It act only as a carrier of propagules of organisms that produce structures adapted to aerial dissemination. However, air dispersal is the most common method of dissemination whether the pathogen is soil-borne or seed-borne, whether it survives through collateral hosts, alternate hosts or on early sown crops. The seeds of phanerogamic plant parasites are carried efficiently by wind. Although viruses are not directly transmitted by air, the speed and direction of blight of their insect vectors is positively determined by wind currents.

Wind is most important source of dispersal of some important plant pathogenic fungi. For dispersal by wind the fungus must have certain characters adapted to conditions in the air. To become air-borne, fungal spores must be able to cross the laminar air flow boundary near the surface of the host. This is accomplished by special mechanisms in certain fungi. The spore discharge occurs with sufficient force (*viz.* ascospores of *Sclerotinia*) to enable them cross this boundary and reach the upper air currents. The role of raindrop splashes carried by wind was mentioned above. The other requirements of spore dispersal by air is their lightness and number. The spores or other propagules that are disseminated by wind are usually small and are produced in enormous numbers (*viz.* urediospores of rust fungi and conidia of powdery mildew fungi). Apart from spores, bits of mycelium are also sometimes disseminated by wind. Among pathogens other than those mentioned above, the dispersal of cysts of nematodes with dust particles in storms also occurs. The cysts are usually present in the upper soil layer and during storms the soil is removed and carried to long distances. Dissemination of the molya disease (*Heterodera avenae*) from Rajasthan into neighbouring areas by this method is possible.

In dispersal of fungus spores by air the altitude reached in the atmosphere, wind current speed, rate of fall and influence of atmospheric conditions on these factors determine the dispersal distance. The urediospores of the wheat stem rust fungus *Puccinia graminis tritici* are disseminated by wind. These spores have been detected at as high as about 4000 meters above the infected wheat fields. Theoretically, if the spores are at an altitude of around 1400 meters their dispersal distance in a 30 miles per hour wind would be 1100 miles and this distance is covered by the spores without losing their viability. Air is so effective agent of dissemination of rust spores that a little inoculum, at least relative to the amount that develops subsequently, increases in various infection centres, being continuously dispersed by wind until it becomes so large as to enable even a new physiologic race of the rust fungus to become the most prevalent and widely distributed in a particular region. In the USA, the urediospores of stem rust are supposed to be blown from the far south (Mexico) into Dakota and Minnesota (in far north) travelling more than 1000 miles in about two days. In Australia the introduction of this rust was suspected to be through wind-borne spores from India or elsewhere. In the plains of northern India the annual recurrence of cereal rusts is solely due to urediospores brought by wind from the source of their survival in the hills in the far north-west or south (Nilgiri and Pulney hills).

8

The Phenomenon of Infection

After landing on the host or having come in contact with the living host surface through dissemination by seed, soil, air or other means the pathogen generally initiates the process of infection. **In strict sense of the term, pathogenesis starts from this point**. The success of this process, however, depends on many factors, listed below and elaborated in this and following chapters:

1) The host should be receptive or susceptible. There must be compatibility at the molecular and genetic level between the host and the pathogen.
2) The host should have disease proneness determined by environments
3) There must be proper aggressiveness in the pathogen
4) The pathogen should be capable of fast multiplication
5) It should have proper inoculum potential
6) Environmental conditions should favour the pathogen in penetration and multiplication.

INOCULUM AND INOCULUM POTENTIAL

The infective propagules coming in contact with the host constitute the **inoculum**. Contact of the inoculum with the host surface does not necessarily ensure infection (establishment of parasitic relationship) for which the above mentioned conditions must contribute their role. Of all these factors inoculum potential is the most important single factor determining the success of infection. **Inoculum potential** has been variously defined as "the resultant of the action of the environment, the vigour of the pathogen to establish an infection, the susceptibility of the host and the amount of inoculum present" or as "the energy of growth of a parasite available for infection of a host at the surface of the host organ to be infected". It includes not only the number of propagules but also the capacity of these propagules to cause penetration. Collective pooling of propagules or spores of many fungi, bacterial cells and fungal hyphae for successful infection is meant to achieve suitable inoculum potential. Under epidemiology (Chapter 9) an equation is given to explain the two components of inoculum potential and their relation to disease severity. Quantification of inoculum in relation to development of epidemics and rate of multiplication of inoculum in different types of diseases is also explained in the same chapter.

Even if environments are favourable for infection, the host has been predisposed and the host cultivar is susceptible, the infection often may not occur and may not be severe unless the pathogen has reached a desired level of propagule density and capacity. Generally, the more specialised a pathogen is less is the requirement of propagule density to cause infection. Thus, while in rusts and powdery mildews very few or even one spore or bacterial cell (as in citrus canker) is capable of causing infection, in non-specialised pathogens like *Pythium, Rhisoctonia, Sclerotium* etc. a high density of inoculum on the susceptible surface is needed for success of infection. It does not mean that a single unit is always incapable of causing infection. Under ideal conditions of experiments where every advantage is given to the pathogen, a single conidium can cause successful infection. However, in those pathogens where enzymatic action determines the success of the propagule(s) to cause penetration the fungus has to develop sufficient quantity of hyphae on the host surface to produce the desired amount of enzymes such as in the rhizomorphic fungus *Armillaria mellea.* In fruit rots where wounds on the surface help infection, the pathogens first feed on dead cells caused by wounds, develop sufficient mass and produce sufficient enzymes and then enter the tissues to cause maceration and rot. Thus, the potential of the inoculum is determined by the nature of the pathogen, its genetic capacity to multiply rapidly and its nutrition. Since these are also affected by the factors of environment the inoculum potential is influenced by environment.

PRE-PENETRATION ACTIVITIES OF THE PATHOGEN ON THE HOST SURFACE

Pathogens such as viruses, mycoplasma-like organisms and fastidious eubacteria (*Liberobacter asiaticum, Xylella fastidiosa, Pseudomonas syzygii* and *Clavibucter xyli*) and protozoa are placed directly into the cells of the host by their vectors, and they are probably immediately surrounded by host cytoplasm and cytoplasmic membranes. Thus, there is no pre-penetration activity on the host surface. Majority of other bacteria and most fungi show pre-penetration activity when brought into contact with the plant surface. These pathogens first secure themselves against being dislodged from the surface and then initiate the attempts to enter the host. The security against displacement is provided in most plant pathogens by surface mucilaginous sheaths on the propagules. The sheath consists of mixtures of polysaccharides, glycoproteins, polymers of hexosamines and fibrillar material which when moistened become sticky and help the pathogen adhere to the plant surface.

1. **Pre-penetration Activity of Bacterial Pathogens:** After landing on or coming in contact with the host surface the bacteria prefer to stick to their preferred sites, for example stomata. The risk of being dislodged by physical forces such as movement of leaves and flow of water on leaf surface is faced by bacteria more than the fungi. Thus, the attachment of plant pathogenic bacteria to the

plant surfaces is the first stage in development of a compatible or incompatible relationship. In bacteria the surface adhesion is provided by structures or molecules arising on the bacterial cell surface. These are filamentous appendages such as fimbriae and pili and polysaccharides such as cellulose fibrils. Small soluble molecules such as cyclic B-1,2- glycan are also present but they mainly act as elicitor inducing receptors for hypersensitive response (HR) in the host. The adhesion to the preferred site initiates the recognition system that determines the compatibility or incompatibility. When there is compatible interaction, the pathogen responds by production of host specific toxins, enzymes, growth regulators, etc.

The entry or ingress of bacteria into the host is mostly through wounds and less frequently through natural openings but never directly. This is a major difference between infection strategies of bacteria and fungi. In the vessel-limited fastidious pathogens the bacteria are placed directly through wounds by the vectors into the xylem or the phloem by the pumping action and liquid flow during feeding. The wounds, lacerations or death of tissues that permit entry of bacteria into the host are caused by wind breakage or rubbing, sand blasting, hail, frost, heat scorching, fire, animal feeding including insects, nematodes, worms, cultural practices including pruning, grafting and trans-planting, leaf scars, root cracks and even other pathogens. The bacteria briefly grow on the dead tissue before they advance into the healthy tissue. During this period they excrete the toxins and enzymes. The enzymes dissolve the middle lamella and create more area for feeding of the bacteria. The toxins kill the host cells in advance of the growth of the bacteria.

Stomata, hydathodes, nectarthodes (in fruit blossoms) and lenticels are natural openings present in all types of plants and provide easy entry to compatible bacterial pathogens. For many bacteria the stomata are the pre-ferred site for aggregation through chemotactic response or by chance. Bac-teria present in the film of water over the stoma easily swim through the aperture into the substomatal cavity where they multiply and start infection. Stomata are open only during the day but hydathodes are permanently open apertures at the margins and tip of the leaves. They are connected to the veins and secrete droplets of liquid containing nutrients. Some bacteria use these openings for easy access to the xylem vessels. Lenticels are openings on fruits, stems and tubers that are filled with loosely connected host cells to allow aeration. They are open during the growing season.

2. Pre-penetration Activity of Nematode Pathogens: Nematode cannot multiply outside the living host or away from the host as they depend for nutrition solely on their host. Adhesion to the host is provided by film of moisture present all over their body and their specific musculature. Their pre-penetration activity mainly involves their orientation towards the root surface and search for suitable location for penetration. In soil the nematodes are attracted by root emanations (exudates and carbon dioxide) which are sensed by the chemoreceptors sense organs located on nematode head. A large number of larvae normally aggregate at the same site. The nematodes first probe the cells in the vicinity of their head without actually puncturing the host. The head is moved from side to side. The object is to locate a weak spot where exudation is taking place. The sensory chemoreceptors located in the lip region help the nematode in this act. The puncture of the host is caused at this location. Body of the infective larva is arched to bring the head and stylet at right angles to root surface before penetration. This provides the pathogen physical strength for penetration which is direct through the intact but weak cuticle.

3. Pre-penetration Activities of Fungal Pathogens: It is only in fungi that complicated activities occur before penetration. The propagules for infection in fungi that come in contact with the host are of varied types. These may be unicellular or multicellular conidia, thick-walled oospores, multicellular sclerotia, etc. Therefore the activities before penetration also differ widely but the structure that approaches the site of penetration is basically a modified hypha which may arise differently in different fungal pathogens. The modified hyphae are specialized structures for the invasion of the plant tissues. Initial events are adhesion to the cuticle and directed growth of the germ tube on the plant surface. Hydrophobic interaction between spore and cuticle provides the initial passive adhesion. This is followed by secretion of a film ensheathing the germ tube and parts of the cuticle in the vicinity of the hypha. These sheaths mediate the adhesion. In some fungi, proteins or glycoproteins present in the extracellular matrix support adhesion and enable the hyphae to sense the surface and to differentiate infection structures (such as appressoria), whereas in others, carbohydrates seem to be involved in adhesion.

Many fungal pathogens first grow on the surface of the host before causing penetration. By this they achieve the proper numerical and chemical strength to bring about breakdown of outer defence barriers of the host. In *Rhzoctonia solani* attacking radish or tomato roots, the fungus, on coming in contact with root surface, first forms **infection cushions** and **appressoria** and from these multiple infections take place by means of **infection pegs.** A similar activity occurs in some species of root infecting *Fusarium.* In *Armillaria mellea* the fungus hyphae must first form the **rhizomorphs** (aggregation of hyphae in rope-like strands) and only these can cause infection because they develop sufficient quantity of necessary enzymes. In other fungi the spore germinates and either the germ tube by itself causes penetration, directly or indirectly, or it first forms an **appressorium** from which infection threads develop and penetrate the host, mostly through the stomata. The conidia of the cereal powdery mildew (*Eysiphe graminis*) regularly produce two germ tubes. The first gem tube is short and does not form appressorium in contact with the host surface. It acts as a probe. Shortly after the emergence of this germ tube, but after the host shows response to its presence, the second germ tube grows out from the conidium. It quickly elongates to its normal length and forms the appressorium. On the

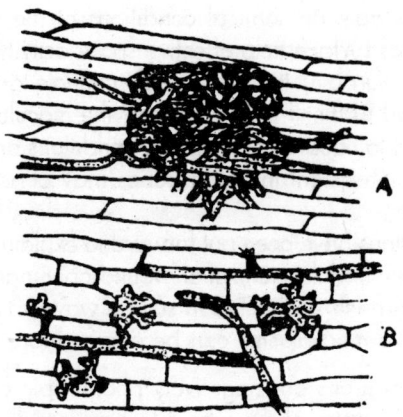

Figure 6. Pre-Penetration Growth of Rshizoctonia Solani on Host Root.

other hand, conidia of *Erysiphe polygoni* produce several germ tubes all of which grow and form appressoria. The penetration hypha from the appressorium or otherwise accumulates components of the cytoskeleton in the tip and secretes a variety of cell-wall degrading enzymes in a highly regulated fashion in order to penetrate the cuticle and the plant cell wall.

The extracellular matrix in which conidia of some fungi are embedded contains many enzymes some of which have been identified as cutinases. These enzymes erode the cuticle, the first barrier against invasion, and facilitate direct penetration. In fungi causing direct penetration and not forming appressoria, the germ tube differs from those that form appressoria. The tip is rich in cell wall degrading enzymes. Enzymes that are either present in the extracellular matrix covering the spores or germ tubes, or released from these structures may contribute to adhesion and preparation of an infection court.

Many factors operate for or against the fungus at this level. Moisture, temperature, pH, oxygen and CO_2 concentration must be proper for determining the time and speed of spore germination or activation of sclerotia. At least in the root zone many biological factors also operate. The antagonistic rhizobacteria are important inhibitors of the pathogens. In the diseases of aerial parts, the favourable conditions for fungal activity are not available throughout the day and night and thus the germination and subsequent activities to cause infection are usually periodical, mostly occurring during hours when there is enough moisture on the leaf surface, stomata are just opening, there is not very strong sunlight, etc. In the downy mildews of maize, sorghum, etc. the sporulation of the fungi is generally considered nocturnal because it is during the later part of night that the temperature and leaf wetness conditions are favourable for formation of sporangia and there germination. In the cereal rusts, infection occurs through the stomata which are open in sunlight. But sunlight quickly dries the leaf surface that inhibits spore germination. Thus, a combination of suitable anatomical conditions of the host and presence of moisture on the leaf surface help effective spore activity. If the environmental conditions are favourable, the plant is susceptible and its surface is in a receptive stage and further multiplication of the inoculum is not required the germ tube proceed to cause penetration. If conditions are unfavourable or the plant is immune or resistant the germ tubes may perish before entering the plant or soon after.

Various suggestions have been put forward to explain why germ tubes enter the host or how they find the natural or wound openings if not present exactly on them. Hyphal growth is directed in such a way as to explore the environment. The operative mechanisms can be one or more of the following:

1) Some germ tubes are negatively phototropic and in such cases this may explain entry of the tube into openings leading to dark interior of the host.

2) Movement of the germ tube towards openings may also be due to response to some chemicals (chemotropic response).

3) Thigmotropic response (contact with hard surface) directs the growth of the germ tube enabling it to recognise an array of anticlinal walls or the stomatal opening.

4) Germ tubes of many fungi (*Puccinia, Uromyces, Colletotrichum*) form appressoria as a result of thigmotropic response, presence of some chemicals in the germ tube or in response to some diffusible or volatile compounds from the host surface including the stomata.

The role of appressoria has been studied in some detail. Appressoria have an adhesion pad at their base. This pad provides firm anchorage to the appressorium on the stomatal opening. In *Colletotrichum* species appressoria formation is so common that even the hyphae of the fungus within the host cells form variously shaped appressoria. The conidia of these fungi germinate quickly and as soon as the germ tube comes in contact with hard surface, which may be even a particle, it forms appressorium. In many obligate parasites, where the entry is through the stomata, appressorium formation over the stomatal slit is due to response to some volatile chemicals emanating from the host. The appressoria may have melanized walls such as in *Colletotrichum* and *Pyricularia grisea*. The melanized walls help increase internal turgor pressure and production of penetration hyphae. Melanization of appressorium is a virulence determinant in *Pyricularia grisea* (the rice blast fungus). Mutants lacking melanized walls are avirulent. Fungicides that show *in vitro* activity against the fungus but control the disease have been shown to inhibit melanization of appressoria.

Pre-penetration Stages in Phanerogamic Plant Parasites: In the total stem parasite dodder (*Cuscuta* spp.) the seeds germinate just like seed of any other plant. The slender young seedlings is raised in more or less vertical position and the growing tip moves in a circle in search of a support. When a support is contacted the parasitic stem twines around it. If it is a living host stem the parasitic stem produces haustoria. During the development of haustoria a sucker-like organ first arises from epidermis of the mother stem and adheres firmly to the host surface. The true haustorium develops later. It originates endogenously mainly from the cortical tissues just outside the pericycle like an adventitious root. Cells of the pre-haustorium dissolve their way into the host tissue partly by pressure and partly by enzymes and provide the space for entry of the main haustorium. In *Dendrophthoe* (semi-parasite of stem) also haustoria are formed for effecting penetration and infection. In broomrape (*Orobanche* spp.) the seed germinates to produce a radicle that grows towards root of the host plants, becomes attached to it and produces a shallow disc or cup-like appressorium which surrounds the host root, penetrates it with a mass of undifferentiated, polymorphic cells that extend to and, occasionally, into the xylem of the host root. Other polymorphic cells become attached to the phloem cells of the host. In *Striga* (witchweed) the endospore nutrients in the weed seed can sustain the seedling only for 3–7 days in absence of the host. If there is no attachment with a host root within this period the parasite seedling dies. If there is a host root within 2–3 mm of the parasite seedling, chemical signals are exchanged that direct the *Striga* radicle to the host, initiate haustorium formation and successful attachment and xylem to xylem connection. The tip of the parasite radicle swells into a conical or bulb-shaped haustorium which presses against the host root. The haustorium dissolves host cells by enzyme secretions and penetrates the host within 8–24 hours.

PENETRATION

Having come in inimate contact with the host and with the formation of structures explained above, the pathogen completes the process of infection by (i) penetration and breakdown of physical and chemical barriers and (ii) establishment of physiological relationship with the host (the infection proper). Entry of pathogens in the host may be indirect through wounds and natural openings or it can be by direct penetration. The infection occurring through wounds is called *trauma or wound infection*. Some of the pathogens entering the host through this channel are called wound parasites because they solely depend on wounds for their initial activity. The pathogens may use wounds just for entry or they may use the wounded tissue first for initial multiplication and creation of suitable inoculum potential before progressing to healthy tissues.

The wounds on plant surface are caused involuntarily by workers' hands during cultural operations, by tools, animals or by such physical agencies as strong winds, sand storms, hail etc. Friction between plant parts (leaves and thorns) also causes mechanical injury. Pruning operation in orchards is a major source of large wounds on the trees. Accumulation of toxic salts around roots or base of the stem under conditions of waterlogging may lead to wounds in the form of killing of superficial cells. Insects and nematodes cause voluntary punctures or wounds on the plant. The wounds caused by these agencies may be minute or large. Depending on season, the minute wounds heal quickly. Wounds sustained early in spring or late autumn take longer to heal and, therefore, expose the injured tissue to infection for a longer period. Usually pathogens are unable to utilize these openings unless carried by insects and nematodes. However, some bacterial diseases are facilitated by even minute wounds such as those caused by sand particles in sand storms. Bacterial cells present on the surface are sucked in by such wounds. The large wounds result in death of sufficient number of cells to form dead substrate on which weak or facultative parasites can establish themselves before invasion of deeper tissues. They may dissolve binding walls of the cells before entering them and in farther tissues or they may grow on the dead tissues to develop suitable inoculum potential. Since obligate parasites (biotrophs) can remain active only in association with living cells they are not benefited by wounds.

The pathogens disseminated by insects and nematodes are helped by wounds caused by these animals. When insect vectors of viruses and vessel-limited fastidious prokaryotes (MLO and some eubacteria) insert their proboscis to suck juice, wounds are caused on the plant parts. The rupture of cells enables the pathogen to come in contact with living protoplasma of the host. The biting or chewing insects externally contaminated by viruses, bacteria and fungi also help in entry of these pathogens.

The ectoparasitic nematodes cause superficial wounds on the root surface thereby helping many root rot pathogens to establish on the exposed root tissues. The endoparasitic nematodes puncture the epidermis and move into deeper tissues. The tunnels created by them help entry of pathogens that prefer the deeper tissues. The vascular wilt fungi (*Fusarium, Verticillium*) and bacteria (*Ralstonia solanacearum*) are usually more aggressive when endoparasitic nematodes such as species of *Meloidogyne* (root knot) have attacked the roots.

In addition to the above mentioned unnatural openings the plants posses various types of natural openings or exposed surfaces that are meant for respiration and transpiration. These include stomata, hydathodes, nectarthodes, lenticels and also ruptures caused by production of adventitious roots. These parts are not covered by any covering such as cuticle and therefore are unprotected and provide direct access to the underlying tissue. Bacteria present in a film of water over these parts easily flow into the tissue. Infection thread of fungi from overlying appressoria and the germ tubes from spores enter the host through these openings in a compatible host-parasite relationship. Often, greater the number and size of stomata per unit area, the greater the suscep-tibility of the organs to pathogens entering through stomata. In leaf spots of groundnut (*Cercosporidium personatum* and *Cercospora arachidicola*) the size of stomata is one of the factors determining resistance of a cultivar. In citrus canker (*Xanthomonas axonopodis* pv. *citri*) level of infection is governed by the developmental stage of the host organ. In young organs, such as leaves, stems and fruits, the front cavity of the stomata has a wide opening because the thin cuticular layer of the epidermis is not enough to elongate the edges. As organs approach maturity and the tissues become harder, the cuticular layer becomes thicker so that the edges develop over the stoma leaving a narrow aperture between them. The slit is so narrow that surface tension prevents entry of rainwater carrying the pathogen into the opening of the mature stoma. In very young leaves, just after emergence, the stoma are immature with no opening and, therefore, only slight infection occurs.

Direct Penetration: The process of direct penetration is more complicated than entry through wounds and natural openings. In this method of entry the pathogen exerts its own physical and chemical efforts to break the host barriers and directly enters through the cuticle or the epidermis without seeking the help of wounds or natural openings. Nature has provided the plants with different mechanisms of defence against the attack of foreign bodies and their entry in the plant system. These mechanisms of defence include structural features of the host, presence of chemical coverings on the cell walls, and anti-infection biochemical nature of the protoplasm. The pathogens that do not enter through wounds or natural openings must be able to break or neutralise these barriers to become established in the host system. The mechanisms of overcoming these barriers by the pathogen, thus, can be divided into two groups, viz., capacity to break the physical or structural barriers and capacity to overcome the biochemical barriers.

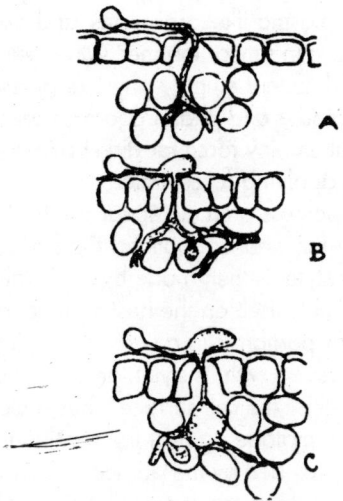

Figure 7. Fungal Penetration Through Stomata without Appressorium (A) and with Oppressorium (B) V = Substomatal Vesicle.

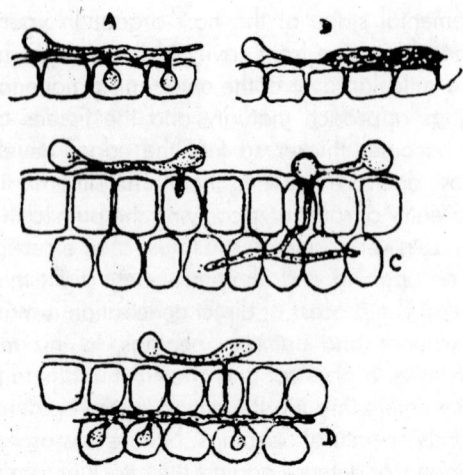

Figure 8. Direct Penetration with and without Appressorium.

Breakdown of Structural Barriers: These barriers in the host are broken down by pathogens through use of physical force. Normally, chemical action of the pathogen is not involved. The methods of entry of viruses are different from other pathogens. These pathogens can establish infection only by direct contact with the living host protoplasm. Viruses have no physical force or enzyme system of their own to overcome the structural or chemical barriers of the host and, therefore, come in contact with host protoplasma only through wounds caused by the vectors. Mycoplasma-like organisms and xylem-or phloem-restricted eubacteria also behave in similar manner. Other bacterial pathogens are mostly weak parasites and cannot employ physical pressure for effecting penetration. Thus, fungi, nematodes and phanerogamic parasites are the only groups of plant pathogens that employ force for direct penetration of the host. However, the mechanisms of applying force differ.

Plant parasitic nematodes possess piercing spear or stylet to penetrate the host surface. They first probe for weaker spots on the surface before finally puncturing it. The insertion of stylet is then made by physical force. As stated earlier the nematode larva orients itself on the host surface in such a manner that atleast its head or anterior portion is in a vertical position to the surface. When the nematode exerts pressure with its stylet, an equal but opposite force (the normal reaction) is exerted by the cell surface. Thus, unless there is some additional force pushing the nematode against the root, penetration can not occur. This additional force is provided by the particular environment in which the nematode is acting. The surface tension of the water film around the nematode body (which helps adhesion also), the normal reactions at the points of contact between the body and the soil particles, and the friction between the nematode and the substratum all may contribute to provide this additional force for penetration.

Due to their structure, size and mode of action, the fungi differ from nematodes in the method of applying force for penetration. Otherwise the stylet of nematodes can be compared with appressorium and infection pegs of fungi. In the case of fungal pathogens, the additional force is not essential because there is very little possibility of opposite force from the host cell surface. The penetration hyphae (germ tubes, infection pegs) accumulate com‑ ponents of the cytoskeleton in the tip and secrete a variety of cell-wall degrad- ing enzymes. Due to influence of thigmotropism the tips of the hyphae or the germ tubes swell into appressoria and infection thread or infection pegs arise from these appressoria. The appressoria provide adherence and turgor pres- sure to facilitate growth and further activity of the infection pegs. These struc- tures exert pressure on the surface of the host. This results in stretching of the epidermis which becomes thin. Then the infection thread easily punctures it and effects its entry. They may become very narrow and pointed while retaining the structural strength.

Breakdown of Chemical Barriers: The chemical barriers in plants present as defence mechanisms include (i) presence of cuticular layer on the epidermis, (ii) lack of suitable nutrients for the pathogen in the host cells, (iii) presence or synthesis of inhibitory or toxic substances in the host cells, and (iv) libera- tion (exudation) of substances toxic to the pathogen or stimulatory for antago- nists of the pathogen.

The outermost layer on the cuticle consists of cuticular wax which may be present as granules or rod-like projections or as continuous layers. The other component of the cuticle is cutin. The outer area of this substance is admixed with the cuticular wax while the inner part that joins the epidermal cells is admixed with pectin and cellulose. Although cutin can be degarded by some fungi and used as sole source of carbon for metabolism, no chemical activity of fungi and bacteria to degrade the cuticular wax is known. It is penetrated only by physical force or with the aid of agencies that can damage this layer. The other barriers, after the cuticular layer, are mainly of chemical nature and the pathogen can develop mechanisms to break them. There are some tissues inside the plant body that have thickened walls or synthesise special antimicro- bial compounds. Pathogens generally avoid these tissues.

The lack of nutrition and presence of toxic substances in the cytoplasm of the host are genetically controlled physiological characters of the protoplasm and account for the mechanism of resistance in the host. The nature of these substance and their role is discussed under effect of infection on the host (Chapter 10) and defence mechanisms (Chapter 12). Exudates from various plant surfaces are well known to play a role in infection by pathogens. The root exudates of plants mainly contain sugars and amino acids which are nutrients for numerous fungi and bacteria. Due to influence of these substances the root region or the rhizosphere contains microbial population much higher than the root-free soil. Among these microorganisms many are deleterious to plant growth while many are plant growth promoting and antagonists of root pathogens either directly killing them or prevent their development. This aspect is further elaborated under biotic environments (Chapter 15). The pathogen can succeed in infection of roots after overcoming this biological barrier. The root exudates also contain substances that are antibacterial, antifungal or antinemic. These are hydrocyanic acid, various other organic acids, antibiotics, etc. The pathogen has to neutralise these compounds or has to be resistant to them before it can enter the roots. Plant leaves also exude chemical substances which encourage high microbial population on their surface (the phylloplane microflora). This is an important biological barrier against leaf pathogens (Chapter 15). There are examples where components of this microflora act as agents of biological control. The glands in leaf hairs of chickpea contain maleic acid which is antifungal and provides resistance to infection by the rust fungus (*Uromyces ciceris arietini*). The protocatechuic acid in red skins of onion is also antifungal.

To overcome the above chemical and biotic barriers the pathogens should possess such chemicals which can render the defence mechanism ineffective or remain unaffected by it. Fungi and bacteria produce various enzymes, toxins, organic acids and growth regulators to accomplish this task. Although these chemicals do not necessarily help penetration, they play major role in further development of the pathogen in the tissues where they help in penetration of cell walls or their disintegration.

It is not essential that a particular pathogen should invariably depend on only one method of penetration. The powdery mildew fungi are ectophytic. The mycelium remains external to the host on the leaf surface where it is anchored by appressoria. Fine infection threads from the appressoria directly penetrate the cuticle and expand in a haustorium. In rust fungi the monokaryon and dikaryon may show different modes of penetration. The infection by the monokaryotic basidiospores occurs, as a rule, by direct penetration by the germ tubes while the germ tubes from dikarytic spores (urediospores and aeciospores) form appressoria over the stomata and as a rule infection threads penetrate the host through stomata. In many other fungi such as *Aletrnaria solani* (early blight of potato and tomato), *Cercosporidium personatum* (leaf spot of groundnut) perennation can be direct through the epidermis or through the stomata, but no appressoria are formed. In some unicellular lower fungi the entire amoeboid thallus enters the host by dissolving a pore in the epidermis. Many pathogens prefer to enter their host through special organs such as floral parts, root hairs, buds, etc. These organs are usually not provided with hard protective layers. Therefore, the pathogen can penetrate them easily.

INFECTION

Through any of the above described methods the pathogen gains entry in the host. This does not always ensure infection and disease development. Even non-pathogens of the host can gain entry in to the host. Although in majority of plant diseases penetration is the first stage in the process of infection, post-penetration stages leading to invasion of tissues and symptom expression are very important. The genetic and molecular basis is discussed later in Chapter 14. The effective infection or host-parasite relationship is possible only when hyphae developing from the infection tube or the bacterial cells that have entered are able to establish biological relationship with the host cells, absorb nutrients and, with further development, cause tissue disintegration, produce toxins and other harmful substances.

As was stated earlier, in viral infections the virus comes in direct contact with host protoplasma, i.e., the virus particles directly reach the cell cavity. Their biological relationship with the cell is only at the genetic level. They do not absorb nutrients. During replication of their genome inside the host cell viruses use the internal cellular environment that is created by the immediate function of genes and consists of such subcellular organelles as nuclei, mito-chondria, ribosome and cytoplasmic components. There is often disruption of these organelles that results in disease. MLOs and vessel-limited eubacteria multiply by reproductive methods and absorb nutrients.

Bacteria and fungi dissolve the cell walls by their enzymes after entry into the host and thus absorb nutrients. The fungi may produce special organs (haustoria) which penetrate cell walls and lie in the cell protoplasm, the absorption of nutrients taken place through osmosis. In many bacteria, toxins are involved in successful infection. Mutants lacking ability to produce toxins are avirulent. In phanerogamic plant parasites the entry of haustoria in the xylem and/or phloem ensures completion of infection.

The attack of nematodes is both internal as well as external. The endophytic nematodes use force as well as enzymes to dissolve or break the cell walls to come in contact with cell protoplasm through their stylet. They may remain sedentary or migratory in the tissues to reach new cells. Infection by the sedentary nematodes is successful only when they have established a nurse cell system (giant cells and synchytia) to draw food. Ectophytic nematodes also use similar methods and remain sedentary or migrate from point to point. They do not bodily enter the host. In each case the enzymes and toxins injected by these organisms cause tissue disintegration and other abnormalities.

Some time elapses between penetration by the pathogen and appearance of effects or symptoms. This period is known as **incubation period** or period of latency and varies from few days to weeks in different diseases and under different environmental conditions. The incubation period is actually manifes-tation of invasion of host tissues. During incubation period the pathogens increase in their quantity to gain suitable concentration for producing their adverse effects on the plant system.

There are many fungal and some bacterial diseases in which propagules of the pathogen become **quiescent** before or after penetration resulting in **latent infection**. Latent, dormant or quiescent parasitic relationship is defined as a condition in which the pathogen spends long periods during the host's life in a quiescent stage until, under specific circumstances, it becomes active (Verhoeff,1974). Quiescence of fungal pathogens is common in post-harvest diseases of fruits and vegetables. It is manifestation of resistant reaction of the host at a particular stage of its growth or it is imposed by environmental conditions. The infection waits for suitable environments, external and internal, to become active and produce disease. Quiescence can occur during spore germination, hyphal development and appressorium formation. Appressoria, germinated appressoria and subcuticular hyphae can all undergo quiescence. Quiescence can be imposed by (i) self-inhibitors in the pathogen, (ii) structural barriers in the host, (iii) presence or synthesis of inhibitory compounds in the host tissue when contacted by pathogen propagules. Self-inhibitors have been identified in some species of *Colletotrichum*. They inhibit their own germination. Structural barriers at different stages of development of plant organs can also enforce quiescence. The thickness of the cuticle and the cell wall can delay activity of germ tubes, appressoria, etc. Hypersensitive or necrotic response of the tissue can also be a cause of induced queiscence.

The more important mechanism that induces quiescence is role of pre-existing or synthesised antimicrobial compounds such as saponin (tomatine in green tomato, avenacin in oats). The saponin, tomatine, is present in high concentrations in the peel of green tomatoes and is inhibitory to infection by *Botrytis cinerea*. In anthracnose of mango (*Colletotrichum gloesporioides*) most appressoria from germ tubes do not germinate immediately but remain firmly attached to the fruit skin as quiescent state of the fungus. The germinating appressoria can cause infection from the start of blossoming until the fruits are more than half grown. Infection pegs from appressoria enter the fruit through pores in the skin but the infection remains latent and the fungus grows only to a limited extent in the epidermal layers of the fruit. The induction of quiescence is reported to be due to presence of mixture of resorcinol compounds in the green fruit skin. During ripening of fruits the level of these compounds declines and the hyphae become active. In anthracnose of banana (*Colletotrichum musae*), in which immature fruits appear resistant, the conidia germinate rapidly on immature fruits and cause quiescent or latent infection through formation of appressoria which remain inactive until the ripening of fruits. During ripening of the fruits the appressoria are activated to form penetration hyphae that colonise the underlying tissue and lead to anthracnose lesions. The unripe fruits contain antimicrobial compounds in their skin which are inhibitory to pectin-degrading enzymes and penetration of fruit tissue. In ripe fruits the concentration of these compounds declines to uninhibitory levels and the fungus gains the ability to spread within the fruit. In anthracnose or ripe rot of chilli green fruits may or may not be infected. When infected the green skin shows only minute black specks. Rotting of the skin develops when the fruits are ripening or are fully ripe. The major antifungal compound in unripe avocado fruit, resistant to *Colletotrichum gloesporioides* is a diene. In bacterial wilt and brown rot of potato (*Ralstonia solanacearum*) temperature-induced latent infection of lenticels in cooler climates is a major source of dispersal of the pathogen. The infection remains dormant until the seed tubers are exposed to high temperature in warmer areas.

COLONIZATION OF THE HOST

After successful establishment in the host tissues, subsequent spread of the pathogen within the host is influenced by many factors including aggressiveness of the pathogen, susceptibility of host tissues, anatomy of the host determining distribution of susceptible and resistant tissues, and environmental conditions affecting host physiology. The lesions are smaller, develop more slowly and produce fewer spores in a resistant than in a susceptible host. The size of leaf spots is restricted by veinlets. Some pathogens become systemic while others remain localised. In many cases of obligate parasitism, hypersensitivity of the tissues may stop growth of mycelium and only very minute necrotic flecks may be formed. Thus, on the basis of post-penetration development the parasites can be grouped in various ways. They may be strictly localised causing small necrotic lesions, they may cause extensive blotches, streaks or stripes, there may be extensive rotting of entire organs, or the parasite may be restricted in growth but produce enzymes, toxins, and other harmful materials which affect organs at a distance from where the parasite is present.

After introduction into the host cell the virus particles must replicate to increase their quantity and spread to other cells. Uptake of virus by protoplasts takes place by charge-dependent, temperature-independent process. Protoplasts are negatively charged. Virus particles with positive charge attach to the protoplasts. For further development the viral genome must be released from its protein coat before it can replicate. The uncoating is probably physical as well as enzymatic. In early stages physical removal of proteins takes place. The naked nucleic acid then replicates. The specific protein(s) are synthesised separately. Then assembly of the nucleic acid and protein coat takes place. Once assembled the particle cannot be disassembled unless the genome has to replicate. Cell to cell transmission within the plant occurs with the cytoplasmic stream through plasmodesmata between cells. Long distance passive movement within the plant is mostly within the phloem where virus particles move along with the carbohydrates to energy requiring organs such as roots and developing shoots. The time required for translocation of a virus from inoculated leaf to other parts of the plant varies from 10 to 21 days for different viruses in different host plants.

The MLOs are confined to sieve elements of the phloem. The distribution and concentration in sieve cells is uneven depending on the amount placed by the insect vectors. Vertical movement of the cells is permitted by pores in the sieve plate. There is no evidence of active or self movement. The MLO cells are translocated with the photosynthates in phloem stream like the plant viruses. In the xylem-restricted eubacterium *Xylella fastidiosa*, causing Pierce's disease of grapevines and phony peach disease, the pathogen is directly placed by its vector in the xylem and the bacteria remain confined to treacheary elements, tracheids or vessels. Only those bacterial cells survive and colonise the plant host which firmly adhere to the xylem vessel walls. The extracellular strands (fimbriae) and extracellular polysaccharides which are more dense at ends of the bacterial cell ensure adherence. The bacteria often form aggregates in xylem vessels that appear to be held together by extracellular strands (microfibrils, fimbriae, etc.) produced by the bacteria. These strands help in uptake of ions of nutrients also. The fibrous network holding the bacterial cells together could also function as protection against host defenses. Although a large number of bacterial cells are killed, those surviving for 3–5 days after inoculation multiply, form microcolonies in the vessels and move rapidly in the leaf veins. In sugarcane ratoon stunting disease (*Clavibacter xyli* subsp *xyli*) colonization of the vascular tissues by the bacterium is most extensive in the lower, more mature internodes than in the upper internodes.

In most diseases caused by eubacteria the target sites after entry are the intercellular spaces and the vascular bundles. The bacteria may immediately start multiplying or there may be delayed activity. In some cases there is even reduction in the initial number of bacterial cells that have entered, as was stated above for *Xylella fastidiosa*. Having multiplied and occupied the spaces, the mass of bacterial cells flows out into other similar locations while continuing their biochemical activities such as production of pectolytic and cellulolytic enzymes, toxins and growth regulating compounds.

In the ectoparasitic fungal pathogens, the main body of the fungus lies on the surface of the host with only the feeding organs (haustoria) penetrating the tissues. The invasion of the host involves the extent of superficial growth and absorption of nutrients through haustoria. In powdery mildews, the mycelium develops extensively on the host surface and only haustoria enter the epidermal cells or few cells beneath the epidermis for drawing nutrition. In *Rhizoctonia solani* the parasitic mycelium invades the internal tissues of potato stem but there is extensive development on the surface also under suitable conditions. The production of external mycelium which grows over the host surface enables invasion to occurs at several different points and thus leads to rapid colonisation, evading possible resistance of certain tissues inside the host.

The endophytic fungi or endoparasites grow subcuticularly (*Diplocarpon rosae*, black spot of rose), in parenchymatous tissues (most fungal pathogens) or in vascular tissues (vascular wilt parasites). The growth in parenchyma is intra- or intercellular. The intracellular growth involves destruction of cell walls and extensive tissue damage. Killing in advance of invasion is common in many necrotrophic fungi. Damping off fungi and fruit rot fungi often cause invasion in this manner. The tissue is first killed by enzymatic action and the mycelium advances into the dead area. In intercellular growth, the fungi only puncture the cell walls to send haustoria in the cell lumen. The hyphal growth continues through intercellular spaces or along the cell walls. The vascular wilt fungus *Fusarium udum* infects the host (pigeonpea) through fine rootlets which are penetrated by the germ tubes. From the laterals the fungus passes into the larger roots. Resistance in pigeonpea to wilt depends upon the extent of colonisation of the vessels by the fungus. The plants wilt when colonisation of the plant is more than 50%. These fungal pathogens produce and release spores within the vessels. These spores are carried in the sap stream and invade vessels far away from the mycelium producing new mycelium which invade new vessels. Growth of mycelium of the fungi in the parenchymatous tissues of the host is indefinite until the plant part or the plant is dead. In some fungal infections, however, while the younger hyphae continue to grow into new healthy tissues, the older hyphae in already infected areas die out and disappear. Many pathogens, mostly those which do not produce mycelium, are **endobiotic** in the sense that the thallus is entirely contained within the host cell as in *Synchytrium endobioticum* (the potato wart disease). It is distributed to the daughter cells during cell division. Release of the pathogen occurs through decay of tissues.

The migratory ectoparasitic nematodes feed from the surface cells with their short stylets for a few minutes or longer and then move on to new locations. Reproduction continues during the period and the nematodes can spread all over the root. *Pratylenchus* species feed for many hours or even days at the same spot on epidermal cells or root hair. The sedentary ectoparasitic nematodes possess long stylets and feed from deeper tissues while the body remains outside. The adult females become slightly swollen and remain attached to the feeding site. Migratory endoparasitic nematodes such as *Radopholus* (burrowing nematode) may invade or leave the root at any stage of their development. They usually feed from cortical cells which are then destroyed by intracellular migration of the nematodes and apparent dissolution of the cells which are ingested by the nematode.

The sedentary endoparasitic nematodes are highly evolved root parasites. They invade the root, partially or completely and migrate to specific sites where they induce the formation of highly specialized nurse cell systems such as synchytia and giant cells. These cells provide a continuous source of nutrients until the parasite dies. Males may not feed at all or feed only in early stages of their development. The females become saccate, and remain completely inside the host tissue. Their posterior portion may partly protrude on the outside and usually produce several hundred eggs. The eggs again hatch, release fresh larvae and colonization of the root continues under favourable conditions of temperature and moisture.

EXIT OF THE PATHOGEN

Rapid multiplication of the pathogens within the host creates a situation when the host tissues cannot support the increasing population. The pathogen must find an exit and disperse to maintain continuity of infection chain or disease cycle and escape death due to overcrowding and depletion of suitable host tissue material.

Viruses can exist only in association with the living protoplasm and their method of exit from the host is solely through other agencies discussed under dispersal. In bacterial diseases, the disintegration of tissues causes release of bacterial cells. Bacterial mass also oozes out from the affected tissues and this ooze is dispersed by insects, water and wind. Nematodes also appear as dormant structures (cysts) and egg masses on the root surface. The migratory forms always try to move to new locations for food.

The fungi have the most elaborate system of exit or production and liberation of secondary inoculum. After growing for some time in the host tissues and when the environmental conditions are favourable most plant pathogenic fungi grow out on the host surface as special asexual spore-producing structures. The spores are produced in vertical series on simple or branched. Spore-bearing hyphae (sporangiophore and conidiophores) or they are produced in aggregations such as sori of smut and rust fungi or spore masses in procumbant chains, special fruit bodies, etc. The spore thus formed are dispersed as discussed earlier and spread of the disease occurs during the same season.

Development of the disease in the population from individuals attacked first follows the third stage, i.e. infection and exit, in the infection chain. The infection is repeated in the same sequence as described so far as long as the host is available and environment is favourable. This is further considered in chapter 9.

FACTORS INFLUENCING INFECTION

The process of infection of plants by pathogens is influenced by many factors including the genetic response of the plant to the activity of the pathogen. The external factors, that is, the factors other than inherent characters, can be broadly called as environments under which the pathogen is trying to establish contact with host tissues. These environmental factors may be related to the plant, the pathogen and the biotic factors which might be interacting with the plant and/or the pathogen. The role of environments in determining survival and dispersal of the biotic infectious agents is described in chapter 15. Infection is a short duration process and is influenced by environments more strikingly than the other stages in pathogenesis. The factors influencing infection and related to the plant are: (i) inherent resistance or susceptibility and (ii) external conditions that increase or decrease the proneness of the plant to attack of the pathogen. Matching factors are associated with the pathogen, viz., (i) the genetic qualities of the pathogen which make it aggressive and (ii) external conditions that favour the pathogen to increase its aggressiveness. The set of factors associated with the plant constitute the **disease potential** and the set of factors associated with the pathogen constitute the **inoculum potential**. These two determine the **disease severity**.

Resistance and Susceptibility of the Host

Genetic response of the host to the pathogen, measured in terms of resistance and susceptibility in a compatible host-parasite system, is an innate plant factor determining success of infection. A genetically susceptible plant is infected readily and rapidly while a resistant plant puts up obstacles in the way of penetration and establishment of infection. However, the genetic character is liable to break down under the influence of several factors such as tissue condition (seedling and adult plant reaction), nutritional status of the host and climate, and presence of aggravating or synergistic biotic agencies such as nematoe-fungus/bacterium complexes.

External Conditions-Predisposition

The term "predisposition" has been used in different ways by different authors. In its strict sense, predisposition is the action of set of environments, prior to penetration and infection, which makes the plant vulnerable to attack by the pathogen. It is related to the effect of environments on the host, not on the pathogen, just before actual penetration occurs. Yarwood (1976) had defined predisposition as "the tendency of treatments and conditions, acting before inoculation or before introduction of the incitant, to affect susceptibility to biotic and abiotic pathogens". Predisposition is an expression of modification of the host response to the pathogen. Other comparable terms for disposition are disease proneness, acquired disposition, induced susceptibility, preconditioning and physiologic susceptibility. Genetic response of the host is directly not involved in predisposition which is increased susceptibility or decreased resistance imposed by external cause. These external causes are numerous and often interrelated. The effect of environments on disease development is discussed in chapter 15. Some of the factors related to disease proneness of the host are briefly listed below:

(i) Plant vigour determined by age and nutrition: Age of the plant (ontogenic disposition) plays a decisive role in infection by many pathogens. In the damping off diseases caused by species of *Pythium* the juvenile stage of seedlings is most susceptible. There is decline in severity of infection as the tissues of the seedlings harden. In some diseases plants become more suscep-

tible as tissue maturity increases (leaf spots of rice caused by *Heliminthosporium oryzae*). However, in such cases of effect of tissue maturity the innate characters of the host may be involved. It may also be due to presence or absence of genetic defence mechanism at the particular stage of plant growth. In many cases susceptibility or resistance may be induced through change in vigour of the plant by the external environments. Thus, specific type of nutrition, climatic conditions determining the uptake of specific nutrients, and tillage operations may increase vigour of the plant which may result in earliness of growth, large leaves, large plants, thin leaves, thin cell walls, turgid cells and active guttation. These conditions, while favouring infection by some pathogens, may increase resistance to others.

(ii) Temperature: Predisposition induced by extremes of temperature may be due to injury to tissues. Absorption of nutrients is also affected by temperature thus change the response to infection. In rice blast, the night temperature (20°C) influences metabolic pattern of the host and increases disease proneness. At still lower temperatures even resistant varieties tend to show susceptibility. Under low temperature conditions reaction of sugarcane cultivars to red rot infection (*Colletotrichum falcatum*) shifts towards resistance while opposite is true at high temperatures.

(iii) Humidity: Atmospheric humidity and soil moisture can enhance disease proneness through their effect on the host as well as on the pathogen, especially the latter. In damping off disease of seedlings (*Pythium* spp.) high moisture helps release and movement of zoospores and at the same time prolongs the juvenile stage of seedlings. The stems remain succulent for longer periods. Accumulation of soluble salts and poor oxygen supply for roots also predispose the stem to infection. In sour rot of citrus fruits (*Geotrichum candidum*) infection is facilitated by high moisture content of the rind. Poor oxygen supply and damage to fibrous roots in citrus predisposes roots to infection by *Phytophthora parasitica* (root rot). The number of zoospores of the fungus is also increased. In sugarcane, even resistant cultivars are prone to attack of the red rot fungus under water-logged conditions. It is due to predisposition of the host as well as changes in the behaviour of the pathogen.

(iv) Light: Holding plants in darkness for 1–5 days before inoculation is a common method of predisposing them to virus infection. Dark treatment before inoculation also increases susceptibility of potatoes to infection of *Macrophomina phaseolina* and Fusarium wilt. These treatments enhance susceptibility through alteration of host metabolism. In pathogens where entry is exclusively through stomata light plays an important role because stomata are open only in light.

(v) Nutrition and pH: The alkalinity and acidity is involved in predisposition through its effect on nutrition of the plant. Numerous experimental results on effect of fertilizers on disease susceptibility suggest that high nitrogen level often favours infection, high potassium often reduces it and phosphorus has variable effect. However, there are many exceptions to this generalization.

Atleast in root diseases the form of nitrogen is more important than the amount of nitrogen. Nitrate and ammonia nitrogen have different effect on different root diseases. The ammonia nitrogen tends to shift pH towards acidity also and thus predisposes the plant to diseases favoured by low pH. Nutrients applied to soil can influence infection through their effect on the plant as well as the pathogen operating from the soil. They have no effect on the pathogens of the foliage. In such cases the role of nutrients is purely through the plant and is related to predisposition. Infection of rice leaves especially those of susceptible varieties by the rice blast fungus, *Pyricularia grisea*, is favoured by heavy nitrogenous fertilization. This is an example of how several factors interact to make the host tissue prone to attack of a pathogen. Greater the accumulation of nitrogen in the leaf tissue the more susceptible the variety is to blast. When plants are grown under nitrogen deficient conditions there is greater absorption of silicon which is reported to impart resistance in leaf tissues. In presence of heavy absorption of nitrogen silicon uptake is reduced and hence there is loss of resistance in the leaf tissues. The nitrogen metabolism is influenced by temperature. At low temperatures under high levels of nitrogen there is very little absorption of silicon and the absorbed nitrogen tends to accumulate in the leaves in the form of soluble nitrogen such as glutamine. The shift in favour of nitrogen metabolism causes exhaustion of carbohydrates which could otherwise help in the production of secondary metabolites imparting resistance. At low temperature such as at 15°C susceptible as well as resistant varieties show heavy infection while at high temperatures such as at 30°C resistant varieties remain free from infection.

(vi) Chemicals: Apart from the major nutrients, the micro-elements such as zinc, boron, manganese, magnesium etc. have also been found to influence susceptibility of the tissues to infection. Deficiencies make the host tissue prone to easy infection. Role of chemicals other than nutrients has also been implicated in predisposition. During decomposition of crop residue phytotoxic chemicals are released in the soil. These substances predispose roots to infection by root parasites. Various pesticides increase susceptibility to different pathogens. for example, 2, 4-D to *Claviceps purpurea* on wheat, DDT and other chemicals like maleic hydrazide to *Puccinia graminis* on wheat.

(vii) Wounds caused by microorganisms or mechanical means: Plants wounded before inoculation often become more susceptible to pathogens, especially the unspecialised types. Wounds on fruit surface predispose the fruit to attack of fruit decay fungi such as *Penicillium, Aspergillus, Rhizopus, Lasiodiplodia* and *Alternaria*. The injury provides avenues of entry and also initial nutrition from dead cells for the necrotrophic pathogen. This initial support to the pathogen increases inoculum potential. The bacterial pathogens cannot enter the host by themselves and need some opening, natural or through wounds. The necessity of slight wounding before inoculation might be a predisposing effect for plant viruses. Injury caused by leaf miners (*Phyllocnistis citrella* and *Thosconyrsa*

citri) aid the citrus canker bacterium. The association of nematodes with many fungal and bacterial diseases is well known. The nematode predispose the host by causing wound through which the pathogens enter the roots. The examples are vascular wilt diseases caused by *Fusarium oxysporum* and bacterial wilt of potato and tomato.

(viii) Plant exudates: Secretions of volatile or non-volatile substances by plant tissues on the surface is a common phenomenon. Roots are particularly active in this respect. The amount and nature of the exudates depend on plant age, nutrition and vigour of the plant or plant organ, growth conditions and treatment of the plant, soil conditions, and specific variety of the host. The exudates are rich in sugars, amino acids and nutrients suitable for parasitic growth. They also attract the infection thread, germ tubes, zoospores and nematode larvae. They can, thus, enhance the chances of infection. The, presence of sugars in wrinkled seeded pea varieties and their susceptibility to seed rot caused by *Pythium ultimum* is an example. Some exudates contain hatching factor for nematode cysts and eggs while others are toxic to pathogens or encourage antagonistic microflora thereby causing inhibition of the pathogen.

(ix) Cultural practices: Frequency, amount and time of irrigation and thick stand of the crop are some of the cultural practices that may pre-dispose the crop to infection of pathogens. These practices provide unfavourable environment for the host and favourable conditions for the pathogen.

Aggressiveness of the Pathogen

The two sets of factors given above were plant related factors constituting the disease potential. In addition to these, the nature of the pathogen and environments influencing activities of the pathogen are also equally important, if not more. Aggressiveness of the pathogen has been described as the capacity to invade the plant, obtain nutrients from and grow and multiply in or on the tissue. It is determined by the number of propagules necessary for infection, percentage of plants attacked successfully, amount of secondary inoculum produced and number of life-cycles completed during the crop period. Most aggressive pathogens are those that cause infection even with a single spore but produce high amount of spores. In less aggressive pathogens a much larger number of propagules or even a short period of ectophytic growth may be required before penetration and invasion.

External Conditions Influencing the Pathogen

Activation of resting structures, germination of resting and propagative spores, growth of the germ tube, its orientation to the host surface or to avenues of entry are influenced by such factors as temperature, light, pH, oxygen and carbon dioxide, and biological stimulants and inhibitors.

9

Epidemiology

Appearance of a disease in a large number of individuals over large areas in relatively short time is an epidemic. Epidemiology deals with outbreak and spread of disease in a population. Epiphytology or epidemiology of plant diseases is essentially a study of the rate of multiplication of a pathogen which determines its capacity to spread a disease in a plant population. From practical viewpoint, this is the most important part of the study of plant diseases. The three links in infection chain (survival, dispersal and infection) discussed in the preceding pages are also, broadly speaking, a part of epidemiology since multiplication of a pathogen in the plant population can occur only when these three stages have occurred. Different aspects of pathogenesis (effect of infection on the host) discussed in the following chapter also are part of epidemiology since they also govern the rate of multiplication of the pathogen in the plant population after infection. In this chapter, epidemiology in its strict sense, i.e., characters of spread of disease and factors affecting the outbreaks of epidemics, are explained. A summary of events prior to spread of the pathogen in plant population is given in Fig. 9.

Aggressiveness, rapid rate of reproduction and rapid long distance dispersal of the pathogen are considered to cause epidemics. But these are not invariably the only cause of epidemics. There are several viral and fungal diseases in which spread of the pathogen is very slow and takes years to reach a high level but the disease may assume epidemic form. Citrus psorosis virus is not transmitted by any known insect vector. Its spread is only through grafting. But epidemics of this disease occur in many parts of the world.

A plant disease is the outcome of interaction between the **plant**, the **pathogen** and the **environment**. In this **disease triangle** the environments favourably or unfavourably influence both the plant and the pathogen. While the plant population in the field influences the environment, pathogens have very little influence on the environment. When favourable interactions between these three components of incidence of a disease continue for long, epidemics occur. The long period or **time**, thus, becomes a fourth component of an epidemic. Often human activities interfere with epidemics through management practices and the epidemics may be encouraged or halted.

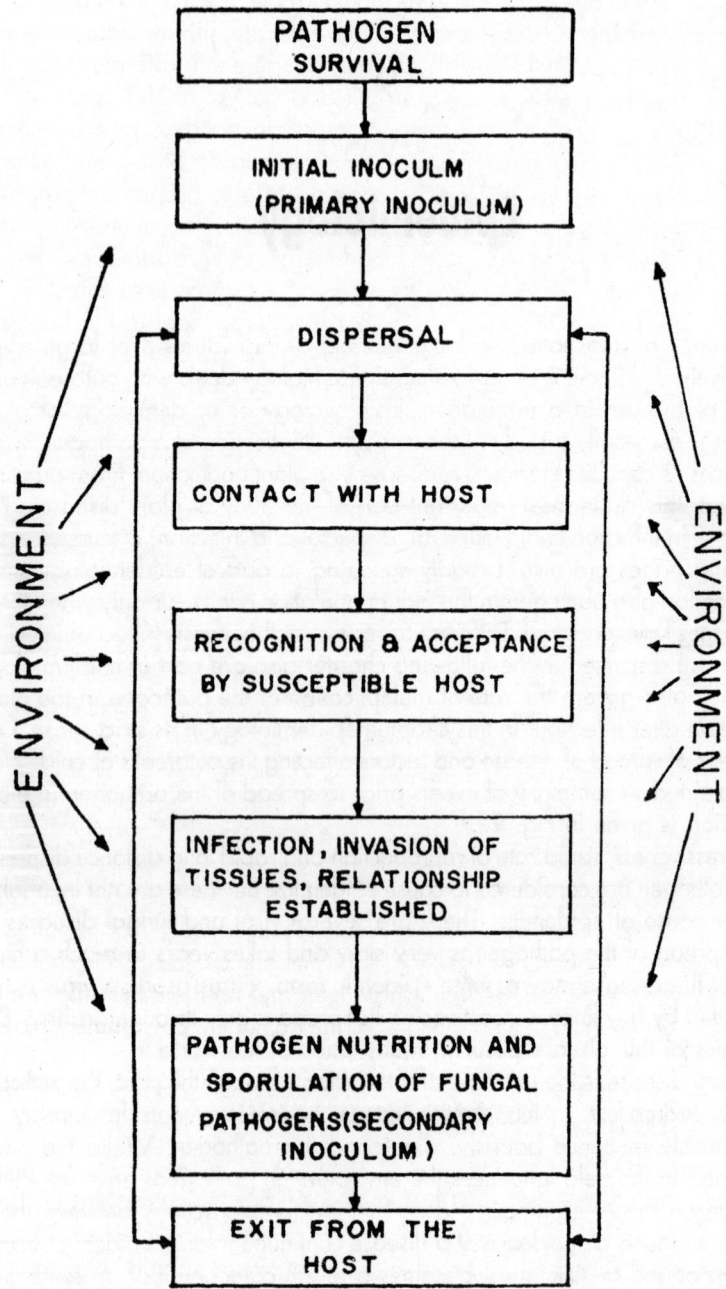

Figure 9. Events Prior to Spread of the Pathogen in the Crop.

The epiphytotics develop in the same manner as the rise in human population. The human population rises due to (i) high birth rate, (ii) low death rate and (iii) low birth rate but still lower death rate or high death rate but still higher birth rate. Epiphytotics can also be attributed to the rate of birth and death of pathogen propagules. In the first category, mostly fungal diseases are included because most fungal pathogens produce enormous quantity of spores of different kinds which not only ensure rapid dispersal but also proper survival from season to season. In the late blight of potato, the initial inoculum is low since only infected seed tubers carry the fungus (*Phytophthora infestans*) and all tubers are not infected. But tubers even with 5% surface area affected can initiate the disease in the field. Once the disease appears in the field and weather is favourable the high rate of multiplication ensures epiphytotic. Penetration of the host occurs within 0.5–2.5 hours and incubation period is only 3–7 days. Thus, many life-cycles occur within a short period. In apple scab (*Venturia inaequalis*), the initial inoculum is very high and the rate of multiplication also is very high. A large number of life-cycles are completed during the apple season in favourable weather.

The high birth rate diseases like late blight, apple scab, rusts and powdery mildews spread very fast but can be effectively managed by fungicidal sprays. In the low death rate category are those diseases in which the pathogen is mostly systemic, living inside the host, and therefore has the advantage of minimum death rate. Such diseases are common among fruit trees and some field crops like sugarcane. The spread is slow. The disease can be checked only by removal of the source of inoculum through sanitary practices and other cultural operations. Such diseases are dangerous because once established it becomes difficult to eradicate the source of primary inoculum. Sugarcane ratoon stunting (*Clavibacter xyli* subsp. *xyli*), citrus greening (*Liberobacter asiqticum*), phony peach disease (*Xylella fastidiosa*) spread slowly but are very persistent unless the plant harbouring the pathogens in xylem or phloem is destroyed. Thus, the build up of an epidemic is actually manifestation of balance between two opposing forces. On the one side are the growth characters and pathogenic potentialities of the pathogen and on the other side are the forces that counteract these capabilities of the pathogen. Environment plays a decisive role.

COMPOUND INTEREST DISEASES

In such diseases the rate of increase is mathematically analogous to compound interest in money. The pathogen produces spores at a very rapid rate. These spores are disseminated by rapid means such as wind. They infect other plants. The incubation period is short and sporulation occurs rapidly. New crop of spores is produced, disseminated and the cycle is repeated. There are, thus, several or many generations of the pathogen in the crop season. These are **polycyclic diseases**. Since the susceptible host area is fixed by the acreage under the crop, amount of uninfected tissues continuously declines as the spread of the disease continues so that the rate of spread in terms of new infections also decreases. Pathogens spreading by means of air disseminated propagules such as rusts, powdery mildews and late blight of potato show this type of spread. However, the spread is continuous only if all the components of the disease triangle are balanced in favour of the pathogen. Often this may not be so and during the crop season the spread of the disease may show several peaks. In bimodal polycyclic diseases different organs (blossoms or fruits) of the plant are attacked at different period.

In India, the cereal rusts (*Puccinia graminis tritici, Puccinia recondita* and *Puccinia striformis*) have no local source of primary inoculum in the plains. The pathogens survive on collateral or alternate hosts in the form of urediospores at remote places. These spores are brought by wind to the main crop. In the beginning only few pustules of rust develop in the crop. Each pustule is capable of producing roughly 50–400 thousand urediospores which are disseminated by wind to other plants in the same field and in neighbouring fields. Since each spore can cause one independent infection, theoretically, 50–400 thousand infections can occur in the crop. Each new infection is capable of developing into a spore bearing pustule within 5–15 days depending on the rust fungus species and its temperature relations. Thus, within a week or a fortnight of appearance of first pustules in the crop hundreds of thousand new pustules are formed which could repeat the process. If the weather conditions remain favourable for only few weeks the entire crop is severely affected by the disease.

SIMPLE INTEREST DISEASES

In the **monocyclic diseases** the increase is mathematically analogous to simple interest in money. There is only one generation of the pathogen during the crop season. The primary inoculum is seed- or soil-borne and secondary infection rarely occurs during the season. All late infections that are noticed in some diseases in the field are from the pre-existing inoculum in the soil. Monocyclic diseases are exemplified by loose smut of wheat (*Ustilago nuda*), covered smut of barley (*U. hordei*) and grain smut of sorghum (*Sphacelotheca sorghi*). Vascular wilt diseases caused by *Fusarium oxysporum* are also monocyclic. In these diseases there may be very slow secondary spread if the pathogen can sporulate on the host surface and other plants are in a position to be attacked. Normally it does not happen since the fungal propagules have to pass through the barrier of soil to reach the roots and most propagules are lost or converted into resting spores during this passage. In loose smut of wheat and barley the pathogen is carried by infected seed, grows in the plant without any external appearance and damage to tissues and finally appears in the ear heads where all flowers are destroyed and mass of resting spores (smut spores) is produced. It is disseminated, causes fresh floral infections but the infected produce carries the disease to the next season. In other smuts, such as covered smut of barley and grain smut of sorghum, inoculum present on seed surface cause infection of seedlings before they emerge out of soil. Further development is as in loose smut. These diseases are mostly systemic in nature, do not produce propagules external to the host during the active season and the dispersal or propagules is restricted by climatic and biotic conditions.

SLOW AND RAPID EPIPHYTOTICS

Simple interest and compound interest diseases can be expressed as slow and rapid epiphytotics. The form an epidemic can take is governed by the nature of the pathogen and its host and the weather which functions as referee in the battle between the host and the pathogen. At one extreme are the epidemics which develop slowly (**tardive**) and at the other end are those that develop rapidly (**explosive**). Many intermediate types may occur. Slow epiphytotics occur among populations of perennial plants such as fruit trees. The causal agent is mostly systemic to varying extents. The infected host survives for several years before dying and the epiphytotic takes an identical time to reach the maximum potential. Most of the characters of a simple interest disease are found in slow epiphytotics. The multiplication rate of the pathogen is slow due to inherent characters including lengthy incubation period or prolonged growth within the host tissues. The movement of the pathogen from plant to plant is much slower than in non-systemic diseases and among herbaceous plants (field crops). They are low death rate pathogens which compensates for the disadvantage of not having high birth rate. Crop sanitation is best method of management of slow epidemics. However, due to complex interactions involved in such diseases the results of control measures are slow.

Rapid epiphytotics are chiefly caused by non-systemic pathogens with very high birth rate which have many generations within a short time. Most annual crops are affected by such epiphytotics. These epiphytotics are more affected by environments than slow epiphytotics. The diseases start early, their increase is rapid, rising to a distinct peak in a short time, then showing sharp or gradual decline, if the rate of increase and decline is graphically plotted the epidemic rate curve will be symmetrical (bell-shaped) as in potato late blight or asymmetrical with the epidemic rate greater in the early season due to greater susceptibility of young leaves (apple scab and most downy mildews) or asymmetrical with the epidemic rate greater in the late season. Unfavourable weather, host tissue becoming resistant due to maturity, dispersal becoming restricted due to crop canopy and many other factors are responsible for decline in the epidemic rate.

ESSENTIAL CONDITIONS FOR AN EPIDEMIC

For establishment of an epiphytotic the following conditions are necessary. These conditions are concerned with the host, the pathogen and the environment that constitute the disease triangle.

Host-related Factors:

(i) *Levels of genetic resistance or susceptibility in the host:* Susceptible hosts lacking genes for resistance provide the ideal susbtrate for establishment and development of new infections leading to epidemics under favourable environmental conditions. Plants with race specific resistance cause maximum delay in epidemic development while plants with horizontal resistance may become infected but the rate of disease and epidemic development depends on the level of resistance and environmental conditions.

(ii) *Abundance and distribution of susceptible hosts:* Epidemics develop in large populations from relatively fewer infected individuals. Continuous cultivation of a susceptible variety in a given area, large areas under a genetically identical susceptible variety or distribution of genetically uniform variety over large contiguous areas are conditions that favour the pathogen to increase the rate of multiplication and use the propagules more effectively to develop an epiphytotic. When a genetically uniform resistance variety is cultivated repeatedly over large areas one of the dangerous effects is development of a new race of the pathogen that can attack the variety which was resistant to earlier races of the pathogen. The wheat cultivar S 227 was resistant to brown rust (*Puccinia recondita*). It had such good commercial qualities that farmers adopted it throughout the wheat belt in India. The repeated cultivation of the variety over large areas resulted in development of a new race of the pathogen and the variety became highly susceptible and was discarded. This aspect is further discussed in the chapter on disease management through host resistance.

(iii) *Distance of the susceptible plants from the source of primary inoculum:* The disease in any area is initiated by the primary inoculum surviving at some source. Longer the distance from the source of survival of the pathogen longer will be the time required for build up of epiphytotic in a susceptible crop. During dispersal in different directions the density of primary inoculum is diluted and as the distance increases fewer propagules are likely to reach the susceptible surface. This can occur in the same field, among different fields and in a larger area of the country.

(iv) *Type of crop:* Epidemics develop much more rapidly in annual crops (cereals, vegetables, cotton, etc.) than in perennial woody plants such as fruit trees.

(v) *Disease proneness in the host due to environments:* Susceptibility is genetically controlled but proneness in the plant to get infected can be induced by environments and other factors. When such conditions exist the host is liable to more vigorous attack and successful infection by the pathogen. A susceptible variety becomes more susceptible and even a moderately resistant variety may tend towards susceptibility when conditions favouring proneness are prevalent. Red rot resistant cultivars of sugarcane tend to become susceptible under water-logged conditions. Under these conditions the pathogen has better chances of multiplying, causing infection and effectively using its propagules for secondary spread. Temperature is known to break resistance to disease in some host-pathogen combinations due to increased susceptibility of leaves or due to interference with resistance gene expression.

(vi) *Presence of suitable alternate or collateral hosts:* This is required only for those pathogens which survive and multiply on wild hosts during absence of the cultivated host. For pathogens spreading through heterogeneous infection chain presence of alternate host is necessary for providing primary inoculum. The amount of inoculum thus available will determine the intensity of primary infection and subsequent secondary spread. Presence of collateral hosts plays the same role for pathogens of homogenous or continuous infection chain. There are numerous grass hosts of many diseases. *Sorghum halepense, S. bicolar* and *Saccharum spontaneum* are grass hosts of *Peronosclerospora sacchari* or *P. philippinensis* (downy mildew of maize). *S. spontaneum* is a host of *Ustilago scitaminea* (sugarcane smut). *Digitaria sanguinalis* is host of *Sclerophthora rayssiae* var. *zeae* (brown stripe downy mildew of maize). In many areas, weeds like *Leersia hexandra* are the main source of survival of the rice blast fungus (*Pyricularia grisea*). Numerous grass hosts of *Rhizoctonia solani* (rice sheath blight) include *Cynodoan dactylon, Cyperus rotundus, Setaria glauca, Paspalum flavidum, Echinocloa curss- galli, Panicum repense, Eriochloa proera* and *Digitaria.* These grow during the rice season and may produce abundance of inoculum aiding in build up of epiphytotics of these diseases.

Pathogen-related Factors

(i) *Presence of aggressive isolate of the pathogen:* Only infectious diseases can take the form of an epiphytotic. For any epiphytotic rapid cycle of infection is essential and successful infection can be caused only by aggressive or virulent isolates of the pathogen.

(ii) *High birth rate of the pathogen:* Among animate causes of plant diseases the high birth rate of the pathogen is an important contributory factor for epiphytotics. The fungi that assume epiphytotic form invariably have the capacity to produce enormous quantity of spores that are adapted to quick and long distance dispersal in a short time so that they can take advantage of the favourable weather during that short period. These spores are asexually formed usually on the exposed surface of the host for dispersal by wind, water and insects. Among fungi the vicissitudes of dispersal by wind, minute size of unprotected spores, and possible chances of falling on wrong hosts are many factors that cause high death rate among the propagules. However, this weakness is offset by extremely high birth rate.

(iii) *Low death rate:* Epiphytotics attributed to low death rate of pathogens are those in which the pathogen is systemic and protected by the plant tissue. Thus, the chances of high mortality of propagules are reduced to a minimum. The chief source of accumulation of inoculum for epiphytotics of such diseases is the diseased plant organ used for propagation of the crop, especially vegetative propagation. Therefore, the build up of epidemic is comparatively slow. When a particular area becomes saturated with diseased planting material chances of occurrence of epiphytotics are very high.

(iv) *Easy and rapid dispersal of the pathogen:* The ability of a pathogen to cause epiphytotics is dependent as much on its dispersal as on high birth rate. The units of propagation produced by the pathogen are dispersed by external agencies which must be available if epiphytotic has to develop. Fungal spores are mostly disseminated by wind while viruses and bacteria are mostly disseminated by insects or, for bacteria, by water. The velocity and direction of wind, moisture, number of suitable insect vectors, their rate of reproduction, feeding habit, etc. determine the degree of epidemics. The fungi have special mechanisms for making their spores wind-borne. Usually the spore discharge is with some force to throw them in upper air currents. Spores held in mucilage are discharged by falling drops of water and then wind-borne. In addition, the fungal spores are light, minute and resistant to adverse conditions encountered in turbulent air.

(v) *Adaptability of the pathogen:* Weeds seldom die because they have the capacity to adapt to adverse conditions. Most of the pathogens causing epiphytotics can be placed in the same category. They can adapt themselves to various conditions listed above. However, exceptions can exist to these requirements. The necessity of adaptability can be substituted with other qualities such as high birth rate.

Environment-related Factors

Meteoropathology deals with the relationship between weather and epiphytotics. The environmental factors that usually influence the progress of an epidemic are moisture, temperature, and cultural practices that modify the moisture and temperature conditions in the plant population and also the biological factors that have restricting influence on the pathogens such as phylloplane and rhizosphere antagonistic microflora. The effect of weather on disease development is discussed in Chapter 15.

Optimum moisture, temperature, light etc. are necessary for activity of animate pathogens. They are as much important as the nutrition for the pathogen. Assuming that a particular fungus meets all the above requirements for causing epidemics; it has high birth rate, high aggressiveness, produces abundance of wind-borne spores and the spores fall on susceptible hosts that are prone to infection; even then infection, invasion and development of epidemic may not occur if weather is not favourable for germination of spores or in absence of light stomata have not opened to permit entry of the infection thread or when the stomata open the moisture is so low that the germ tube has dried. The weather also affects the activity of the pathogen on the host surface. It may not permit sporulation on the host surface thus reducing amount of inoculum for secondary spread. The spores may be washed from leaf surface by rains before penetration has occurred. A stormy weather coupled with showers may create too many injuries on the host surface to permit large number of infections thus large amount of secondary inoculum.

Relation of Epidemics to Human Activities

The activities of the grower in managing his crop have also close relationship with possibilities of epidemic development. Selection of low lying and poorly drained fields, especially when near the source of inoculum favour initiation of epidemics. Farmers using infected or infested seed, nursery stock and other propagative materials create conditions for epidemics but when proper precautions have been taken to ensure seed health the chances of epidemics are greatly reduced. Improper use of fertilizers, sprinkler irrigation system, poor sanitation are contributory factors for an epidemic. Delay in disease management practices also is a cause of epidemic which can be prevented by timely and effective use of fungicides.

DECLINE OF THE EPIDEMIC

No epidemic remains forever in the population. After development of the epidemic a stage is reached when it shows decline by itself. This stage is very common in epiphytotics of annual crop plants. The causes of decline in epiphytotics are as follows:

(i) *Saturation of the pathogen in the host population:* In an epidemic the pathogen infects and colonises a very large number of individuals. These individuals are killed or destroyed. In human populations the unaffected individuals migrate to safer areas or get themselves chemically protected. Plants can not migrate and, hence, the available host tissues goes on decreasing. The non-availability of more host plants limits further spread of the pathogen. This results in production of less inoculum, fewer secondary infections and finally, no new infections. The plants that escape infection are those that possessed resistance or in which resistance developed during the epidemic. Thus, one of the advantages of an epidemic is that it eliminates susceptible individuals and permits only the resistant individuals of the population to survive and breed.

(ii) *Decline of proneness in the host:* Most diseases attack the plant at a particular stage of its growth. When the plant has crossed that stage its proneness for contacting infection is reduced or completely lost. Under these conditions the epidemic will automatically decline. When the plant is receptive for infection throughout its life and its population has been affected by an epidemic, the weather conditions may not remain always favourable for disease development. As a result further spread of the disease in the population will be checked and the epidemic will decline. Brown and stripe rusts of wheat in northern India are favoured by the low temperature prevailing during December–February. Epidemics of these rusts decline when the temperature starts warming up in March and later.

(iii) *Reduction in aggressiveness of the pathogen:* Due to above mentioned and other causes the aggressiveness of the pathogen may be reduced. When all susceptible individuals are destroyed by the pathogen, it may try to parasitise the remaining resistant individuals of the same species. In these adverse conditions it may lose power of successful infection, its reproduction may slow down and, thus, it may not remain as aggressive as when the conditions were favourable.

DISEASE MEASUREMENT

There are three aspects of disease measurement, each of which is significant in its own way. Measuring **disease incidence** is quick and easy. It is a measurement of prevalence of the disease in the population and reflects the number or proportion of plant units diseased. Thus, out of 100 plants when 10 plants show disease the incidence is 10%. But, it does not fully reflect the damage done to the crop. In diseases like vascular wilts, rice neck blast and cereal smuts, measurement of disease incidence has direct relationship to the **severity and loss**. But in many other diseases such as most leaf spots, root lesions and rusts disease incidence has little relationship to the severity of the disease which is expressed as the percentage or proportion of plant area destroyed by the pathogen. It is closer to the yield loss caused by the disease. The yield loss is positively correlated with **economic loss**, the loss which occurs due to damage to plants resulting in less yield and due to disease management cost. The level of disease (amount of plant damage) at which incremental control costs just equal incremental crop return is called **economic threshold**. The economic threshold of a crop-pathogen system varies with the tolerance level (**damage threshold**) of the crop. The damage threshold of the crop depends on the growth stage of the crop when attacked, crop management practices, environment, shift in virulence of the pathogen and control practices. Change in market price of the produce also affects the economic threshold.

DISEASE SEVERITY

The conditions for development and decline of epiphytotics can be expressed by the following equation. In this equation **disease severity** has been expressed in terms of balance between qualities and weaknesses of the host and the pathogen.

Disease severity = Inoculum potential × Disease potential
= [Inoculum density × capacity] × [Proneness × susceptibility].

Inoculum potential of the pathogen is a function of the number of infective propagules [inoculum density] and their pathogenic capacity [inoculum capacity]. The number of propagules alone do not determine inoculum potential. The propagules should have the energy or capacity to cause infection. A fixed number of well-nourished strong spores can cause better infection than the same number of ill-nourished and weak spores. Freshly hatched larvae of *Meloidogyne javanica*, the root knot nematode, cause better infection than those which have wandered in soil for some time and have lost much of their energy reserves. During movement, the larvae lose their lipid reserves, the source of energy that sustains them until they find their host, and lack the power of thrust required for insertion of their stylet into the host. Thus, even if density of propagules is high but capacity low, disease incidence will be adversely affected. On the other hand, if the capacity is high even fewer propagules may cause high incidence of the disease. If both these factors are high the pathogen is in ideal condition for causing an epiphytotic unless the opposing forces in the host are strong enough to counteract its aggressiveness. The conditions of the pathogen listed earlier necessary for development of epiphytotics determine the inoculum potential.

Disease potential is concerned with the condition of the host. Due to unfavourable environments, poor or unbalanced nutrition, susceptible stage of growth, etc., the host may be predisposed to attack by the pathogen. This proneness is different from the inherent susceptibility which is genetically determined. High proneness and high inherent susceptibility are ideal for development of an epidemic in the plant population.

Both the above factors, inoculum potential and disease potential, are affected by environment which may turn them favourable or unfavourable for development of epiphytotics. If all the above factors are of high degree and environment is favourable development of epiphytotic is unavoidable. If intensity of any one of these factors is reduced there is corresponding decrease in disease severity. If majority of these factors are unfavourable either the epidemic will not develop or it will show the stage of decline.

DISEASE PROGRESS CURVE AND MATHEMATICAL MODELS

The disease incidence and disease severity as influenced by time and human interferences can be expressed in progress curves or as mathematical models of the disease development. When the pattern of an epidemic in terms of the numbers of lesions, the amount of diseased tissue or the numbers of diseased plants is plotted against time over which these effects have occurred it gives a **disease-progress curve**. The amount of disease is maximum near the source of primary inoculum appearing in the field as infection foci. It decreases as the distance from the infection foci increases. This is shown by **disease-gradient curve**.

The terms compound interest and simple interest for explaining rate of increase of pathogens in plant population during the crop season were introduced by Van der Plank, in 1963, who for the first time put plant disease epidemiology on a mathematical basis. The model suggested by him still is widely recognized. This model is largely based on the infection rate r for polycyclic pathogens (compound interest disease) and r_m for monocyclic pathogens (simple interest diseases). It is the rate at which the population of the pathogen increases. The r is an average estimated from successive estimates of the population of the pathogen as proportion X of infected plants in case of systemic diseases or of infected susceptible tissue in case of local lesion diseases. Van der Plank had suggested that at least in its initial stages the epidemic can be described by the following equation:

$$X = X_0 e^{rt}$$

where X = proportion of disease at any one time
         X_0 = the amount of critical inoculum
         r = average infection rate
         t = time during which infection has occurred

The value of e in the equation is the base of natural logarithm

$$e = \left(1 + \frac{1}{n}\right)^n$$

Thus, we are dealing here with an exponential function and the basic assumption is that at any given time the rate of disease increase is proportional to the amount of disease present at that moment. In early stages of the epidemic, when there is plenty of susceptible host tissue available for colonization by the pathogen, this assumption is true. As the epidemic progresses and available susceptible tissue declines, the rate of increase of disease is determined not only by the amount of the disease present (x) but also the proportion of the susceptible tissue left, i.e., ($1-x$).

The item of interest in the above equation of the model is r. It gives an overall measure of the rate at which the epidemic is progressing. It can be used to compare epidemics of the same disease in different localities and in different cultivars. The comparative r is derived by taking log and transposing:

$$\log e^x = \log e^{xo} + rt$$

$$rt = \log e^x - \log e^{xo}$$

$$r = \log e \frac{1}{t} \log e \frac{x}{xo}$$

The r can be assessed at any time during the epidemic but its use is simpler in early stages of the development of the epidemic and this stage is more important than when the epidemic has established. In order to find out the rate of increase in r in a varietal trial or in a locality two readings or assessments are essential for comparison and computation of the rate. Thus, after the initial measurement has been taken, another assessment is made after some time to calculate the average rate of increase of the disease during the time elapsing between first and second assessment. By convention, the first measurement is x_1 at time t_1 and the second is x_2 at time t_2. Average infection rate is then calculated as:

$$r = \frac{1}{t_2 - t_1} \log e \frac{x_2}{x_1}$$

When the disease has reached very high intensity, the formula for calculating the rate is more complicated because the factor $1-x$ is introduced. In that case:

$$r = \frac{1}{t_2 - t_1} \log e \frac{x_2(1-x)}{x_1(1-x)}$$

ANALYSIS OF EPIDEMICS

Epidemiological studies have made rapid progress since Van der Plank proposed the mathematical models for different aspects of epidemics. **Epidemic is a system.** A system is regarded as an interlocking complex of processes characterized by many reciprocal cause-effect pathways, OR, any phenomenon, structural or functional, having two or more separable components and some interaction between these components. Each component of the system can be studied separately. However, in epidemics we are concerned with interaction of these components and the effects of these interactions. Analysis of a system with two or few non-variable components is easy, but analysis of a system with too many variable and non-variable components with their own subcomponents and all interacting with each other, is a tedious and time consuming job.

As a system the epidemic exists in a relatively very small compartment of a much larger system, the **agroecosystem**, which has developed over a long period of time from the **ecosystem**. The epidemics interact with other subsystems of the agroecosystem such as cropping system or crop management system, pest management system, associated biological systems other than the epidemic, etc., which interact among themselves also (Fig 10). These systems and their components are influenced by physical parameters of the agroecosystem and, in turn, influence the latter also. The different interacting systems, with their

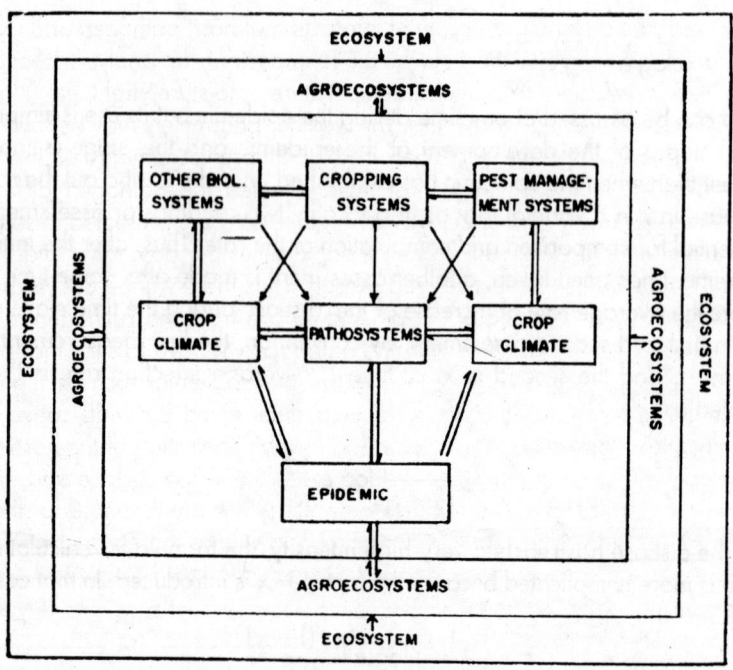

Figure 10. Position of Epidemic within the Ecosystem.

interacting components give a picture of the complexity of data for analysis of epidemics.

Analysis of epidemics is a modern approach in the study of epidemiology. Although the mathematical equations for r tells us the rise and fall of an epidemic over a period of time during the growing season of the crop and about comparison of the epidemic between varieties, fields, or regions, it does not answer how the different components of the epidemic behaved during that time. The picture remains incomplete unless all the possible interactions are integrated. What factors of weather, host or pathogen and at what time contributed how much to the development of the epidemic? Such questions are difficult to answer by calculations on a piece of paper with the help of mathematical equations proposed from time to time. But answers to such questions are important not only to understand about the epidemic that has occurred but also from the viewpoint of applied epidemiology or **disease forecasting** discussed later in Chapter 15. Although no direct application of analysis of an epidemic has been possible, since no prediction of behaviour of the multiple variables involved in epidemics can be made, the answer to some of these questions can at least pin point why and how a particular forecasting system, based only on physical parameters of the environment in relation to the pathogen, has failed.

Since plant disease is outcome of interaction of host, pathogen and weather over a period of time, the time factor is important for analysis because of variation in weather conditions (temperature, moisture, light etc.,) over a period of 24 hours, days or weeks. The weather parameters may be favourable or unfavourable for the pathogen and at different segments of the entire period of the epidemic they must have affected the pathogen differently. Therefore, for analysis of epidemic the weather data are presented in segments of the epidemic period, say, 3-hour slices, 8 slices per day, for a total period of, say, 100 days of the active crop season. Thus there will be 800 segments.

Along with weather data, the other data are from the pathogen and the host. To find out the variable with the pathogen we have to slice that part of disease cycle which is directly involved in the epidemic, such as flight and landing of spores on the host surface as determined by wind speed, their germination, appressoria formation (if formed), penetration of the host, lesion development, production of conidiophores on the lesion surface and, finally, more spore production. The weather variables which are recorded for the time segment are those which directly or indirectly influence these stages of disease cycle, such as, temperature, rainfall, relative humidity, leaf wetness and dryness, day light or night, wind speed, etc. Let us presume there are 8–9 variables associated with the pathogen and 8 with weather. Not taking into account the host variables, except for those determined by weather such as leaf wetness or dryness, the number of interactions will be approximately $8 \times 8 \times 800$, i.e., 51200. This incredible number of facts is to be used for analysis of epidemic and only a computer can do this. It can remember these facts without error, sort them out on command, calculate all possible correlations, arrange them in sequence of the calendar and draw a curve of the epidemic for the year under test.

Computers are extremely quick and precise provided a proper programme or model is available and all the basic data for the epidemic have been properly simplified and placed on the punch card or other units with the help of which the computer does the calculations in the sequence in which data have been arranged. Thus, when the time slices with weather data and also the data on conidial germination, appressorium formation, penetration etc., have been fed to computer according to the programming, the computer can be commanded to answer any question regarding how the pathogen behaved at a particular time. However, it must be noted that a computer is just a tool or equipment like a laboratory microscope. If light and magnification in the microscope have not been adjusted properly, it gives a blurred image. Same is true for a computer except that it is more sturdy and more exacting. It will give the answer according to information fed to it and the way the question is asked. The limitation is the completeness of the required information and its accuracy. Owing to the possible weaknesses in programming and to intrinsic variations in the pathogens, computers work out only the probabilities.

124

Waggoner and Horsfall (1969) had developed the first computer simulation programme called EPIDEM by modelling each stage of life cycle of a pathogen as a function of environment. EPIDEM could simulate, imitate or mimic an epidemic of early blight of potato and tomato (*Alternaria solani*). Subsequently, computer simulations were written for *Mycosphaerella* blight of chrysanthemums (MYCOS), for southern corn leaf blight caused by *Helminthosporium maydis* (EPICORN) and for apple scab caused by *Venturia inaequalis* (EPIVEN). Such programmes are available now for many other diseases.

10

Effects of Infection on the Host

Host tissues show different types of responses to activities of the pathogen after it has established infection and when it starts parasitic and pathogen activities. In the initial stages of penetration there may be a striking increase in protoplasmic strands and the nucleus of the cell may move to the site of penetration. Cytoplasmic particles, in rapid Brownian movement, appear followed by granulation of the cytoplasm and appearance of many more particles in Brownian movement. Later, the cell contents become yellow and finally dark brown when Brownian movement ceases and the cell is dead.

Normal physiological activities of the host cell are disturbed and anatomical and morphological changes (*morbid anatomy*) appear as visible changes. In pathogenesis, the first stage after infection is the manifestation of these responses of the host cells which appear in the following forms:

1) Structural changes: In diseased plants usually abnormal structures are seen. Examples are overgrowth (hypertrophy), sterile flowers, phyllody, hairy root, witches' broom, bunchy top, crown gall, root knots, etc. These abnormalities are discussed in the present and the following chapters. However, the appearance of abnormal structures on the sick plants is not due to physical but to chemical reactions occurring within the plant body and are, therefore, expression of physiological malfunctioning of the host cells.

2) Physiological changes: Harmful effects of infection on physiology of the host are the main causes of symptom expression, damage and loss. In this chapter, brief account of following harmful effects is given:

i)   Disintegration of tissue by enzymes of the pathogen
ii) Effect of pathogenesis on growth of the host plant due to hormonal imbalance
iii) Effect on reproduction of the host
iv) Effect on uptake and translocation of nutrients and water
v)   Abnormal respiration of the host tissues
vi) Reduction in photosynthesis

The role of toxins in disturbing the physiology of the host including effect on permeability of cell membranes is discussed in the following chapter.

TISSUE DISINTEGRATION

Among diverse symptoms of plant diseases the most common are those caused by disintegration of tissues. There are very few diseases that do not show tissue disintegration and death of cells at some stage after infection by the pathogen. These symptoms are so prominent that they had attracted the attention of man in ancient times. Usually the word "rot" is used for such symptoms. The word is derived from "ret" connected with "retting", the process of soaking fiber-bearing tissues in water and macerating the tissues by biological action to separate the fiber. Thus, in tissue disintegration the cells and tissues of the host plant are separated from each other resulting in the condition known as rot. This condition is present even in many necrotic sports. The material binding the cells to form tissues is destroyed by the pathogen to enable it to reach the host protoplasm. Such pathogens are mostly facultative parasites or facultative saprophytes. However, obligate parasites or pathogens and even non-parasitic causes of disease can induce tissue disintegration through indirect effects. As a result of tissue disintegration the symptoms known as blight, canker, anthracnose and soft rot appear on the plant.

Fungi, bacteria and nematode bring about tissue disintegration through the action of enzymes secreted by them. These enzymes are of different types for action on different tissues and chemical constituents of the cell wall.

i) Cuticular enzymes: The epidermis of plants is protected by cuticle. The propagules of the pathogen first contact the host on this surface. The major chemical substance in the cuticle is a cutin framework with waxes embedded in it and extruded from its surface to give a water-repellent surface (the cuticular wax). Leaf hairs also provide similar water-repellent surface. The central region of the cuticle consists of cutin, a polyester of hydroxylated monocarboxylic acids each containing 16 to 18 carbon atoms and 2–3 hydroxyl groups. On hydrolysis, the polyesters yield fatty and hydroxyl fatty acids. The wax portion consists of complex mixtures of long chain paraffins, alcohols, ketones, esters and acids. Paraffin and esters predominate on the outer surface. Small quantities of other substances such as proteins, carbohydrates, pigments and occluded pectin and cellulose may also be present. Thickness of cuticle varies with plant species. The amount of wax also similarly varies with plant type. Thick cuticle often, but not necessarily, provides resistance to penetration by the pathogen.

While pathogens do not produce enzymes that can act on the cuticular wax, other components of the cuticle can be degraded by fungal enzymes. The enzymes suggested to be involved in dissolution of the cuticle are (i) cutinase which catalyzes the breakdown of cutin, hydrolysing it into cutin acids, (ii) those enzymes which degrade other cuticular substances such as proteins, cellulose, pectin, pigments, etc. Germinating spores of may fungi such as *Colletotrichum gloeosporioides* (on orange leaves), *Sphaerotheca pannosa* (on rose), *Venturia inaequalis* (apple scab) and *Helminthosporium victoriae* (on oats) are reported to degrade the cuticle.

ii) Pectic enzymes: After the cuticle has been penetrated the pathogen comes in contact with cell wall protecting the protoplasm. The main components of the cell wall are pectin or pectic substances, cellulose, hemicellulose, lignin and a small quantity of protein. Fungi, most bacteria and nematode which feed on the protoplasm have to degrade these substances to dissolve the middle lamella and cell walls so that protoplasm can be reached.

Pectin or pectic substances are major chemical constituents of middle lamella which binds the cells together. Fungi, bacteria and nematodes are known to secrete pectic or pectinolytic enzymes and use these enzymes against this group of materials. Pectic substances are polysaccharides mostly consisting of chains of galacturonan molecules interspersed with a much smaller number of rhamnose molecules and small side chains of galacturonan and some other sugars. Degradation of pectic substances is brought about by two groups of pectinolytic enzymes: the *pectinesterases* and *polygalacturonases* (PG). Pectinesterases are widely distributed in plants and microorganisms. Those in fungi have a lower optimum pH for their activity. The pectinesterases or pectin methylesterases (PME) catalyse the hydrolysis of methyl ester groups of pectinic acids to methyl alcohol and pectinic acid of reduced methoxy content. Eventually, pectic acid is formed. This is attacked by polygalacturonase group of enzymes which includes glycosidases and lyases (pectin lyases or transeliminases). These enzymes break the links between adjacent galacturonic acid units. Polygalacturonase (PG) attacks pectic acid and polymethylgalacturonase (PMG) attacks pectin. When these enzymes are activated the pathogen absorbs some molecules and increases its production of enzymes.

The presence and amount of pectinolytic enzyme differ in different fungi and are governed by many factors such as pH. The fact that many fungi produce these enzymes in culture does not necessarily mean that same fungi will produce the same enzymes in the host also and the enzymes will play some role in pathogenesis. Under certain conditions the pectinolytic enzymes are inactivated or rendered ineffective. Phenolic compounds or their oxidation products common in darkened tissues at sites of injury inactivate pectic and other enzymes. Indole acetic acid also inhibits certain pectic enzymes.

Pectin degradation results in liquefaction of the pectic substances in the middle lamella leading to **maceration** (softening and loss of coherence of plant tissues and separation of individual cells which eventually die). The degradation of pectic substances provides nutrients for many fungal pathogens and due to weakening of the cell walls facilitates inter- and intracellular invasion by hyphae. The cell protoplasm is not affected by pectic enzymes. Loss of strength of the cell wall makes it unable to support the osmotically fragile protoplast and the protoplasts burst. These enzymes are apparently of primary importance in diseases characterized by soft rotting of tissues such as those caused by *Botrytis cinerea, Sclerotinia sclerotiorum, Sclerotium rolfsii, Rhizoctonia solani,* species of *Pythium, Phytophthora* and *Rhizopus, Erwinia carotovora, E. chrysanthemi* and *Pseudomonas* species. However, more specialized pathogens like *Puccinia graminis* also produce pectic enzymes during spore germination. This facilitates movement of hyphae in between the host cells. The activity of pectic enzymes may be aggravated by synergistic action of other enzymes and metabolites. In the invasion of *Sclerotium rolfsii*, oxalic acid production by the fungus plays a similar role which is crucial for pathogenesis and tissue disintegration.

iii) Cellulolytic enzymes: Cellulose is the major component and basic unit of structural framework of plant cell walls. The cellulose content of tissues varies from 12% in the nonwoody tissues of the grasses to about 50% in mature woody tissues to more than 90% in the cotton fibers. Cellulose is polysaccharide but contains chains of glucose molecules. The glucose chains are held to each other by a large number of hydrogen bonds. Relatively less is known about the significance of cellulolytic enzymes (cellulases) and enzymatic degradation of cellulose in plant diseases. Not all fungi are cellulolytic although they do cause tissue disintegration. However, even in such pathogens cellulases are present but they cannot degrade the natural cellulose of plant cells.

Some believe that a single enzyme converts cellulose into glucose by random cleavage of the molecules. Others have suggested that native cellulose is converted to disaccharide cellobiose by one enzyme (C_1 enzyme) and thence to glucose by a second one. In noncellulolytic pathogens, perhaps cellulolytic saprophytes provide help in the degradation of the natural cellulose. Still others suggest that a series of steps are involved in the conversion of natural cellulose into glucose. The native cellulose molecules are released or loosened from the linear cellulose chains by C_1 enzyme. Enzyme C_2 also attacks native cellulose and breaks it into shorter chains. The loosened molecules then take up water and are hydrolyzed by other enzymes (C_x enzymes) to soluble low molecular cellosaccharides and finally to cellobiose and glucose. These are then attacked by β-glycosidases to form glucose.

The degradation of pectic substances and native cellulose not only helps the pathogen in invasion of tissues and disintegration, it causes other effects also. The degraded products are used by the pathogen as food. Large molecules released by degradation often cause plugging of vessels thus partly contributing to development of symptoms of wilt in many plants.

iv) Hemicelluloses: In addition to pectin and cellulose, the plant cell walls contain the complex water-insoluble polysaccharides known as hemicelluloses which are attached with cellulose and lignin. The hemicelluloses are important constituents of mature and thickened cell walls. They contain complex mixtures of such pentosans as xylans, mannans, galactans and arabans. Many parasitic and saprophytic microorganisms produce the enzymes hemicellulases for hydrolysis of these polysaccharides and produce simple sugars. Certain components of hemicelluloses are degraded by cellulolytic enzymes also. The degradation of hemicellulose exposes the cellulose and lignin of the cell walls for action of fungal enzymes.

Reports implicating hemicellulases in plant diseases are few. Xylanase and arabinase have been found in the hypocotyl of sunflower attacked by *Sclerotinia sclerotiorum*. *S. fructigena* produces arabinofuranosidase in cultures. *Sclerotium rolfsii* produces exogalactanase, endomannase, galactosidase and endoxylanase in cultures.

v) Lignolytic enzymes: The most complex chemical compounds in plant cell walls are the lignins. Pectic materials and cellulose are major constituents of herbaceous agricultural crops which contain lignin mostly in the walls of their xylem vessels and in fibrous tissues. Such structures usually remain intact after degradation of the cell walls and disintegration of tissues. However, woody plants, especially the perennials, contain relatively larger amounts of lignin. In mature woody plants the lignin content may vary from 15 to 38%. In woody cells lignin constitutes almost all of the middle lamella and forms its own framework in the walls so that the latter are strengthened and remain intact even after cellulose and hemicellulose are removed. Deposition of lignin is usually followed by death of the cells and the fungi that attack lignified tissues are usually perthotrophs.

Lignin is an amorphous, three-dimensional polymer that is different from both carbohydrates and proteins in composition and properties. It is formed by the oxidative polymerisation of three substituted cinnamaryl alcohols: *p*-coumaryl alcohol, coniferyl alcohol and snapyl alcohol. The amount of these alcohols in lignin differs with plant species.

Although enormous quantities of lignin are decomposed annually, only a small group of microorganisms is capable of degrading lignin. Most of the knowledge about lignin degradation is based on the study of wood rotting fungi, chiefly white rot fungi (higher Basidiomycotina) although the foot rot of cereals in which fairly high amounts of lignin are present indicates that lignolytic enzymes might be involved in this disease also. Of about 500 fungi known to degrade lignin about one fourth cannot utilize it. The wood rotting fungi produce polyphenol oxidases and through their action assimilate and metabolise lignin. However, many wood rotting fungi degrade lignin but cannot utilize it. Among other fungi reported to cause partial degradation of lignin are *Alternaria*, *Cephalosporium*, *Chaetomium*, *Xylaria*, *Pestalotia*, *Fusarium* and *Penicillium*. Among phytopathogenic bacteria, the species of *Pseudomonas* and *Xanthomonas* can cause degradation of lignin.

vi) Other enzymes involved in degradation of cell contents: After breakdown of the cell wall the pathogens come in contact with the host protoplasm. They get the required food mainly from this source. The protoplasm contains mainly proteins, starch and lipids. In addition, phosphorus, potash, sulphur and iron are also present. Nucleic acids and proteins are constituents of the nucleus.

Plant cells contain a vast variety of proteins which play diverse roles as catalysts of cellular reactions (enzymes) or as structural material (membranes). This makes the degradation of proteins very important effect of pathogenesis. All pathogens are capable of degrading many kinds of protein molecules. Mechanisms of protein breakdown by plant pathogens is same as found in plants and animals. The enzymes involved are called proteinases or proteolytic enzymes. They degrade protein to form polypeptides. The enzymes peptidases attack and break down the polypeptides into lower peptides and amino acids which are utilised by the pathogen.

Starch is the main reserve polysaccharide found in plant cells. It is synthesised in the chloroplasts and, in nonphotosynthetic organs, in amyloplasts. Starch is a glucose polymer and exists as amylose and amylopectin. Most pathogens utilise starch and other reserve polysaccharides. The degradation of starch is brought about by the action of enzymes called amylases to produce glucose used by the pathogens as nutrients. Lipids occur in plants as oils and fats in seeds, wax lipids on epidermis, and as phospholipids and the glycolipids which along with protein are the main constituents of cell membranes. All lipids contains fatty acids which may be saturated or unsaturated. Many fungi, bacteria and nematodes degrade lipids with the help of enzymes called lipases and phospholipidases. Many species of *Pseudomonas* produce lipase, phospholipase and alkaline phosphatase. The breakdown of lipids releases fatty acids which are presumably utilized directly by the pathogens.

The nucleic acids (ribonucleic acid or RNA and deoxyribonucleic acid or DNA) are present in the cell and are liable to be acted upon by fungal or bacterial enzymes. DNA is mainly present in the nucleus in the chromosomal beads. In minute quantities it is present in the chloroplasts and mitochondria also. Chromosomal DNA is not degraded but the cytoplasmic DNA can be degraded. RNA is present throughout the cell. These acids contain linear chains of alternating molecules of phosphate and a sugar (ribose in RNA and deoxyribose in DNA). Enzymatic hydrolysis of these acids in the cell cytoplasm results in the release of mononucleotides. Non-specific phosphatases hydrolyse nucleotides into inorganic phosphate and nucleosides. Further action by deaminases separates the amino groups from the nucleosides and makes them available to the parasite.

From the above description it is obvious that tissue disintegration is a function of enzymes secreted by the pathogenic organisms. The destruction of cell walls leads to plasmolysis of protoplasts and death of the cell and utilization of cell contents by the parasite. These degrading processes result in rots, blights and cankers etc. The tissues thus broken down are only those present at the site where the parasite is active. Sometimes disintegration of tissues occurs at a distance from this site. It is due to translocated toxins produced by the pathogen (chapter 11). Necrosis caused by many obligate parasites and pathogens including viruses is not due to enzymic action of their own but due to indirect effect. This may include hypersensitive response of the host cells including programmed cell death, toxins, starvation of the cells and non-availability of materials required for synthesis of cell walls. In deficiency diseases the necrosis representing tissue disintegration is due to shortage of elements required for cell wall synthesis and inhibition of other cellular activities. Boron helps in utilization of calcium for the formation of cell walls. Deficiency of boron in fruits (viz., apple) is known to cause disintegration of the internal tissues.

EFFECT ON GROWTH OF THE HOST

The growth of plants is controlled by a group of naturally occurring compounds in the plant body that are called growth regulators. In some diseases this control system is disturbed and various structural abnormalities appear on the host. The regulatory substances are of two types: growth promoting which include specific hormones and inhibitory substances which also play equally important role. Auxins, gibberellins and cytokinins are the known growth promoting compounds. Growth regulators act in very small concentration and even a slight deviation from the normal concentration induces striking difference in the pattern of plant growth The growth promoting compounds generally continue rising in the plant body but at a particular stage the hormone-inhibitory system of the plant suppresses them so that only normal growth continues. The phenomenon of rise and suppression of regulators is genetically controlled.

In a broad sense, gibberellins and cytokinins are also auxins but in strict sense only indole acetic acid (IAA) is known as auxin. Dormin, ethylene, etc., are growth inhibitors or induce such reactions in the plant organs that lead to premature ripening of fruits and untimely formation of abscission layers leading to fall of leaves, fruits and flowers. The growth inhibitors can inhibit the action of growth promoters or the former can be rendered ineffective by the latter. This depends on the condition of the plant and its response to infection by a pathogen. The pathogens (fungi, bacteria and nematodes) also produce growth regulators. When such compound are produced during the period of infection the host cells are induced to show growth responses. These responses at wrong place and wrong time adversely affect the plant growth. Pathogens also produce metabolites that are not hormones but affect the regulatory mechanisms in the plant thereby causing unrestricted production of growth regulators by the plant. Production of growth promoting compounds in excess of normal requirements of the plant causes overgrowth of cells and tissues. In addition to their own effect the growth inhibitors produced by the pathogen can render the growth promoting substances of the plant ineffective or inhibit their production thus causing growth retardation or stunting of the organ or the entire plant. The imbalance in growth promoting and growth inhibiting substances causes appearance of symptoms known as hypertrophy and atrophy. Hypertrophy may appear as galls, tumors, knots, witches' broom, etc.

Auxins: The naturally occurring auxin in plants is indole-3-acetic acid (IAA). It is continually produced in growing plant tissues and moved from young green tissues to other tissues but is constantly being destroyed by the enzyme indole-3-acetic acid oxidase to maintain the balance. Many plant pathogens produce small quantities of IAA or induce the plant to produce more of this compounds. Although the amino acid tryptophan is most important precursor of this auxin, it appears that different organisms have evolved different pathways of IAA synthesis involving precursors other than tryptophan. It is also possible that some auxin other than IAA and gibberellins may be involved in pathogenesis.

The functions of IAA in the plant are numerous. It is required for cell elongation and differentiation and also affects cell wall permeability. Due to effect of IAA on oxidative enzyme system of the plant there may be abnormal increase in respiration of the tissues. Possibility of this auxin affecting the genetics of the plant was also reported. The conversion of tryptophan into IAA takes place in following steps: Tryptophan — Indole pyruvic acid — Indole acetaldehyde — IAA or Tryptophan — Tryptamine — Indole acetaldehyde — IAA.

The increase in the amount of IAA has been noticed in many diseased conditions of plants. These diseases can be due to any infectious cause such as fungi, bacteria, nematodes and viruses. The fungi causing clubroot of brassica (*Plasmodiophora brassicae*), late blight of potato (*Phytophthora infestans*), smut of maize (*Ustilago maydis*), Panama disease of banana (*Fusarium oxysporum* f.sp. *cubense*), downy mildews of maize and pearl millet (*Peronosclerospora sacchari, P. philippinensis, Sclerospora gramnicola*), and the root knot nematodes (*Meloidogyne* species) not only induce the plant to synthesise more IAA but also themselves produce this auxin. Excess IAA detected in diseased plants, thus, could be due to excessive production of IAA by the plant and due to production of auxin by the pathogen. It could also be due to reduced destruction of IAA in the diseased tissue by IAA oxidase. The pathogen inactivates this enzyme by its metabolites and thus the level of IAA continues rising. This condition has been proved in maize smut and wheat rust (*Puccinia graminis*). Hyperauxinity (excess of IAA) has been detected in many other diseases such as rust of *Euphorbia cyparissias* caused by *Uromyces pisi*, powdery mildew of wheat (*Erysiphe graminis tritici*) and white blister of *Brassica napus* (*Albugo candida*).

The production and activities of auxins have been studied in some detail in bacterial plant diseases. Two examples are cited here. In bacterial wilt and brown rot of potato (*Ralstonia solanacearum*) the bacterium grows in vascular bundles of the tubers and stems and causes vascular browning, rot and wilt. The diseased plants contain 100 times more IAA than the healthy plants. At the same time phenolic compounds (scopoletin) also increase 10 times. Phenolics are known to suppress the activity of IAA oxidase. Thus, one of the explanations for increased level of IAA in diseased potato plants can be that the rise in phenols in the diseased tissues suppresses the IAA oxidizing enzyme that regulates the production and accumulation of the auxin. However, the level of the auxin rises in the beginning of pathogenesis also suggesting that the plant also synthesises the auxins to some extent. After the death of the plant, rise in level of IAA continues. This suggests that the bacterium also synthesises the auxin in dead tissues. Thus, all the three possibilities of the cause of hyperauxinity exist in this disease.

The high level of IAA in the plant increases plasticity and permeability of the cell walls. This makes the pectin, cellulose and proteins easily available to the pathogen. Although phenols help in lignin synthesis which makes the tissues resistant, this does not occur in the diseased plant because IAA interferes with lignification of tissues. The bacterium gets more time for tissue disintegration. Increased respiration and transpiration are other effects produced by IAA through altered cell wall permeability.

The other example is that of the crown gall disease caused by *Agrobacterium tumefaciens*. This bacterium attacks more than 100 plant species and produces galls on root crown, stems and petioles. Crown gall has served as a classic model for investigations of the role of auxins, IAA and related indole compounds, in plant pathogenesis. The bacterium is not present in the galled tissue.

The gall or tumour is initiated in two phases. The first is conditioning phase in which a fresh wound is required without the presence of the bacterium. Then the bacterium enters the host from the soil and second phase starts. A piece of the tumour inducing plasmid DNA of the bacterium (T-DNA) is introduced into the cell and it is integrated with the host DNA. The conditioned cells are transformed into tumour cells. This phase occurs only at temperatures below 29°C. Afterwards, there is no role of the host and the bacterium in the development of the gall. The tumour cells multiply and develop the galls. The process cannot be checked by killing the bacterium.

The tumour cells contain more than normal quantity of IAA. In addition they contain cytokinins. The bacterium can also produce IAA and cytokinins. Genes coding for IAA and cytokinins are present in the Ti plasmid of the bacterium. The presence of higher level of IAA in tumour cells in absence of the bacterium suggests that cells themselves are capable of synthesising this auxin and the cytokinins. Since the enzymes capable of oxidizing IAA are in identical amounts in normal and tumour cells it is evident that excess auxin in tumour cells is due to its synthesis rather than due to lack of its degradation. The increased levels of IAA and cytokinins are sufficient to cause autonomous enlargement and division of the tumour cells. However, IAA and cytokinins alone cannot cause the transformation of healthy cells into tumour cells. There are indications that several metabolic systems are gradually, but permanently, activated during the transition from a normal cell to fully altered tumour cell. The other substances involved in the process are not fully known.

In the knot disease of olive and oleander, caused by the bacterium *Pseudomonas savastonoi*, the pathogen produces IAA which induces gall formation. The more IAA a strain of the bacterium produces the more severe the symptom it causes. Strains that do not produce IAA fail to induce the disease. The genes for IAA synthesis are in a plasmid carried in the bacterium. Some IAA synthesis is also carried out by a gene in the chromosomal DNA of the bacterium.

Gibberellins: The role of gibberellins in pathogenesis, although well-recognised, has been studied in relatively few plant diseases. The gibberellins were discovered by Japanese workers investigating the "bakanae or foolish seedling disease" of rice caused by *Gibberella fujikuroi*, an ascomycetous fungus with its imperfect stage in *Fusarium moniliforme*. The disease is characterised by abnormal elongation of stem due to excessive elongation of internodes. Gibberellins, a group of chemically related compounds, were isolated from these seedlings. At least 38 of these compounds are reported. They are different from each other in their structure and/or biological activity. The best known gibberellin is gibberellic acid (GA 3).

Gibberellins are considered normal constituents of green plants and are also found in microorganisms. They perform numerous functions. One of the important functions is their role as chemical signals which activate cell extension, dormancy breaking and flowering. These chemical signals activate various enzymes in the plant body. Thus, during seed germination the embryo secretes the hormone which activates the cells of aleurone layer to secrete hydrolytic enzymes for liquefaction of reserve starch. It also promotes enzymes that aid in digestion of endosperm cells and softening of seed coat. Other enzymes are also helped by gibberellins. The proteinase enzymes synthesised under the signals from gibberellins cause degradation of proteins, thus releasing amino acids. Among these acids tryptophan, the precursor of IAA, could be one. Thus, there is close relationship between gibberellins and IAA. Tissues treated with gibberellic acid develop higher concentration of IAA. The latter often helps the functions of gibberellins and both substances seem to work synergistically for maximum stem elongation. It is possible that gibberellic acid neutralizes some growth inhibiting systems in the plant such as IAA oxidase.

Gibberellins have strong growth promoting qualities. They speed elongation of dwarf varieties to normal sizes, promote flowering, stem and root elongation, and growth of fruit. Symptoms of growth inhibition can be reversed by application of gibberellic acid. These compounds are suspected to be operating in many other host-pathogen systems such as downy mildew of sugarcane (*Peronosclerospora sacchari*), rust caused by *Uromyces pisi* on *Euphorbia cyparissias* and smut of *Bromus* caused by *Ustilago hypodytes.*

The site of activity of gibberellins in the cell is close to the nucleic acid system. It activates genes that have become inactive and in this process synthesises new messenger RNA which direct the synthesis of various enzymes for different functions.

Cytokinins: The best known cytokinin is kinetin (6-furfuryl-amino-purine). These growth regulating substances are derivatives of adenine, a constituent of DNA and RNA. As such they are essential for growth and differentiation of cells and tissues. The type of tissues and plant organs is determined by the amount of cytokinin in the primordial tissue. Low cytokinin activity causes root formation while high levels induce bud formation. In presence of auxin, these compounds induce rapid cell division. They prevent degradation of protein and nucleic acid thereby delay senescence. Amino acids and other chemicals flow towards points where cytokinin activity is high. The mode of action of cytokinins is similar to that of gibberellins and they also often function synergistically with auxins.

The significance of kinins in plant pathogenesis is uncertain, although their role in many pathosystems has been suspected. These include bean rust, root knot, Victoria blight and some bacterial diseases. Increase in the level of cytokinin-like substances in leaves of bean (*Phaseolus vulgaris*) attacked by *Uromyces phaseoli* and leaves of *Vicia faba* attacked by *Uromyces fabae* is reported to induce formation of "green islands" around the infection centre. Nutrients accumulate in the green tissues. The tissues with low cytokinins thus become senescent.

Growth and differentiation inhibitors: Many chemically different compounds works as growth inhibitors in the plant and are synthesised by them. Excess of these compounds causes such effects as inhibition of cell division, induction of dormancy, formation of abscission layer, epinasty, etc. Growth inhibitors interact with growth promoting substances in the plant and render them ineffective. Thus, normal growth of the plant or plant organs is arrested. Dormin and ethylene are two such compounds.

Dormin or abscission II induces dormancy by converting developing leaf primordia of a bud into bud scales. The inhibitory effects of dormin are counteracted by presence or application of gibberellins. On the other hand dormin can function as antagonist of gibberellins. Dormin can also mask the effect of IAA which cannot be reversed by application of additional IAA although gibberellins can offset this effect of dormin on IAA activity. At least some plant pathogens produce compounds that induce abscission layer at the wrong time. Role of formation of abscission layers in many fruit diseases is known.

Ethylene ($CH_2=CH_2$) is the earliest known and a highly active growth regulator best qualified for a primary role in pathogenesis. It is produced by plants independently or under the influence of pathogenesis. A number of fungal and bacterial plant pathogens also produce ethylene. It is biologically active in even as low concentration as 0.1 part per million. Some prominent effects of ethylene are epinasty, tissue proliferation, marked increase in the rate of respiration, premature senescence and shedding of leaves and stimulation of root proliferation. Ethylene is highly mobile in plants and does not accumulate in tissues. However, when its movement is blocked such as when vascular occlusions and stomatal closure occur in bacterial and fungal wilts, ethylene may accumulate and symptoms of epinasty become apparent.

In the case of Fusarium wilt of tomato (*Fusarium oxysporum* f.sp. *lycopersici*) ethylene production of the fungus is sufficient to account for the epinastic symptoms of the disease. Banana plants attacked by *Ralstonia solanacearum* (bacterial wilt or Moko disease) show premature ripening and yellowing of fruits which is linked with high level of ethylene in the yellowed fruits. Many species of *Pseudomonas* and some species of *Xanthomonas* and *Erwinia* also produce such effects. In most cases the production of ethylene is by the damaged tissues. In a number of viral diseases, leaves with necrotic local lesions produce more ethylene than those from systemically infected plants without necrotic lesions. Necrosis induced by toxic chemicals also results in increased ethylene level. This suggests that in such cases ethylene is the product rather than cause of tissue damage.

From the above account of the effect of pathogenesis on plant growth it can be concluded that changes in growth pattern are caused by imbalance in production, accumulation and translocation of growth regulators in the plant. The normal plant synthesises growth promoting substances in quantities just enough for its normal growth. The plant also produces growth inhibitors to regulate the activity of growth promoters and other chemical substances. In pathogenesis this regulatory mechanism is disturbed or destroyed and as a consequence there is unregulated synthesis of growth hormones and other substances and therefore the changes in growth habit are seen (see Table 5).

TABLE 5: Examples of Plant Diseases in which Symptoms Suggest Overactivity of Growth Regulators.

Disease and symptoms	Host	Parasite
I. Galls caused by cellular proliferation		
Crown gall	Many	*Agrobacterium tumefaciens*
Smut	Maize	*Ustilago maydis*
Wart	Potato	*Synchytrium endobioticum*
Rust	Cedar apple	*Gymnosporangium juniperivirgineanae*
II. Increase in length of stem		
Downy mildew	Sugarcane	*Peronosclerospora sacchari*
Bakanae disease	Rice	*Gibberella fujikuroi*
Rust	*Euphorbia cyparissias*	*Uromyces pisi* (aecial)

Rust	Wheat	*Puccinia graminis*
Rust	Safflower	*Puccinia carthami*

III. Suppression of abscission layer

Leaf blight	Cherry	*Gnomonia erythrostroma*

IV. Stimulation of abscission layer

Leaf spot	Coffee	*Omphalia flavida*

V. Excessive abnormal branching

Witches' broom	Various trees	Many fungi
Witches' broom	Sweet pea	*Rhodococcus fascians*
Downy mildew	Maize	*Peronosclerospora sacchari*

EFFECT ON REPRODUCTION

From practical viewpoint the loss from disease is due to reduction in reproductivity of the plant. Reproduction in plants is determined by its age, nutrition, environments (light, temperature, moisture) and normal physiological activities. Disturbance in one physiological activity initiates a chain reaction that affects other activities thus influencing growth and reproduction. Inaminate causes and unfavourable environments mainly reduce the reproduction of the plant. Infectious diseases reduce or completely suppress reproduction of the host. As was discussed in the preceding section, under normal conditions the abscission layers are formed when the fruit has ripened and contains viable seeds or when the leaf has reached senescence. Pathogenesis induces formation of abscission layers in immature fruits and there is loss of reproduction.

The infectious diseases, localised or systemic, affect physiological activities of the host. When pathogenesis reaches a particular stage reproductive system of the plant is also affected. These effects can be direct or indirect. The direct effects usually lead to partial or complete destruction of reproductive organs or the produce (fruits, seed, etc.). The indirect effect, which is universally associated with almost every disease, results from weakening of the plant or loss of the crop resulting from seeds produced by diseased plants. The process by which pathogens reduce reproduction in plants is mediated by the physical and chemicals means given in the preceding sections.

In loose smuts of cereals there is direct and complete loss of ears and no grains are formed. In many downy mildews also the infected plants produce ears which are partially or completely devoid of grains. In many plant diseases the host produces normal fruits and viable seeds. However, the inoculum gets mixed with the seed and reaches the next crop where it harms the young plant or destroys the reproductive organs of the mature plant. In wilt disease of tomato caused by *Fusarium oxysporum* f.sp. *lycopersici* the fungus is often associated with the viable seeds and harms the seedlings from such seeds. In seed rot, seedling rot and stalk rot of maize (*Fusarium moniliforme, Cephalosporium* spp.) the fungi are present in and on the seed. When the seed is planted these pathogens, depending on the environments, may cause rotting of seeds and seedlings. *Fusarium moniliforme* causes direct loss of seed also when it attacks the cobs and induces cob rot. In loose smut of wheat (*Ustilago segetum tritici*) the seed itself is not damaged by infection of the flowers but the plant developing from it in the next season produces ears completely devoid of grains and full of smut spores. The green ear or downy mildew of pearl millet (*Sclerospora graminicola*) is carried by seed and soil and the infection becomes systemic in the plant developing from the contaminated seed or on contaminated soil. The plants turn yellow, remain stunted and may never reach the flowering stage. If ears develop they bear leafy structures in place of grains. Since the infection is systemic in such diseases there is complete loss of productivity in them. Downy mildews of maize caused by *Peronosclerospora sacchari* and *Sclerophthora macrospora* also often cause complete loss of seed formation. In downy mildew or green ear disease of pearl millet, the cob is partly or wholly converted into leafy structures formed from the floral parts. There is no grain setting.

The direct infection and loss of floral organs and seed occurs in such localised diseases as smut of pearl millet (*Tolyposporium penicillariae*), Karnal bunt of whet (*Neovossia indica*), and ergot of pearl millet (*Claviceps fusiformis*). The grains are partially or completely damaged although majority of grains remain unaffected. In false smut of rice (*Ustilaginoidea virens*) the individual grains in the ear are converted into smut balls. In these two disease the flowers adjacent to the site of infection are usually sterile. Destruction of individual flowers or clusters of flowers are often most striking symptoms of many fruit tree diseases. In fire blight of apple and pear (*Erwinia amylovora*) floral clusters become water-soaked and are blighted. In powdery mildew of apple (*Podosphaera leucotricha*) there is reduced blossom bud production and blossoms are aborted. In powdery mildew of mango (*Oidium mangiferae*) the major damage is through blighting of the flowers. Sterility induced by many viruses and MLOs is well known.

The above examples are only of those diseases where the loss of reproductive parts or the produce is conspicuous and forms part of symptoms. But, loss of reproduction occurs in all diseases because of interference in physiological system resulting in general weakness of the plant. The powdery mildews of legumes do not show serious effect on reproductive organs but pod development may be significantly reduced due to disturbed physiology of the host including serious loss of photosynthesis. The seeds formed in these pods are weak. The increased transpiration in plants affected by rusts also results in shrivelled grains of low viability.

Thus, the effects on reproductivity of the diseased plants are the result of direct consumption of floral or reproductive parts by the pathogen (smuts caused by species of *Ustilago*) or due to altered physiology such as low uptake of water and nutrients (root diseases), low rate of translocation of water and nutrients (wilt diseases), tissue disintegration before flowering, lack of normal photosynthesis (powdery mildew) or photosynthetic area (leaf spot diseases).

EFFECT ON UPTAKE AND TRANSLOCATION OF WATER AND NUTRIENTS

Living cells of plants need sufficient water and nutrients (organic and mineral) for their normal activity. If there is interference with the availability of these requirements the cells fail to perform their physiological functions. Minerals and water are absorbed by roots and translocated by the xylem vessels of the stem upward towards leaves. From vascular bundles of petioles and leaf veins the water and nutrients enter the leaf cells. The minerals and a part of water is utilised by the leaf cells for synthesis of various essential substances. However, a major portion of water reaches the intercellular spaces and diffuses into the atmosphere through stomata and lenticels in the process of transpiration. The organic nutrition of the plant is mostly synthesised in the leaf by photosynthesis and translocated through phloem vessels downward upto the roots. The excess organic nutrients, in various forms (amino acids, sugars, organic acids) are exuded out into the soil as root exudates. It is, thus, obvious that if due to effects of pathogenesis uptake and translocation of minerals and water or photosynthetic process in leaves are checked the plant tissues will starve and when due to starvation the physiological activities of these tissues are affected there will be deficiency of substances produced by these tissues for the entire plant. In this way the entire plant will be sick. For an example, if water is not absorbed by roots or there is obstruction in its translocation the leaves cease to be active and photosynthetic activities decrease or stop. The organic nutrition will not be available to roots. Thus, not only leaves but roots also will be adversely affected. The effect of pathogens on photosynthesis is discussed elsewhere. Here only the uptake and translocation of minerals and water is being considered.

Effect on absorption of water by roots: The water uptake capacity of roots can be affected in three ways: root development is checked and root mass is reduced, roots are injured, and permeability of root cell walls is altered. Some vascular parasites reduce the number of root hairs thus decreasing the effective surface area and therefore reduced uptake of water. Many pathogens such as the fungi causing root decay or root rot and damping off, most phytopathogenic nematodes and some viruses cause sufficient injury to the roots before visible symptoms appear on aerial parts. The deleterious rhizobacteria also contribute to obstructed root functioning. Injuries or wounds kill the finer roots and reduce the root mass. Water uptake is proportionately decreased. Aggravation of effect of roots loss can occur when the plant is unable to produce new roots to replace the damaged roots. In gummosis and fibrous root rot of citrus (*Phytophthora parasitica, P. citrophthora* and *P. palmivora*) when the soil remains continuously wet the pathogens are favoured but new root formation is inhibited.

Roots absorb water and mineral nutrients dissolved in water through the process of osmosis. If the pathogen or its growth regulators, toxins and enzymes alter cell wall permeability of roots osmosis is affected. The role of enzymes and growth regulators was discussed in the preceding sections. Role of toxins is discussed in the next chapter. The reduced osmotic activity causes decrease in water uptake. If the root cells are killed, toxins generated by plant debris or pathogens can enter the roots and affect physiological activities of the plant.

Effect on translocation of water by xylem vessels: The fungal and bacterial pathogens involved in damping off, root rot and cankers can enter the xylem. If the plant is young these vessels can be disintegrated. In affected vessels organs of the pathogen or biochemical substances produced by the pathogen or resistance structures formed by the plant also cause obstruction. Disintegration as well as obstructions both reduce the water carrying capacity of the root system. The crown gall bacterium (*Agrobacterium tumefaciens*), the clubroot fungus (*Plasmodiophora brassicae*) and the root knot nematodes (*Meloidogyne* spp.) develop galls on roots and stems due to overgrowth of cells. The xylem vessels adjacent to these proliferating tissues are crushed or dislocated and thus lose their normal water conducting capacity. The tomato mosaic virus causes necrosis in the stem. Cells adjacent to the necrotic area show hyperplasia and the vessels that come in the path of these overgrowths are rendered non-functional due to pressure. A common example of malfunctioning of the xylem vessels is the group of wilt diseases caused by *Ceratocystis, Fusarium, Verticillium, Pseudomonas* and *Erwinia*. Although these pathogens can obstruct water flow by their physical presence, normally it can happen only when the entire system is colonised. Host reaction resulting in formation of tyloses, gels, etc., and biochemical activities of the pathogen through its enzymes, polysaccharides, toxins etc., as well as punctures in cell wall permitting entry of air bubbles resulting in vascular embolism are more important causes of obstruction in translocation.

The production of pectic and cellulolytic enzymes in vascular infections by wilt causing pathogens is known. The enzymes degrade the middle lamella and release pectic acid and other substances which form gels and gums. The pectic substances form plugs which block the septal pores and membrane pits. Thus, upward and lateral water movement is checked. Browning of vessels is a common feature in vascular wilts. This is due to the pigment called melanin. The pigment is formed by the action of enzymes. The pectinolytic enzymes of the pathogen disintegrate the host cells and start oxidation of phenolic compound. The oxidised products form molecules of the pigment. Since the action of enzymes and growth regulators alters cell wall permeability the molecules of the pigment enter the cells easily. Coloured deposits form a coating of vessel walls which prevents lateral movement of water. These materials may be produced by the host also due to stress.

Fungi and bacteria are capable of producing polysaccharides that can induce wilt symptoms *in vitro*. In many wilt diseases the symptoms have been attributed to these complex compounds. The pathogenic polysaccharides are macromolecules that can not pass through openings in the cell walls. Their entrapping by the cell wall pores causes obstruction in the movement of water from one vessel to another and laterally to other cells. Increase in viscosity and thereby decrease in rate of flow of tracheal fluid is also attributed to polysaccharides and partly accounts for wilt symptoms. In bacterial brown rot and wilt of potato the amount of polysaccharides produced by the pathogen is proportionate to the severity of the symptoms.

Abnormal development of xylem vessels, even in areas not yet invaded by the pathogen, often occurs in vascular wilt diseases. The walls of the new vessels are thinner than normal and they are usually flattened instead of being circular and appear collapsed. In many fungal, bacterial and viral vascular wilt diseases the presence of the pathogen, its toxins and permanent deficiency of water causes development of tyloses in the xylem vessels. Tyloses are outgrowth of the parenchyma adjacent to the xylem and appear as peg-like structures. They obstruct passage of water in the same manner as gels and gums. When formed in the leaf area the tyloses and gels account for chlorosis and necrosis.

Thus, obstruction to water transport by xylem vessels can be caused by one or more of the following:

i)   Vascular occlusion by the pathogen.
ii) Vascular occlusion by host reaction to infection: tyloses, gels.
iii) Vascular occlusion by pathogen induced changes. Vessel lining materials, vessel wall degradation, vascular embolism due to wall penetration and entry of air bubbles.
iv) Impaired water absorption from xylem.

Effect on transpiration: Pathogens affect transpiration also in addition to uptake and translocation of water. Increased transpiration is noticed in most leaf diseases (rusts, powdery mildews, apple scab). The main cause of this increase in transpiration or loss of water from the plant body is disintegration of the cuticle. Increased cell wall permeability and stomatal malfunctioning are also contributory factors. In rust diseases, a major portion of the leaf surface is exposed due to rupture of epidermis by rust pustules. This causes unrestricted loss of water. If there is no translocation of water from roots in proportion to water lost in transpiration symptoms of wilt may appear. The increased suction tension caused by increase in transpiration may result in collapse of vessels or formation of tyloses.

In some diseases, such as blights, death of cells reduces the number of healthy and active cells. This decreases the suction tension. Therefore, the transport of water to leaves by the xylem is also reduced. Reduced permeability of cell walls also produce similar effects. On the other hand, in some wilt diseases the related toxins (fusaric acid, lycomarasmin, etc.,) enhance cell wall permeability. In this situation also the loss of water is increased.

Physiological and pathological wilting: Wilt symptoms usually indicate water deficiency in the plant. This deficiency may occur in plants without infection due to non-availability of water. No pathogen is involved in these cases. The wilting thus caused is known as physiological wilting in contrast to pathological wilting in which non-availability of water is directly or indirectly associated with some pathogen. Soil moisture generally has no relationship with pathological wilting.

Effect on translocation of nutrients and their deficiency in plants: The mineral nutrient from soil enter the roots as solutions in water and are translocated through the xylem. Therefore, the abnormalities that obstruct water uptake and translocation affect the uptake and translocation of mineral nutrients also.

The organic nutrients synthesised in leaf cells (photosynthesis) enter the phloem vessels through plasmodesmata. Due to difference in osmotic pressure they move down the sieve tubes and during this downward movement they continue passing again into the adjacent non-photosynthetic cells through plasmodesmata. These cells then use the nutrients for their activities or store them. Plant pathogens can interfere with this process of organic nutrition at every stage, i.e. synthesis, entry into phloem, downward translocation and utilization by non-photosynthetic cells. Thus, the sick plant is starved of carbohydrates and does not get proper energy for its various activities. Since effect of pathogenesis on photosynthesis is discussed separately, only aspects concerned with translocation of photosynthetic products are given here.

The obligate fungal parasites such as rust and powdery mildew fungi are known to cause accumulation of carbohydrates and minerals at the site of infection. However, there is less photosynthesis and more respiration at these sites. This suggests that increased quantity of photosynthetic products at infection sites is due to their translocation from healthy cells to the diseased area. Two explanations are available to account for accumulation of nutrients at the infection site. First, the nutrients are unable to move out of the diseased area and, second, the accumulation is due to cytokinin activity. The level of cytokinins is higher in diseased cell than in healthy cells. Cytokinins are known to attract water and nutrients. In infection of facultative parasites and saprophytes no such accumulation of nutrients has been observed.

Obstruction in phloem vessels may cause accumulation of carbohydrates in the leaves where they are synthesised. In some viral diseases level of starch in leaves is much higher than in healthy leaves. One of the reasons given for this is that the virus causes necrosis of the phloem. Thus, starch does not flow down the phloem and accumulates in the leaves. This accumulation of starch induces many deformities of the leaf surface. Starch in translocated after hydrolysis by enzymes into smaller translocable molecules. Some viruses render the necessary enzymes ineffective. Since hydrolysis of starch does not occur it remains in the leaf.

EFFECT ON RESPIRATION OF THE HOST

The respiratory process: Respiration is the process by which the cells, through enzymatic oxidation (aerobic or anaerobic) of organic materials (energy rich carbohydrates and fatty acids), produce energy for various activities (maintenance, growth and synthetic processes) and carbon skeletons. In plants the process of respiration takes place in two major steps:

(1) In the first step known as Embden-Meyerhof glycolytic pathway or **glycolysis**, hexose sugars such as glucose are degraded by enzymes of the cell to form pyruvates. The process can occur under aerobic as well as anaerobic conditions but is usually known as anaerobic respiration. To some extent the pentose pathway also helps in this process.

(2) In the second step, the terminal phase, pyruvates are oxidised to produce carbon dioxide, water and energy. The series of reactions participating in this process constitute the Krebs cycle or tricarboxylic acid cycle and its companion, the glyoxalate shunt. Krebs cycle is strictly aerobic and accounts for major portion of the energy produced.

Thus, the two steps in respiration can occur simultaneously only in presence of oxygen. In presence of oxygen one molecule of glucose produces six molecules of carbon dioxide, six molecules of water and about 678,000 calories of energy. However, in absence of oxygen, only the first step occurs and the products of glycolysis are fermented by enzymatic action into lactic acid and alcohol.

A part of the energy produced under aerobic conditions by the above mentioned two steps is lost but about half of it is converted into high energy bonds of adenosine triphosphate (ATP). Actually, respiration does not liberate energy as such but forms these chemical bonds and distributes them throughout the plant. Glycolysis yields 2 molecules of ATP per mole of energy. The rest and major portion of energy is produced by Krebs cycle and glyoxalate shunt. In absence of oxygen this cycle cannot occur and therefore under anaerobic conditions much less energy is produced. In order to meet the energy demands of the cells the rate of glycolysis is increased. This is known as Pasteur effect.

In the course of oxidation of sugars the liberated energy is trapped by inorganic phosphate in the cell. In presence of a transphosphorylating enzyme the energy rich phosphate bond is attached to adenosine disphosphate (ADP) to form adenosine triphosphate (ATP). In this process oxidative energy is spent. The conversion of ADP into ATP is known a phosphorylation and when respiratory oxidation is attached to it the process is known as oxidative phosphorylation. ATP is the molecular form of energy and means of its distribution and storage. The cells requiring energy reverse the process of phosphorylation and breakdown ATP into ADP and inorganic phosphate. This releases energy. Due to increase in the level of ADP and PO_4 in the cell the rate of respiration is increased. If somehow ATP is not fully utilised, less or no

ADP is formed in the cell and rate of respirations is decreased. Thus, the level of ADP and PO_4 in the cell is determined by the rate of utilization of ATP or energy and, in turn, determines the rate of respiration.

The energy produced during respiration is not always directly converted into ATP bonds. Often it is entrapped by reduction of certain coenzymes such as nicotinamide adenine dinucleotide (NAD), nicotinamide adenine dinucleotide phosphate (NADP), flavin and cytochrome. On oxidation these coenzymes release energy to aid in formation of ATP. The oxidation of coenzymes is known as terminal oxidation. In this two hydrogen ions of the reduced NAD take the place of oxygen to form water and enough energy is available to produce 3 molecule to ATP from each $NADH_2$ mole.

In Krebs cycle the pyruvates formed by glycolysis or pentose pathway are

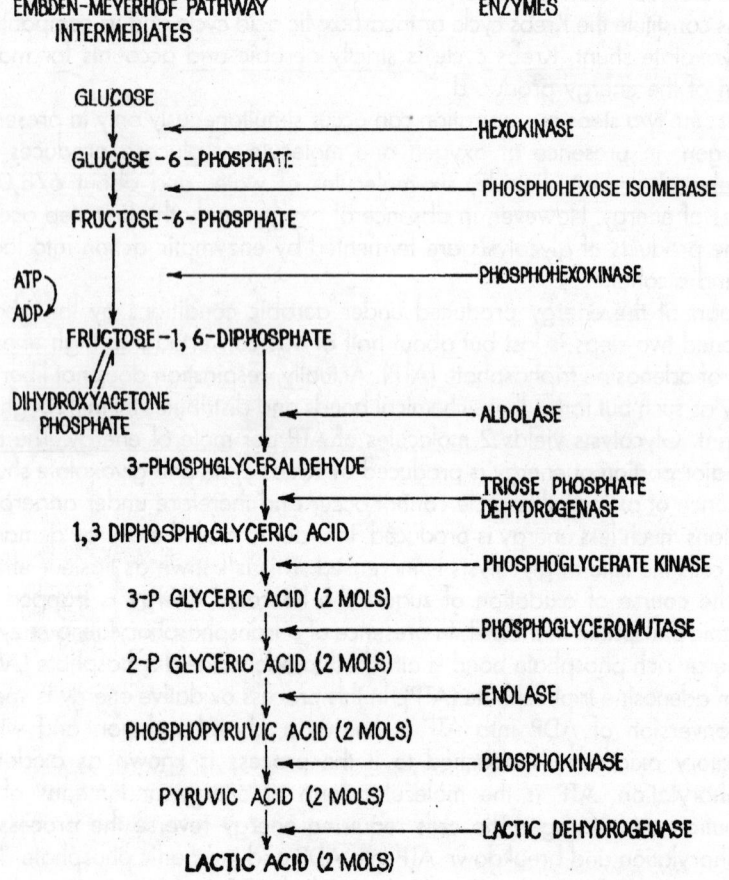

Figure 11. The Glycilytic Scheme (Embden Meyerhof Pathway) with Involved Enzymes.

oxidised and decarboxylated in presence of NAD to produce acetate and $NADH_2$. The acetate reacts with a coenzyme (CoA) forming acetyl CoA. The energy of the acetyl CoA bond is used in a further step-reaction with oxalacetate-to form citric acid and carbon dioxide. The citric acid is again decarboxylated and oxidised resulting in formation of oxalacetic acid. With the revolution of the cycle one molecule of pyruvic acid is converted to 3 moles of $CO_2 + H_2O$. In the process a large amount of energy is released and conserved as ATP.

The energy available from these reactions is used by the plant cell for various activities such as flow of protoplasm, synthesis and translocation of various substances, synthesis of proteins, organic phosphate, phenols, enzymes, cell growth and differentiation, etc. The complexity of respiratory cycles and its effect on cell activity and life account for the fact that in a sick plant respiration is the first to be adversely affected.

Respiration of the diseased plant: When the plant is attacked by an infectious pathogen it generally shows increased rate of respiration and slight rise in

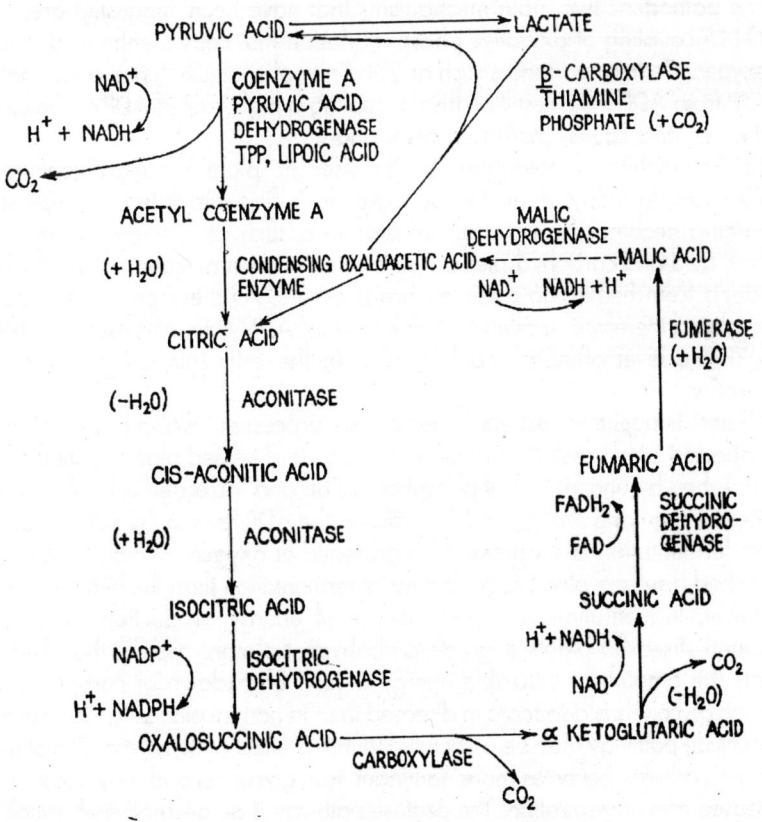

Figure 12. The Krebs Cycles and Involved Enzymes.

temperature of the affected parts. Both obligate and facultative pathogens as well as mechanical and chemical injuries can induce an increase in respiration suggesting non-specific nature of the reaction. However, abnormal respiration due to abiotic mechanical and chemical injuries is temporary and respiration becomes normal after some time.

The increase in respiration starts soon after inoculation and rises to a maximum rate coincident with the sporulation of a fungal pathogen and then declines to normal or even subnormal level. In resistant plants the rate of increase in respiration is very rapid but declines soon while in susceptible plants respiration rises slowly but lasts for longer time. Increased respiration has been noticed in cereal rusts, powdery mildews, blast of rice, late blight of potato and many other diseases. This increase significantly alters metabolic processes in the host. Level of many enzymes associated with respiratory processes is increased in the diseased plants. Accumulation and oxidation of phenols also increases with increase in respiration. It has been suggested that increased respiration in diseased plants is due to respiration of the host as well as the pathogen. Two main mechanisms that have been suggested are:

(1) Uncoupling of oxidative phosphorylation that causes enhanced uptake of oxygen. Certain substances such as 2,4-dinitrophenol (DNP) prevent formation of ATP from ADP. As a result of this uncoupling a high level of ADP accumulates in the cell and causes increased respiration.

(2) Stimulation of metabolism in the diseased plant is a more convincing mechanism. In many plant diseases, growth is first stimulated, protoplasmic streaming becomes faster, synthetic processes such as synthesis of proteins, nucleic acid and carbohydrates are stimulated and translocation of synthesised products from healthy to diseased areas occurs. All these processes require energy and therefore accelerate breakdown of ATP. With increased utilization of ATP the level of ADP and PO_4 rises in the cell. This causes increased respiration.

Changes in respiratory patterns: The various processes involved in respiration are affected by diseased condition of a plant. In diseased plants Pasteur effect is abolished by uncoupling of phosphorylation and increased rate of synthetic processes requiring energy and thus raising the ADP level in the cell. In Pasteur effect fermentation is suppressed by presence of oxygen. When this effect is abolished diseased plants carry out more fermentation than the healthy plants. However, fermentation is a poor source of energy production. Therefore, although diseased plants degrade carbohydrates more rapidly than healthy plants the amount of utilizable energy is low. Breakdown of carbohydrates through glycolysis is also more in diseased than in normal plants. The breakdown by pentose pathway may be associated with this effect of glycolysis. Sometimes, pentose pathway becomes more dominant than glycolysis and may account for increased rate of respiration. The pentose pathway is an alternative of glycolysis and helps in metabolism of carbohydrates. In plants reaching old age or under conditions of wounds and infection glycolysis is replaced by pentose pathway. However, since pentose pathway is not directly linked with formation of ATP the increased respiration through it does not supply as much energy as glycolysis. The pentose pathway is a major source of phenol synthesis and, therefore, when this process is accelerated the diseased plants contain more phenols which may serve as a source of biochemical barriers to spread of invasion by the pathogen.

The oxidation of pyruvates (Krebs cycle) formed by glycolysis is also affected to some degree in a diseased plant in some cases. The adverse effect on Krebs cycle has been supported by the fact that substances inhibiting certain steps in the cycle in a healthy plant have no effect in diseased plant suggesting that this inhibition had already occurred due to pathogenesis. On the other hand, there is also evidence of activation of Krebs cycle in infected plant through increase in concentration of coenzyme A (CoA).

In healthy plants terminal oxidation is a major source of energy providing 10 time more ATP than any of the other pathways. In this process, the $NADPH_2$ produced during glycolysis and Krebs cycle is oxidised through action of flavin, cytochrome and cytochrome oxidase to free oxygen with formation of water. During this oxidation ATP is formed in the cells. The oxidation process is similar in healthy and diseased plants but due to domination of pentose pathway in a diseased plant the released energy is in $NADPH_2$ molecules. The mechanism of oxidation of these molecules to release energy is probably through some non-cytochrome system such as through the role of ascorbic acid oxidase and polyphenol oxidase.

EFFECT OF PATHOGENS ON PHOTOSYNTHESIS

Derangement of the photosynthetic process resulting in chlorotic symptoms is a common phenomenon in plant diseases affecting the green parts. Leaf spots, blights, rusts, mildews and most viral diseases are common examples. Dependence of most of the cellular activities on organic nutrition was emphasized in the preceding sections. Carbohydrates are the major constituents of nutrition because the entire energy for cellular activities is derived from their breakdown. In many crops, such as those which store food as carbohydrates in structures like tubers, derangement of photosynthesis directly affects the yield. In late blight of potato, when 75% of the foliage is killed no photosynthates are available for the developing tubers and they stop growing.

Carbohydrates are synthesised by chloroplasts in green parts of the plant through the process of photosynthesis. This is a basic function of all green plants that enables them to convert light energy (from sun) into chemical bonds of energy for utilization by the cells. This process is related to respiration but the reactions are opposite. While in respiration carbohydrates are degraded to release carbon dioxide, water and energy, in photosynthesis carbon dioxide, water and light energy combine in the chloroplasts to form carbohydrates.

In the basic reaction of photosynthesis atmospheric carbon dioxide and water taken from soil combine in the chloroplasts and in presence of energy from light react to form glucose and release oxygen. Glucose provides the substrate for reactions that follow in the synthesis of other compounds in the plant. The energy released by breakdown of glucose is stored as ATP molecules. The pigments, enzymes and cofactors of photosynthesis are confined to chloroplasts of the green plants. Only few stages of the photosynthetic process require light (**light reactions**). Rest of the reactions (**dark reactions**) can occur in absence of light. Light helps in generation of two essential substances, ATP and $NADPH_2$, for photosynthesis. If these substances can be provided fixation of carbon dioxide in leaves can occur in darkness.

Since photosynthesis provides the basic material for synthesis of all the organic compounds in plant tissues it is apparent that any condition that obstructs this process is pathogenic. In a large number of plant diseases the symptoms on leaves are clear enough to suggest that photosynthetic processes are impaired due to infection. Such symptoms include chlorosis, necrotic lesions, reduced growth, leaf, spots, etc. The effects of pathogens on photosynthesis can be attributed to causes grouped under two categories: (i) the destruction of chlorophyll including chloroplasts, and (ii) decreased efficiency of the photosynthetic process per mole of chlorophyll.

Leaf blights, spots and similar diseases that cause tissue disintegration as a major activity of their pathogens reduce photosynthesis through reduction of the assimilative surface. The main action of these necrotrophic pathogens is not on the chlorophyll and chloroplasts but on the cell wall. However, the non-necrotrophic pathogens, including rusts, mildews, wilt causing fungi, viruses affect photosynthesis through their chemical action on chlorophyll, chloroplasts or the enzyme system associated with the process of photosynthesis, and also respiration.

The reduction in photosynthetic area is not always correlated with loss in photosynthesis. When barley leaves are inoculated with the powdery mildew fungus there is initially an increase in photosynthesis but after 4 days photosynthesis rapidly declines. The reduction obtained 7–10 days after inoculation does not always relate to the amount of leaf surface covered by the mildew: below 30% coverage the reduction is more than what can be explained just in terms of loss of leaf surface. Similar changes have been noted in leaves infected with rusts when reduction in photosynthesis is linked especially with sporulation of the pathogen. With senescence of leaf, sporulation decreases, re-development of chlorophyll occurs and pustules appear on green islands on the yellowing leaves.

Characteristic symptoms of virus diseases often include chlorosis (loss of green colour and appearance of yellow colour). The cause of chlorosis in the cells is either loss of chlorophyll or breakdown of chloroplasts, or both. The ultimate result is impaired photosynthesis. The breakdown of chlorophyll is brought about by the enzyme chlorophyllase which breaks chlorophyll into chlorine rings containing magnesium and chain alcohol. Degradation of chlorophyll is followed by degradation of chloroplasts. On the other hand, the symptomless tissues with less virus content also show reduced photosynthesis per mole of chlorophyll. Often there is increased amount of chlorophyll as is evident from dark green islands on the leaf surface. This suggests that components of the photosynthetic process, other than chlorophyll, are affected first. Many studies have emphasised the effect of virus infection on enzymatic processes of photosynthesis. Starch metabolism is also deranged. Either there is low level of starch in the cells or its quantity is higher than normal. The low level is obviously due to nonfunctioning of the photosynthetic and starch synthesising apparatus. The accumulation of starch is due to inactivation of enzymes hydrolysing starch for translocation and nonfunctioning of phloem due to necrosis and collapse.

Some bacterial infections are characterised by chlorosis. For example, species of *Xanthomonas* producing leaf spots, show yellow halo around the necrotic lesions. This limited chlorosis can be the effect of toxins of the pathogen on chloroplasts or due to reduced rate of photosynthesis. Pathogens that induce occlusions in the vascular system indirectly affect photosynthesis through water stress and closing of stomata and resultant inhibition of carbon dioxide intake.

Like bacteria and viruses, fungal pathogens also obstruct photosynthesis by affecting chloroplasts in the same manner. Genetically, the primary reactions with which light is essentially associated are not affected. The toxins of fungi can affect chloroplasts and rate of photosynthesis. Occlusions of vascular elements in wilt diseases caused by species of *Fusarium* and *Verticillium* cause water stress, induce closure of stomatal openings thus reducing carbon dioxide intake and also affect the enzyme system responsible for photosynthesis. On the other hand, there are some fungal diseases that enhance photosynthesis of portions of the plant that are not directly invaded by the pathogen.

EFFECT ON PERMEABILITY OF CELL MEMBRANES

Membranes function as permeability barriers that allow passage into a cell only of substances that cell needs and inhibit leakage of essential molecules that the cell needs. Change in permeability of cell membrane is often the first detectable response of the cell to infection by pathogens, to host specific and host-nonspecific toxins and to certain air pollutants. Loss of electrolytes is the most important effect of change in membrane permeability.

THE LOSS OF YIELD

As discussed in the preceding sections, changes induced by pathogens cause damage to structures and functions and are bound to prevent the plant from yielding to its normal potential. However, it cannot be said that the loss in potential yield is enough to warrant controlling the disease with inputs that will cost money. A good example is that of late blight of potato. The disease can occur in the plains of northern India any time from November to February. As was stated earlier, tuber growth ceases when 75% of the foliage is destroyed by the blight. If this point is reached before tuberization starts there will be no yield and the loss will be 100%. Thus, epidemics starting in late November or early December are most dangerous and need all efforts for control. If 75% foliage is lost in later December, there will be heavy loss warranting control measures. If the 75% blight point is reached late in the season when tubers are reaching maturity the loss in yield will be nominal, not warranting any treatment of the crop. Same can be true for many other diseases like rusts and powdery mildews. In the Sigatoka disease of banana (Cercospora musae and C. fijiensis) the loss of foliage is directly responsible for loss in yield. It takes at least 12 healthy leaves on mature banana plant to carry the fruits to maturity. Due to destruction of most of the mature leaves by the disease, only few leaves are left which are insufficient to support maturity of the fruits. If more than 4 leaves are destroyed on a plant, the fruits produced are small and susceptible to various types of rots.

11

Role of Toxins in Plant Pathogenesis

The role of toxic metabolites of pathogens in infectious diseases of man and animals was suspected even during the 19th century when the etiology of such diseases was not yet fully understood. Later, the discovery of endotoxins of *Clostridium tetani*, *C. botulinum* and *Corynebacterium diphtheriae* established the role of microbial metabolites in these highly toxic diseases of man. The significance of such toxins in plant pathogenesis is of much later realisation but the subsequent advances in the study of toxins as substances injurious to plants have been very rapid. In 1954, Ernst Gaumann had asserted that a microorganism cannot be pathogenic unless it is toxigenic. In other words, unless the organism produces toxins or microbial poisons that penetrate the host tissues and damage the cells, it cannot cause disease. At that time these toxic metabolites included all injurious substances produced by the pathogens, including the enzymes. With better information available on the origin, selectivity and mode of action of these substances it became possible to designate them separately as toxins, phytoaggressins, etc. It also became evident that in most diseases of plants some toxin was invariably associated. These are often called antimetabolites because they suppress the metabolic activities of the host.

PATHOTOXINS, VIVOTOXINS AND PHYTOTOXINS

A toxin can be defined as a microbial metabolite excreted (**exotoxin**), or released by lysed cells (**endotoxin**), which in very low concentration, is directly toxic to cells of the suscept. In plant pathology, the term toxin is used for a product of the pathogen, its host or pathogen-host interaction which even at very low concentration directly acts on living host protoplasm to influence the course of disease development or symptom expression.

The above definition emphasises that a toxin is injurious or lethal to only macroorganisms (plants and animals). This differentiates toxins from antibiotics which also are microbial metabolites. Some toxins are structurally similar to antibiotics. However, antibiotics are toxic only to microorganisms (fungi, bacteria, mycoplasma) at low concentrations. They can be toxic to man and animals and plants at excessive dosages only. The effect of viral infections is usually not included in toxic effects because the viruses do not directly affect the protoplasm but function at the genetic level inducing malfunction. Enzymes, causing rot or tissue disintegration are also not toxins because they do not affect the protoplasm and only disintegrate the cell wall, using the degradation products of middle lamella as substrate for their activity. Such substances have been termed as **phytoaggressins**. In strict sense, a toxin has low molecular weight, is mobile in the plant and does not affect structural integrity of the host tissues.

The toxin hypothesis was first presented with experimental evidence obtained from the studies of the metabolites of *Cochliobolus* (*Helminthosporium*) *victoriae*, the fungus which causes victoria blight of oat. It postulated that (a) the toxin will produce all symptoms characteristic of the disease, (b) sensitivity to toxin will be correlated with susceptibility to the pathogen, and (c) toxin production by the pathogen will be directly related to its ability to cause disease. Except **victorin**, the toxic metabolite of *C. victoriae*, the vast majority of toxins associated with plant diseases fail to exhibit all the above characters. Nevertheless, they are associated with disease syndrome. It has been now accepted that in most cases not one but more than one toxins function together to exhibit the syndrome.

The toxins which have been shown to "play a major causal role in disease" fulfilling the above three criteria are called **pathotoxins** (Wheeler and Luke, 1963). These compounds produce all or most of the essential symptoms of the disease in a susceptible host as a convincing evidence of their causal role and can be host specific, such as victorin, or nonspecific. Substances for which a causal role in disease is merely suspected rather than established can be called **phytotoxins**. These are products of parasites which induce few or none of the symptoms caused by the living pathogen. They are nonspecific and there is no relationship between the toxin production and pathogenicity of the disease causing agent. Alternaric acid, produced by *Alternaria solani*, is a typical example of phytotoxin in this restricted sense.

A list of pathotoxins, proposed by Wheeler (1975) and others is given in Table 6.

Vivotoxin is a substance produced in the infected host by the pathogen and/ or its host which functions in the production of the disease but is not itself the initial inciting agent of the disease. Three criteria were suggested for a vivotoxin: reproducible separation of the toxin from the sick plant, purification or chemical characterization, and induction of atleast a part of the disease syndrome by placing the toxin in a healthy plant. Production by the causal organism of the disease was suggested as an additional criterion by Graniti (1972). Vivotoxins are usually nonspecific. Typical examples are fusaric acid, lycomarasmin and pyricularin.

According to specificity the toxins can be : (i) nonspecific which can affect the protoplasm of many unrelated plant species in addition to the main host of the pathogen producing the toxin, and (ii) host specific toxins which adversely affect only the specific host of the pathogen. The latter are very active and can produce their effect even in extremely low quantities.

TABLE 6: Some Pathotoxins

Specificity	Toxin	Pathogen
	Pathogen produced toxins	
Selective	Victorin	*Helminthosporium victoriae*
	T-toxin	*H. maydis* race T
	HC-toxin	*H. carbonum*
	HS-Toxin	*H. sacchari*
	AK-toxin	*Alternaria kikuchiana*
	PC-toxin	*Periconia circinata*
	PM-toxin	*Phyllosticta maydis*
Non-selective	Fumaric acid	*Rhizopus* spp.
	Tentoxin	*Alternaria tenuis*
	Marticin	*Fusarium* spp.
	Tabtoxin	*Pseudomonas syringae* pv. *tabaci*
	Syringomycin	*P. syringae* pv. *syringae*
	Phaseolotoxin	*P. savastonoi* pv. *phaseolicola*
	Fusicoccin	*Fusicoccum amygdali*
	Oxalic acid	*Sclerotium* and *Sclerotinia*
	Zinniol	*Alternaria* species
	Ceratoulmin	*Ceratocystis ulmi*
	Ophiobolins	*Helminthosporium* spp.
	Cercosporin	*Cercospora* spp.
	Plant or plant x pathogen produced toxins	
Selective	Amylovorin	*Erwinia amylovora*
Non-selective	Juglone	*Juglans nigra*

EFFECT OF TOXINS ON PLANT TISSUES

It is difficult to separate the effect of toxins from those of other biochemicals on plant tissues during pathogenesis. This is because many effects, such as increased respiration, are common to many type of tissue injury. The possible mechanisms of phytotoxicity include the following:

i) Changes in cell wall permeability: All diseased tissues show changes in cell wall permeability. Some show this effect even before any visible symptom of the disease appears. This change may be caused not only by toxins but also by other biological, chemical and physical agents. The event is the first step in functional disorder caused by pathogenesis.

Some toxins kill plant cells by altering the permeability of the plasma membrane thus permitting loss of water and electrolytes and also unrestricted entry of substances including the toxin. Potassium ions are major loss from leaking cells. The plasma membrane consists largely of lipids and proteins. Toxins probably either destroy them or interfere with the mechanisms involved in their synthesis. Lycomarasmin, fusaric acid, picolinic acid, victorin and other toxins are known to affect permeability of the cell wall. In the case of victorin, the toxin causes increased permeability in susceptible tissues even at very low concentrations making the cells leaky. These cells cannot accumulate salts and other substances. Similar effects can be obtained with resistant tissues but only with high concentration of the toxin.

ii) Disruption of normal metabolic processes: Due to changes in cell wall permeability many physiological activities of the host cells are disrupted. The disturbed salt balance in the protoplasm causes increase in respiration. The loss of water and other substances causes malfunctioning of the enzyme system finally resulting in death of the cell. Coagulation or hydrolysis of protoplasmic protein occurs through blocking of the enzyme system or through action of the toxin as an antimetabolite to some vital metabolite, such as obstruction of metabolism of methionine by tabtoxinine. Pyricularin inhibits polyphenol oxidase system and victorin acts by uncoupling of oxidative phosphorylation. Fusaric acid and picolinic acid can form chelating complexes with heavy metal ions and could inhibit enzyme systems which require these metals for their activity. Toxin resistant tissues inactivate the toxin or have alternative enzyme systems for normal functioning.

iii) Other mechanisms: Interference with growth regulatory system of the plant may cause stimulation of growth of plant parts. Some toxins inhibit root growth. *Fusarium moniliforme* produces a thermostable toxin even in soil around the roots which causes browning of the roots and their restricted development. Stomatal dysfunction has also been reported for certain pathotoxins. Physical blocking effect of large molecules is also a mechanism of phytotoxicity by toxins.

THE SELECTIVE (HOST-SPECIFIC) TOXINS

A host specific toxin (HST) is a metabolic product of a pathogenic microorganism which is selectively toxic only to the susceptible host of the pathogen. These compounds have a very high degree of toxicity to the suscept in low concentrations at which they do not affect other organisms. They produce all the essential symptoms of the disease when placed in the healthy susceptible host. Host specific toxins are known from at least nine fungal species in four genera (*Helminthosporium, Periconia, Alternaria, Phyllosticta*). Among the bacterial pathogens, amylovorin, a toxin produced by *Erwinia amylovora*, is also classified as selective toxin.

Victorin or **HV-toxin** is one of the most potent and selective pathotoxins. It is the first well-documented and widely recognised determinant of host selectivity discovered in 1946. This toxin is produced by *Cochliobolus victoriae* (*Helminthosporium victoriae*), the fungus that causes victoria blight of oat. The disease is characterised by necrosis of the root and stem base and blighting of leaves. The agronomic background of this disease and the pathotoxin involved in it are quoted as an example of how introduction of a new plant gene rather than a pathogen, and how genetic uniformity of a crop can create serious economic problem by large scale devastation of the crop.

The problem of Victoria blight of oat began with the efforts of the plant breeders to control crown rust (*Puccinia coronata* f.sp. *avenae*) through genetic resistance. A gene for resistance to the rust was located in the oat cultivar Victoria from Argentina and was incorporated into several oat cultivars. These were widely planted in 1945. Soon their cultivation had to be abandoned because they were devastated by a new disease, later identified as blight caused by *Helminthosporium victoriae*. This fungus was already present in the USA as a weak parasite of grasses and as saprophyte and was endemic in the country. The gene carrying resistance to most races of the rust fungus (Pc) conferred unusual susceptibility to *H. victoriae* and its toxin. The toxin was so specific in its effects on susceptible genotype that it was used by plant breeders to screen for disease resistance. The gene (Pc) from cv. Victoria for rust resistance is probably identical with and tightly linked to the gene (Vb) for victoria blight susceptibility. Crown rust resistance and victoria blight susceptibility in oats are fully dominant.

Cochliobolus victoriae (*H. victoriae*) can produce the toxin victorin or HV-toxin in culture. A single gene in the fungus, called Tox 3, controls victorin production. The toxin is heat resistant, soluble in methanol and butanol and is adsorbed on alumina. It is a low molecular weight, chlorinated, partially cyclic pentapeptide. It is linked to a tricyclic amine named victoxinine (victalanine). Empirical formula of victoxinine is $C_{17} H_{29} NO$. The peptide and the amine can be separated under mild alkaline conditions. The peptide is non-toxic and contains aspartic acid, glutamic acid, glycine, valine and leucine. It is host-specific. Free victoxinine is mildly toxic but has no host specificity. Biological effect on the oat plant results from a combination of the two. Many nonpathogenic strains of *H. victoriae* and other species of *Helminthosporium* (*Bipolaris*) produce victoxinine but not in combination with the peptide. Further, only the fungus isolates that produce victorin in culture are pathogenic and those that do not are nonpathogenic.

Victorin is highly mobile in the plant. When produced at the stem base it can move up to the leaves inducing blight. Early responses of the sensitive plant cells to victorin include depolarization of the plasma membrane potential, changes in permeability, rapid leakage of electrolytes, lysis of protoplasts, and extracellular synthesis of callose, the last is synthesised by protoplasts in response to victorin. Respiration is increased due to uncoupling of oxidative phosphorylation. Carbon dioxide fixation in dark and free amino acid contents are also increased. Root growth, transpiration and auxin-induced cell elongation are inhibited. The toxin is detected on the cell surface but there is no proof that it is firmly bound to the cell membranes. A highly active preparation of victorin causes complete inhibition of root growth of susceptible seedlings at 0.2 µg/ml while resistant plants can tolerate more than 400000 times higher concentration (Scheffer and Yoder, 1972).

T-toxin or ***Helminthosporium maydis race T-toxin (HMT-toxin)*** is produced by the fungus *Helminthosporium (Bipolaris) maydis (Cochiobolus heterostrophus)*. It has a similar history as the victorin. The pathogen causes leaf blight of maize. The disease is best known for its epidemic in maize in 1970 in the USA. Two factors were responsible for the epidemic: (i) a change in the maize population of the 1960s from a variety of genetic types to virtually homogenous Texas male-sterile cytoplasm (Tcms) because of economics of seed production and (ii) the appearance in the pathogen population of strains able to produce T-toxin which Makes *C. heterostrophus* highly virulent on Tcms maize, resulting in lesions several-fold larger than normal. Strains of *C. heterostrophus* that produce T-toxin are known as race T. All other strains are known as race O. Resistance and susceptibility to *H. maydis* race T and its toxin are inherited maternally (in cytoplasmic genes). The ability of the pathogen to produce T-toxin and its virulence to Tcms maize are controlled by one and the same gene. The susceptibility of Tcms to T-toxin is 25 times more than normal resistant plants. The toxin is sufficiently host specific to be useful as an aid in selection for disease resistance.

T-toxin is a family of linear polyketols that act by disrupting the function of the mitochondria of Tcms maize. Following the toxin treatment, the susceptible mitochondria are swollen, oxidation and phosphorylation are uncoupled and the inner membrane is disrupted (Scheffer, 1976). Dysfunction (closure) of stomata and disruption of photosynthesis are also suspected although chloroplasts are not affected.

HC-toxin or ***Helminthosporium carbonum toxin:*** The fungus causes leaf spot of maize. Certain maize inbred lines and their hybrids are highly susceptible to *H. carbonum* and its toxin. The fungus species has four races, 1, 2, 3 and 4. Only race 1 produces the HC-toxin. The toxin is produced in synthetic media also. It is a cyclic substituted polyamide with the formula $C_{32}H_{50}N_6O_{10}$ (mol wt 679). The HC-toxin is highly selective and produces physiological effects similar to those in tissues infected by the pathogen. The effects are somewhat similar to those of victorin. Root growth is inhibited by 50% at as low concentration of the toxin as 0.2 µg/ml in susceptible hosts and 20 µg/ml in the resistant hosts. There is increased respiration and electrolyte leakage. At low concentrations the toxin causes increased growth and synthesis by maize seedlings but cells exposed to higher concentrations become leaky and die. The toxin, however, has no effect on mitochondria or non-dividing mesophyll protoplasts, chloroplasts and on protein synthesis. A single gene diffecence between races 1 and 2 of *H. carbonum* controls the exceptional virulence of the race 1. The gene is now called Tox 2. It is involved in the synthesis of HC-toxin.

HS-toxin or ***Helminthosporoside:*** The eyespot disease caused by *Helminthosporium sacchari* is especially severe in seedlings of certain sugarcane cultivars. The host specific toxin produced by the fungus induces reddish brown stripes when injected into susceptible leaves. The toxin brings about changes in chloroplasts and cell wall permeability and is reported to bind a single protein in the suscentible host but not in the resistant host. The protein in the resistant lines has no affinity for the toxin. The structure of the toxin is probably 2-hydroxycyclopropyl-α-D-galactopyranoside.

PC-toxin or *Periconia toxin:* Milo disease or Periconia blight, caused by *Periconia circinata* attacks only milo type of grain sorghum (*Sorghum vulgare* var. *subglabroscens*). The pathogen is soil-borne and invades the roots and the lower internodes of the culm. In older roots the cortex decays, the central cylinder turns red and dies, followed by reddening and death of the culm base. Leaves of older plants, although not invaded by the pathogen, may roll, wilt, and turn yellow and show the blight symptom. When young seedlings are attacked, the foliage has a scalded appearance. The effects on the leaves are due to the toxin produced by the pathogen at the stem base. Only virulent strains produce the toxin. The culture filtrates of the fungus are capable of reproducing the symptoms of the disease in susceptible cultivars but not in resistant cultivars. The purified toxin from the culture filtrates completely inhibits root growth of susceptible seedlings with as little as 0.1 µg/ml but does not affect resistant plants at 26 µg/ml. The toxin is a low molecular weight polypeptide. There is evidence that it contains three or more related and selectively toxic compounds with different chemical properties. Acid hydrolysis of one of these compounds has yielded alanine, aspartic acid, glutamic acid and serine in 6 : 4 : 2 : 2 proportion. In structure and biological activity it appears equivalent to T-toxin. The toxin causes rapid loss of potassium ions and other materials through leakage of the plasma membrane of susceptible but not resistant tissues. It does not affect activities of isolated mitochondria, chloroplasts and nuclei and has on direct effect on protein synthesis.

AK-toxin or *Phytoalternarin:* This host specific toxin is produced by *Alternaria kikuchiana*, the fungus causing black spot of Japanese pear (*Pyrus serotina*). Only the cultivar Nijisseiki and its derivatives are susceptible to the disease. The disease became serious only when these cultivars were put to widespread cultivation in Japan and Korea. Three selective toxins designated as phytoalternarin A, B and C have been obtained from culture filtrates of the fungus. Only one has been chemically defined. Drops of culture filtrate applied to leaves cause disease symptoms on susceptible hosts. Mechanism of action of AK-toxin is similar to that of victorin. Cell wall permeability of susceptible pear cells is changed within minutes after exposure to the toxin. There is direct damage to the plasma membrane leading to a rapid loss of electrolytes (potassium and phosphate) from susceptible leaf tissues within 5–30 min of the exposure (Park, 1976). The plasma membrane develops prominent invaginations and cell walls show conspicuous degradation. The initial toxic effect of the toxin occurs at the interface between the cell wall and the plasma membrane. Similar drastic increase in loss of electrolytes is evident during infection of the host by the pathogen. The spores of *A. kikuchiana* do not contain AK-toxin but synthesise and release the toxin during germination. By 4 hours after inoculation with spores, one germinating spore yields about $10^{\times 6}$ µg toxin which is sufficient to disturb function of about 100 host cells. This occurs before actual invasion. During this process loss of electrolytes is increased. Then the invading hyphae produce more toxin and disrupt host membranes causing a second high increase in electrolyte leakage in 9 hours. Visible symptoms appear 11–12 hours after inoculation as tiny, dark, necrotic spots.

AM-toxins or **Alternaria mali toxins:** Blotch of apple is caused by *Alternaria mali (A. alternata)*. Highly virulent strains of the fungus produce at least 7 toxins which are collectively known as AM-toxins. Each of these toxins has a high degree of host specificity. The two major toxins of the group have been isolated in crystalline form. These are: AM-toxin I or alternariolide ($C_{23}H_{31}N_3O_6$) and AM-toxin II ($C_{32}H_{29}N_3O_6$). Alternariolide is a depsipeptide and AM-toxin II is its demethoxy derivative. These toxins cause veinal necrosis in apple leaves. Unlike the AK-toxin, the resistant cultivars are not immune to AM-toxin but can tolerate approximately 10000 times higher concentration than susceptible cultivars. The toxin acts on plasma membrane causing leakage of electrolytes and also on chloroplasts resulting in reduced photosynthetic carbon dioxide fixation. Due to presence of so many toxins in *A. mali*, the fungus does not lose its pathogenicity easily because. If one toxin is not present others take its place thus maintaining pathogenicity. This is not true for *A. kikuchiana* which loses pathogenicity rapidly because only one toxin is responsible for pathogenicity and gene controlling the synthesis of the toxin may be lost.

Alternaria alternata is generally a saprophytic fungus sometimes becoming an opportunistic pathogen. In the course of its evolution, as a result of mutation or other genetic changes, some strains of the fungus have acquired an ability to produce HSTs and have become highly specialized pathogens.

THE NON-SELECTIVE (NON-SPECIFIC) TOXINS

These toxins, produced by pathogens, plants or plant-pathogen interactions, have no specificity and can affect the physiology of those plants also that are normally not infected by the pathogen. The non-specific toxins can be **pathotoxins** (tentoxin, tabtoxin, marticin), **vivotoxins** (fusaric acid, lyconarasmin, pyricularin) or **phytotoxins** (alternaric acid, colletotin, diaporthin, cochliobolin, skyrin, polysaccharides and ethylene, etc.) and, rarely, **mycotoxins**, the toxins produced by fungi, mainly on foodstuff, which are toxic to mostly vertebrates (moniliformin, aflatoxins, etc.)

While the selective pathotoxins, discussed in the preceding section, are considered as "primary determinants of disease" the non-specific toxins are often termed "secondary determinants of disease" However, many of these non-selective toxins seem to be necessary for successful infection of the host and expression of full syndrome, while others play only a secondary role.

When injected into a host or nonhost plant the non-specific toxins produce few or none of the typical symptoms of the disease.

Three main lines of evidence have been used to show involvement of non-selective toxins in plant disease. These are: reproduction by the toxin of distinctive early disease symptoms, correlation of toxin production and pathogenicity, and recovery of the toxin from diseased plant in quantities sufficient to account for symptom development. Non-selective pathotoxins are those for which at least two of these three lines of evidence have been provided. However, such evidence is usually not very conclusive.

Tabtoxin or ***Wildfire Toxin:*** Five structurally distinct classes of toxins that cause either chlorotic or necrotic symptoms in infected plant tissues are produced by pathovars of *Pseudomonas syringae* or *P. savastonoi*. Chlorosis inducing toxins are most common. These include tabtoxin, phaseolotoxin, coronatine and tagetitoxin. The necrosis inducing toxins are restricted to pathovars of *P. syringae* only.

Tabtoxin is produced by *Pseudomonas syringae* pv. *tabaci*, the causal bacterium of tobacco wildfire disease and also by *P. syringae* pv. *coronafaciens* (halo blight of oat), and pv. *garcae*. In tobacco wildfire disease the necrotic lesions on the leaves are surrounded by a yellow halo. The cell-free culture filtrates of the bacterium can produce similar symptoms not only on tobacco but also on leaves of many non-host plants. Most of the toxins produced by plant pathogens are pleiotropic, that is, they have more than one effect on the host cells, but most bacterial toxins, including tabtoxin, are monotropic, having a single effect.

The wildfire toxin is a dipeptide with the chemical formula of $C_{10}H_{16}O_6N_2$. In the host and/or in the bacterium the toxin is hydrolysed to **tabtoxinine-β-lactam** by non-specific peptidases. It is this compound, not the intact tabtoxin, that is responsible for the biological effects of the toxin. Tabtoxinine is a toxic amino acid attached to lactic acid by a lactone ring. Strains of *P. syringae* pv. *tabaci* lacking tabtoxinine do not produce chlorotic halo around the leaf lesions. Tabtoxinine is structurally similar to methionine, an essential amino acid in plant cells. However, methionine does not protect tobacco leaves from the effect of the toxin. Tabtoxinine inhibits the enzyme glutamine synthetase leading to accumulation of toxic concentrations of ammonia which in turn uncouples the carbon and energy fixation components of photosynthesis and destroys the membrane of chloroplasts. The toxin is a virulence determinant for the bacterium.

Phaseolotoxin: *Pseudomonas syringae* pv. *phaseolicola*, renamed *P. savastonoi* pv. *phaseolicola*, causes halo blight of bean (*Phaseolus vulgaris*) and some other legumes. The chlorotic halos are accompanied by ornithine accumulation in the tissues. The functional toxin in the plant is **octicidin** which is more potent than the intact phaseolotoxin. Octicidin blocks conversion of ornithine to citrullin, a precursor of arginine. The resultant chlorosis is due to depleted arginine pool in the developing leaves. The chlorotic effect of phaseolotoxin can be reversed by application of arginine to the affected tissues.

Syringomycins: This group of non-specific toxins is exclusively produced by *Pseudomonas syringae* pv. *syringae*, a major pathogen of stone fruit trees causing leaf spots and cankers. These toxins are cyclic lipodepsinonapeptides that are members of polypeptide class of antibiotics produced by some bacteria. Such as *Bacillus* spp. **Syringomycin** contains serine, diaminobutyric acid, arginine, phenylalanine and some amino acids in combinations. Syringostatin and syringotoxin are amino acid analogs of syringomycin. **Syringostatin** has homoserine, ornithine and threonine substituted for the arginine, phenylalanine and one mole of serine in syringomycin. **Syringotoxins** differs from syringostatin by substitution of glycine for one of the residues of diaminobutyric acid. These compounds are wide-spectrum antibiotics that are phytotoxic. They cause rapid, detergent-like lysis of cellular membranes resulting in necrosis.

Cornatine: This toxin is implicated in diseases caused by *Pseudomonas syringae* pv. *atropurpurea*, pv. *tomato*, pv. *morsprunorum* (canker of cherry and plum), pv. *maculicola* and pv. *glycinea* (*P. savastonoi* pv. *glicinea*, the causal agent of bacterial blight of soybean). All naturally occurring strains of *P.s.* pv. *tomato* produce coronatine. The toxin is composed of an alpha amino acid, coronamic acid, linked by an amide bond to a polyketide moiety, coronafacid acid. The toxin causes a light-dependent chlorosis which may eventually become necrotic and cause stunting of plant tissues. In addition, the toxin induces hypertrophic growth of potato tuber tissues and increases activities of enzymes which result in loosening of the cell wall.

Tagetitoxin: This toxin is produced by *Pseudomonas syringae* pv. *tagetis*, a pathogen of several members of the Compositae family including marigold and sunflower. The toxin causes a striking apical chlorosis as well as necrotic leaf spots, sometimes accompanied by chlorotic halo. It acts by inhibiting chloroplast RNA polymerase. Plastid ribosomes are completely absent.

Tentoxin is produced by *A. tenuis* (*A. alternata*). The toxin and the pathogen cause striking variegated chlorosis in seedlings of cucumber, cotton, citrus and other plants. The toxin is effective even at a concentration of 2 ppm. It is a cyclopeptide with molecular formula of $C_{24}H_{32}N_4O_4$. Acid hydrolysis of the toxin gives leucine and N-methylalanine. The toxin causes large reduction in chlorophyll content, inhibits cyclic photophosphorylation and induces stomatal closure in sensitive plants.

Fusarial toxins: Species of *Fusarium* produce several toxins, both patho- and vivotoxins. They are well known for their wilt inducing character. Common examples of these wilt diseases are wilt of cotton (*Fusarium oxysporum* f.sp. *vasinfectum*), wilt of tomato (*F. oxysporum* f.sp. *lycopersici*) and wilt of pigeon pea (*F. udum*). The main symptoms of these diseases are epinasty, browning and plugging of xyiem vessels, necrosis of tissues, wilt and death of the plant. Some of these symptoms such as epinasty can be produced by metabolites of the pathogen on many plant species.

Marticin is a pathotoxin produced by *Fusarium* species of the Martierella group. The pea pathogen, *F. solani* f.sp. *pisi*, produces large quantities of marticin in culture. Marticin has been extracted from diseased plant tissues in quantities sufficient to induce wilting and general necrosis. Marticin and similar red pigmented compounds are naphthazarin toxins in naphthaquinone derivative group.

Fusaric acid is the most studied pathogen produced wilt toxin classified as a non-specific vivotoxin. It does not produce all the symptoms of wilt. Chemically this toxin is 5-*n*-butyl- picolinic acid. The toxin is produced by many *Fusarium* forme species of the elegans group. These include *Fusarium oxysporum* f.sp. *batatis*, f.sp. *conglutinans*, f.sp. *cubense*, f.sp. *lini*, f.sp. *lycopersici*, f.sp. *vasinfectum*, *Fusarium udum* and *F. moniliforme*. Production of the toxin by some species in the rhizosphere is also reported. The toxin is active at 20–200 mg/kg fresh weight. Sometimes, another toxin, dehydrofusaric acid, is associated with fusaric acid which is easily converted into the latter. It mainly causes interveinal chlorosis. The role of fusaric acid in the plants is said to be of many types. It causes chelation of iron and copper in the host cells and alters the cell wall permeability. This disturbs the ionic balance of the cell. It also affects the enzymatic processes in the cell. By chelating the enzymes or by rendering respiratory enzymes ineffective it alters respiratory pattern of the plant. Wilt syndrome in plants is produced by a combination of several toxins and metabolites. Phytonevein, a toxin present in culture filtrates of *F. oxysporum* f.sp. *niveum*, causes wilting, fusaric acid causes leaf necrosis and pectic enzymes cause faccidity.

Lycomarasmin is another vivotoxin produced exclusively by *F. oxysporum* f.sp. *lycopersici* (tomato wilt). It is a dipeptide with empirical formula of $C_9H_{15}O_7N_3$. The toxin inactivates the growth factor strepogenin. If this growth factor is applied to the infected plants symptoms are reduced. Purified lycomarasmin acts as a chelating agent and after application to tomato cuttings it forms a water soluble, unstable chelate with iron which is translocated to leaves. Iron is set free in the leaves and causes typical symptoms. Iron increases toxicity of the toxin while copper reduces it.

Pyricularin: The blast disease of rice, caused by *Pyricularia grisea* (*Magnaporthe grisea*), is characterised by spotting of leaves and collar and culm rot. Culture filtrates of the fungus produce some of these symptoms on young and old rice plants. The symptoms have been attributed to two toxins, one of which (alfa picolinic acid) is non-specific and is produced in quantity at later stages of the disease. The other, pyricularin, is more active and semi-specific since it affects susceptible varieties more than the resistant varieties. At low concentrations, the toxin is toxic to conidial germination of the pathogen but the fungus synthesises a pyricularin-binding protein which destroys the fungitoxicity but not the phytotoxicity. The toxin induces increase in polyphenols and oxidases. The toxicity is counteracted by chlorogenic acid. The toxin also increases respiration and growth at low concentrations and inhibits them at high concentrations.

Fusicoccin (FC): Fusicoccum amygdali causes twig blight of almond and peach. The phytotoxin produced by it induces symptoms of blight when introduced into the xylem. The toxin is a carbotricyclic terpene glycoside. The effects of FC on plant cells are somewhat similar to those of plant growth regulators. It affects cellular transport systems and increases uptake of several anions, sugars and amino acids by affected cells. The toxin is regarded as the first clear example of wilt toxin because it has been shown to reproduce physiological cause of wilt, not just the wilt symptoms. It also stimulates stomatal opening, respiration and cell enlargement.

12

Defence Mechanisms in Plants

Resistance of plants to harmful effects of other organisms seems to be the rule while susceptibility to such effects the exception. In spite of many kinds of diseases attacking a single plant species and hundreds and thousands of fungal spores or bacterial cells reaching the host, the species have survived. The attacked individual may suffer partial or complete damage but other individuals in the population either tolerate the attack of the pathogen or remain unaffected. This fact suggests that plants are not passive hosts to the constant on slaught of microorganisms with which they interact in the environment. They have some built-in mechanism(s) of defence or can respond to attack with suitable post-infection activities that enables them to survive in presence of so many diseases around them. The mechanisms might have developed in the plant during its coevolution with the pathogens through changes in the genetic make up or through activation of dormant, pre-existing resistance genes. This pathogen-host relationship has been known to man since the early days when he started cultivation of crop plants and has provided the basis for selection of resistant individuals for cultivation and also the reason for understanding these mechanisms so that natural methods of disease management could be developed. These pathogen-host relationships are important from viewpoint of epidemiology because the defence mechanisms in the plant population determine the course of an epidemic to a great extent. This natural basis of resistance in plants to attack of pathogens is the subject matter of this chapter. The topic was briefly mentioned in the chapter on phenomenon of infection under breakdown of host barriers and will again be discussed in the chapter on disease management. Here, some principles are explained.

A plant can be successfully invaded by a pathogen or not is determined by the genetic characters of the plant and the pathogen. Cereal rusts do not attack mango, chickpea or pea because cereals are genetically different from these plants which do not provide basic compatibility. The genetic qualities deciding the makeup of the cells of these plants may not provide the genetically controlled requirements of the pathogen or may react in such a manner that the pathogen is repelled or expelled. If a cereal rust fungus attacks wheat crop, the degree of invasion and type of reaction will vary among different varieties of the crop. In other words, with genetic differences among the varieties there have been differences in pathogenesis. Obviously, whether a plant is resistant or susceptible and the degree of these qualities is determined by genetics of the plant and the pathogen. The plant must recognise and accept the parasite (compatibility) for the association that leads to the condition known as disease. This is accomplished, in most host-pathogen systems studied, through a system of matching genes in the host and the pathogen and may be mediated through complicated biochemical systems.

Disease resistance in plants is a condition in which the plant, when challenged by a pathogen, suffers little or no injury. Susceptibility is the condition when the above quality is lacking in the plant. Natural resistance or susceptibility is hereditary character of the plant. What determines this hereditary character is a topic for principles of genetics. Only a brief introduction is given here. The plant body is made up of tissues which are composed of numerous units known as cells. The nucleus in the cell is considered the controller of all cellular activities and, therefore, all the activities and qualities of the plant. The genetic information of the organisms including plants, is encoded in DNA molecules present as **genes** contained in chromosomes in the nucleus. This is chromosomal DNA. The mictochondria and chloroplasts also carry DNA in the plant. The genes store and transfer the information about characters of the plant from parents to progeny. Nature of the plant, its structure, cellular activities, synthesis, quantity and degradation of various chemical components, reaction of synthesised substances, etc. all are determined and controlled by the information coded in the genes. All genes are not active or expressed. Some genes remain inactive or "turned off". These are "turned on" or activated on certain occasions under the influence of chemical, physical and biological stimuli.

In describing the events in plant pathogenesis, this book has followed the disease-cycle and steps that occur sequentially before, during and after successful infection. On the same pattern the defence mechanisms are also described in the following order: (i) the pathogen arrives on the host and faces the pre-existing physical and chemical barriers which form part of the hereditary characters of the host, (ii) the pathogen succeeds in crossing the above barriers and tries to establish parasitic relationship with the tissues, the host reacts to the foreign material and as a result new physical and chemical barriers are created. These two conditions can be stated as resistance to penetration and resistance to disease development and spread and invasion of the tissues. A brief account of these two barriers was given in the chapter on phenomenon of infection. On the following pages defence mechanisms in plants are explained under following groups for clarity and sequence.

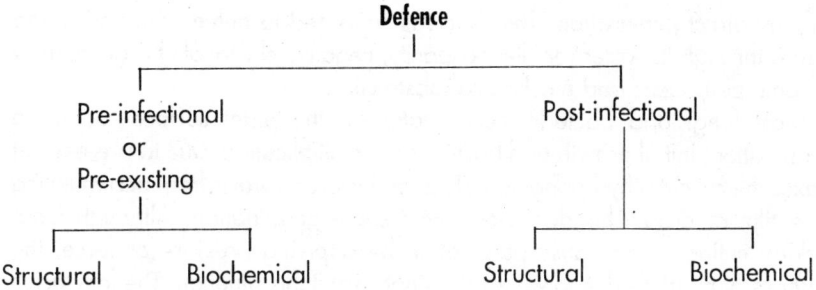

Defence

Pre-infectional or Pre-existing — Post-infectional

Pre-existing: Structural, Biochemical

Post-infectional: Structural, Biochemical

PASSIVE OR PRE-EXISTING DEFENCE

The passive or pre-infection defence mechanisms involve mechanical or structural barriers such as waxy cuticle, thick epidermal cell walls, structure of natural opening, etc. or biochemical barriers such as strategically positioned reservoirs of antimicrobial compounds.

Pre-Existing Structural Defence: The first line of defence in plants is present on its surface. Many characters of the plant surface function as barriers to penetration which the pathogen must break to enter the host. The pathogens enter the plant host by penetrating the epidermis either by force or through natural openings or wounds and insect punctures. The structure of the epidermis along with the overlying cuticle and cuticular wax, and structure and number of natural openings (stomata, lenticels, hydathodes, etc.) which are present before the attack of the pathogen can obstruct penetration. If penetration succeeds, the pathogen encounters pre-existing internal structural barriers. The external and internal structural barriers present before the attack of the pathogen are also called **pre-existing defence structures** or **static anti-infection structures.**

1) The Cuticle: The surface of the plant cell walls directly exposed to the air is covered by the non-cellular, many layered, biopolyster membrane known as **cutin.** This layer is generally attached to the pectinaceous layer of the cells and is often embedded in waxes. Cutin and waxes together constitute the cuticle. Presence of waxes makes the cuticle in most plants highly water repellent. Thickness of the cutin and the amount of waxes present may determine the efficacy of the cuticle in its function as the first line of defence against pathogens causing diseases of leaves and fruits.

No parasite is known to produce enzymes that can degrade the cuticular waxes. However, fungi such as *Fusarium solani* f.sp. *pisi* produce the enzyme cutinase, a glycoprotein (Shaykh, *et al*, 1977). This enzyme breaks down cutin. Many fungi can utilise cutin as sole source of carbon in their nutrition (Kolattukudy, 1980). But this capacity is effective only where cutin is not protected by a layer of waxes. Thus, the major defence structure is the wax layer in the cuticle. It increases resistance against fungi which enter the host only by direct penetration. The cuticle is supposed to defend the underlying tissues through its water repellent capacity, negative electrical charge, toxicity to some pathogens and mechanical obstruction.

Most fungi and bacteria need water on the plant surface for spore germination, initial growth and bacterial cell multiplication. Due to presence of waxes the cuticle is hydrophobic surface and prevents water from accumulating as a film or drop. This does not permit spore germination. Although fungi lacking cutinase can cause penetration by applying pressure or force, the absence of moisture due to waxes does not permit this situation. The fatty acids of the cutin impart the surface a negative electrical charge. Spores that carry a negative charge are, therefore, repelled by the surface while those with positive charge are held on the surface. Angtifungal substances have been isolated from the cuticle of many plants. The presence of these antifungal compounds can be better discussed under pre-existing biochemical defence. Mechanical obstruction by cuticle has, perhaps, been studied in much detail and there is some agreement that thickness of the cuticular wax may obstruct the path of infection tube as well as the exit of the pathogen from inside the host. This results in less inoculum production and slow spread of the disease in the population. Waxes are important in the prevention of post-harvest decay of some fruits. They are even externally applied not only to provide protection to the fruit skin but also as carrier of fungicides. In addition to obstructing penetration by germ tubes, presence of wax on the fruit skin prevents high loss of water from inside.

2) Structure of the Epidermal Cell Walls: Many plant pathologists, after reviewing work done on resistance due to cuticle, had concluded that contribution of the cuticle in defence mechanism is not much although in some studies chemical inhibitors in the cuticle were shown to play significant role. The external walls of the epidermal cells are more important and may directly impede entrance of the pathogen. The degree of this impediment depends on thickness and toughness of the walls. Even if the walls are of similar thickness their toughness due to relative lignification or deposition of silicic acid may differ.

The degree of infection or susceptibility in species of *Berberis* to same strain of *Puccinia graminis* had been found to vary with the species of the host as early as 1927. The young leaves of the species *B. canadensis* and *B. vulgaris*, with minimum wall thickness were highly susceptible while those of *B. thunbergi* with maximum thickness of epidermal cells were resistant. *Pyricularia grisea*, causing leaf blast of rice, directly penetrates the epidermis. The entry is through the motor cells and guard cells. More than half of the germ tubes enter through motor cells. This preference for specific cells for entry is attributed to the fact that lignification of outer cells occurs rapidly while the motor cells are either not lignified or it takes place very late. The fungus can easily penetrate these cells.

TABLE 7: Formation of Appressoria and Penetrating Hyphae of *Pyricularia Grisea* on Different Cells of Rice

(Ito and Shimada, 1937)

Epidermal cells	Per cent formation	
	Appressoria	Infection hyphae
Motor cells	53.0	63.7
Long cells	16.2	7.9
Short cells	12.8	6.1
Hairs	1.7	-
Stomata	6.0	-
Short cells contacted with stomata	7.7	51.1
Middle lamella between stoma and short cell	2.6	6.1

The severity of rice blast disease is related to the amount of silicic acid in leaves. Application of calcium silicate slag as a source of plant available silicon in silicon deficient soils reduces blast (*Pyricularia grisea*) as well as brown sport caused by *Dreschlera oryzae* (Datnoff, et al, 1992). The accumulation of silicic acid in leaves increases with age of the plant and occurs in walls of motor cells also. This is how the susceptibility of rice to leaf blast decreases with increasing age. Transplantation method of rice cultivation and deficiency of nitrogen increases uptake of silicon by roots and so impart resistance. At low temperatures under conditions of high level of nitrogen very little silicon is absorbed by roots and thus susceptibility is increased. This is an example where metabolism of the plant helps in development or otherwise of a structural defence mechanism.

The strength of the epidermis is due to toughness of the polymers of cellulose and hemicellulose that make up the cell walls. Potato tubers resistant

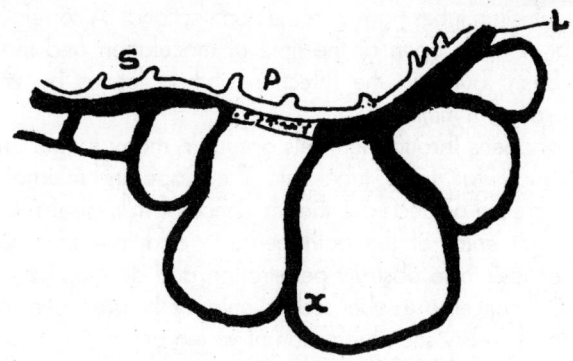

Figure 13. The Cross Section of Motor Cells of Rice Leaf. L = Lignified Cell Wall, P = Lectin Layer (Dotted Area), S = Silicic Acid Deposition, X = Wall of Water Cell on which Silicic Acid Deposits Easily (Yoshii, 1936).

to *Pythium debaryanum* have a higher fibre content than susceptible varieties. If the pathogen can overcome the toughness barrier in the epidermis, the underlying tissues may be quite susceptible. In many plants the epidermis is replaced by more resistant tissue–the pericycle. This occurs during the secondary growth. The cells comprising the periderm arise from the cork cambium. This layer, which is renewable as the tree grows, provides strong mechanical resistance against pathogens.

3) Structure and Number of Natural Openings: Many fungi and bacteria can enter the host only through the natural openings such as stomata and lenticels. This is especially true for bacterial pathogens since fungi have alternative mode of penetration also and can dilate the stomata by pressure if they are small. Therefore, the location, number and structure of these openings often determines pathogenicity of many pathogens. Infection of peach by *Xanthomonas orboricola* pv. *pruni* (*X. pruni*) occurs only when the bacterial suspension is sprayed on the under surface of the leaf where most stomata are located. The Szinkum variety of citrus or Mandarin oranges are resistant to *Xanthomonas axonopodis* pv. *citri* (*X. citri*) causing citrus canker because the stomata are small and surrounded by raised, broad-lipped structures which prevent entry of water drops containing the bacterial cells. In later studies it has been found that greater the number and size of stomata per unit area the greater is the susceptibility of the organ. The stomatal invasion by the bacterium is governed by the developmental stage of the organ as was explained in chapter 8.

In cereal rusts the infection tube penetrates through the stomata. In some species of the rust fungus, the pathogen can force open the stomata by pressure. But in the stem rust of wheat the fungus (*Puccinia graminis tritici*) cannot do so and must find open stomata. A functional defence mechanisms has been observed in some wheat varieties (such as the cultivar Hope) in which the stomata open late in the day when moisture on leaf surface has dried and the infection tubes have become nonfunctional. A correlation between the number of stomata open at the time of inoculation and the amount of disease has been found in the infection of hop leaves by zoospores of *Pseudoperonospora humuli.*

Entry of pathogens through lenticels occurs in many fungal and bacterial diseases affecting twigs, fruits, tubers etc. The shape and internal structure of lenticels can increase or decrease the incidence of fruit diseases. Lenticels of small size restrict entry of the pathogens. In addition, the suberin layers beneath the lenticels also obstruct penetration and development of infection tubes. When the lenticels are suberised the entry of the pathogens is prevented. Lenticels are the primary site of infection of young potato tubers and stems by *Streptomyces scabies* (common scab disease). However, infection by the bacterium occurs in a brief stage: when unsuberised tissue of the young lenticels is exposed subsequent to the loss of guard cells. After suberisation resistance develops.

Hydathodes are natural openings on the edges of leaves and serve to excrete excess water from the interior. They are early entry points of bacterial pathogens such as *Xanthomonas campestris* pv. *campestris* (black rot of cabbage). Similar to hydathodes are the nectarthodes in inflorescence of many plants. They secrete sugary nectar and this serves as barrier to those organisms that cannot tolerate high osmotic pressure. *Erwinia amylovora* (fire blight of apple and pear) can tolerate high osmotic pressure and can enter through nectaries. Leaf hairs on glabrous leaves and on nectaries also resist entry of pathogens. High hairiness of leaves and pods in chickpea is resistance character against *Ascochyta rabiei*. Groundnut varieties showing resistance to Cercospora leaf sports have thick epidermis-cum-cuticle and compact palisade layer, fewer and smaller stomata and high frequency of trichomes on the abaxial surface of leaf (Kaur and Dhillon, 1988; Mayee and Suryavanshi, 1995).

4) Internal Structures: The thickness and toughness of walls of internal tissues is important in invasion by some pathogens. If the thickness and toughness is increased, growth of the pathogen is slowed. In many leaf diseases the spots or necrotic lesions are often surrounded by leaf veins. This indicates that the pathogen was unable to disintegrate the tissues of the veins which obstructed its spread. The sclerenchyma tissues are made up of thick and tough-walled cells. The amount and distribution of these tissues varies in different species of plants. In wheat, varieties containing these tissues in relatively high proportion in the stem possess certain degree of resistance to stem rust (*Puccinia graminis*). The leaf smut of rice (*Entyloma oryzae*) produces dull black linear spots on leaves which are restricted by leaf veins. If due to some reason, such as use of high levels of fertilizers, these mechanical barriers are few or become weak, the resistance of the plant to invasion by the pathogen is also reduced.

Species of *Pythium* invade only juvenile tissues in which secondary thickenings have not yet developed. Seedlings become resistant as they advance in age and secondary thickenings are formed. There may also be involvement of some biochemical factors in this age based resistance. Lignification, suberisation, deposition of carbohydrates, etc. to strengthen the internal tissues are also mechanical barriers to infection.

Pre-existing Biochemical Defence: There is much more evidence to suggest that physiological or biochemical defence is more important and common than structural defence resisting invasion by plant pathogens. Through biochemical conditions and reactions the plant inactivates the pathogen or its toxins or kills it before the infection spreads and the disease becomes serious. These biochemical mechanisms may be present in the plant before it is attacked by a pathogen but more commonly they develop as a result of activation of defence responses when the attack occurs (post-infectional biochemical defence). In the latter case resistance in plant—pathogen interaction is accompanied by rapid multicomponent responses that include hypersensitive reaction, antimicrobial phytoalexins, hydrolytic enzymes and structural defensive barriers. These responses are explained later in this chapter. In this section only the pre-existing biochemical mechanisms are explained.

Pre-existing and post-infection chemical compounds that provide resistance have been variously classified. Ingham (1973) classified the antifungal compounds in plants into four categories: prohibitins, inhibitins, post-inhibitins and phytoalexins. Prohibitins and inhibitins are normal constituents of the plant involved in constitutive or semiconstitutive resistance. These are the pre-existing biochemical defence materials. Post-inhibitins are formed by minor alteration of pre-existing compounds while phytoalexins are post-infectionally produced metabolites inducing resistance. Bell (1981) had called the compounds pre-existing in the plant as constitutive antibiotics and those which are formed in response to wounds as wound antibiotics.

1) Antifungal and Antimicrobial Compounds Released by the Plant in its Environment: During the growth and accompanying activities of higher plants there is a continuous exchange of materials with the surrounding environment. Plants not only take in water and nutrients from the soil and carbon dioxide and oxygen from the atmosphere but also liberate gases as well as organic substances from roots and leaves. These leaf and root exudates contain those biochemicals which are produced during metabolic processes of the plant cells, such as amino acids, sugars, glycosides, organic acids, enzymes, alkaloids, nucleotides and flavonones, inorganic ions and also certain growth factors and toxic materials. They have a profound effect on the nature of the surrounding environment including the phyllosphere and the rhizosphere microflora and fauna. These substances may accumulate in minute drop on leaf surface or diffuse in the moisture of the atmosphere around leaves and roots. The nature, quantity and distribution of these exudates vary with plant species and varieties.

Although these substances are ideal nutrients for microbes and help in growth of many saprophytes and parasites, a number of inhibitory compounds are also contained in the exudates. These inhibitory substances directly affect the microorganisms or encourage certain groups to dominate the environment and function as antagonists of the pathogen. Antibiotic producing microflora in the phyllosphere or the rhizosphere also form a part of the enhanced microbial population due to nutritive effect of the exudates. There are numerous examples of the resident microflora enhanced by exudates which are potential biocontrol agents of fungal and bacterial pathogens (Kalita et al, 1996). The phylloplane microorganisms are also responsible for induced systemic acquired resistance (Sticher et al, 1997). The root exudates favour development of plant growth promoting rhizobacteria (PGPR) which not only suppress pathogens but also suppress the activity of growth inhibiting bacteria or deleterious rhizobacteria and fungi (Schippers et al, 1987; Weller, 1988).

Tomato leaves are known to excrete chemicals that provide resistance to attack of *Botrytis cinerea*. Cowpea leaves resistant to *Cercospora* leaf spots possess toxic substances that inhibit germination of conidia. In leaf spot of sugarbeet (*Cercospora beticola*) low incidence of local lesions on the leaves of a resistant variety has been correlated with the presence of a diffusible inhibitor from healthy leaves. Spore germination of the fungus is inhibited by resistant leaves, their water washings and dew deposits on such leaves. These water extracts also inhibit growth of the germ tube suggesting their antibiotic nature. Certain powdery mildew resistant varieties of apple exude toxic waxes on leaf surface which prevent germination of conidia of *Podosphaera leucotricha*. Apple varieties producing low amount of this wax are generally susceptible to powdery mildew. Epidermal excretions play an important but indirect role in the infection of grapevine leaves by *Plasmopara viticola* (downy mildew). The excretions depend on the living condition of the guard cells of stomata and attract the germ tubes of the fungus. If the guard cells are dead no such excretions occur and germ tubes fail to reach the stoma for infection. Red scales of onion (red varieties) contain protocatechuic acid and catechol which may exude in drops and impart resistance to attack of *Colletotrichum circinans* (onion smudge disease). This phenolic substance inhibits spore germination of the fungus. If the red scales are removed the onion bulb becomes susceptible to the disease. Spore germination inhibitory property of catechol against some species of *Alternaria* is also reported.

The role of roots exudates in defence of plants against root disease pathogens has been extensively studied. Root exudates in certain conditions and certain types of plants enhance infectivity of the pathogens while in others they inhibit the pathogen. The wrinkled seeded varieties of pea are susceptible to seed and root rot because the seeds exude very high quantities of sugars which encourage the growth of *Pythium ultimum*. However, indirectly the root exudates may suppress the growth of pathogens by encouraging other microorganisms to grow and compete with the pathogen for nutrition or produce antibiotics.

Sometimes the root exudates contain compounds that are directly toxic to a pathogen. Certain varieties of linseed (flax) resist wilt caused by *Fusarium oxysporum* f.sp. *lini* through the presence of hydrocyanides (HCN) in their root exudates. The compound is highly toxic to the wilt pathogen and reduces its infectivity around the roots (Trione, 1960). The substance is present in the roots also and imparts resistance. In addition, HCN does not affect the development of *Trichoderma* spp. which are highly antagonistic microorganisms. Thus, HCN encourages biological control of the disease also. Presence of HCN in roots of maize and sorghum is also reported. HCN in sorghum roots exudates might be responsible for reduction of pigeonpea wilt (*Fusarium udum*) when sorghum and pigeonpea are grown as mixed crops. Observations similar to those for linseed wilt have been reported for pea wilt caused by *F. oxysporum* f.sp. *pisi* also.

Enemy or antagonistic plants produce some toxic compounds that destroy plant parasitic nematodes in soil around the roots. Some grasses, varieties of mustard, marigold (*Tagetes erecta*), species of *Crotalaria* and asparagus have been listed as enemy plants. Reduction of 96% in the population of *Pratylenchus* by growing marigold is reported. These enemy plants reduce the populations of *Tylenchorhynchus* and *Meloidogyne* also. The antinemic property of marigold is attributed to the presence of the polyenes, terthienyl and derivatives of bithienyl in the roots and root exudates. *Asparagus officinalis* also produces a toxic material (asparaguisic acid) against the nematode *Trichodorus christei*. Bitter cucumber produces a compound cucurbitacin which is repellent for nematodes.

2) **Inhibitors or antimicrobial compounds present in the plant cell:** Plant produce thousands of naturally synthesised compounds many of which are unique to specific taxonomic groups and are toxic to animals including insects and to microorganisms (Bell, 1981, Swain 1977, Webster, 1974). The compounds may be completely synthesised and stored in vacuoles, lysigenous glands, ducts, heartwood and periderm. Alternatively, less toxic precursors of these antimicrobial compounds may be stored in vacuoles, with the final antibiotic being formed rapidly by the action of hydrolases and oxidases located in other parts of the cells. The antimicrobial substances pre-existing in the plant cells include unsaturated lactones, cyanogenic glycosides, sulphur containing compounds, phenols and phenolic glycosides and saponins (tomatine, solanine, avenacin).

The unsaturated lactones and derivatives of hydrocyanic acid (HCN) are present in intact tissues of many plant species as glycosides. Mechanical or pathogenic injury to the tissue brings these glycosides in contact with separately stored enzymes. Action of hydrolyzing enzymes on cyanogenic glycosides instantly releases HCN which is extremely toxic at low concentrations. The concentration of glycosides in roots, stems, leaves, flowers and seeds of different plants may vary from 1 to 5 mg per g fresh weight. The compounds may be released into the outer environment also as discussed in the preceding section. However, once the primary conditions of infection are fulfilled without obstruction in the outer environment, the pathogen reaches the host tissues where it may encounter these toxic compounds. In many diseases these pre-existing toxic substances in the cells form the basis of immunity or resistance. In a resistant variety these substances will be more abundant while in susceptible variety they may be less or completely absent.

Extracts of plant parts and plant oil have been extensively studied in recent past for their antimicrobial activity and have been demonstrated as good plant disease control agents (Singh, 2000). The diseases in which extracts and oils have been found effective include fungal diseases as well as virus diseases. Oil from American elm seed contains 55% capric acid which is toxic to *Ceratocystis ulmi* (Dutch elm disease). Oil of sunflower, olive, maize and rapeseed have provided more than 99% control of the powdery mildew of apple (*Podosphaera leucotricha*) under controlled conditions. Rape oil or mustard oils are esters of the isothiocyanic acid and occur in the plants as glycosides. When tissues are crushed or damaged the glycoside is acted upon by the enzyme myrosinase separately stored in myrosine cells. This produces isothiocyante. In some plants action of isomerase converts this compound into thiocyanate. Due to these compounds mustard oil has antifungal, antibacterial and antinemic properties. Mechanically emulsified rape seed oil is comparable to the fungicide dinocap (Karathane) in the control of powdery mildew of apple (Northover and Schneider, 1993). Azam et al, (1998) also have reported significant control of grapevine powdery mildew (Uncinula necator) with rape oil derivatives. Similar generation of antimicrobial compounds due to wounds (crushing) is reported from allyl sulphoxide in onions and garlic, glucosinolates in cabbage, cyanogenic glucosides in sorghum, lima bean and peach, para-hydroquinone glucosides in pear and walnut, and benzoxazines in maize, wheat and rye. In many of these hosts phytoalexins have not been identified to be related to resistance and these compounds may explain the resistance.

Aqueous extracts of plant parts are known to possess germicidal properties and have been used for the treatment of human and cattle diseases. In recent years screening of plants extracts for control of fungal, bacterial and viral diseases has been extensively done (Datar, 1999; Sindhan, et al., 1999). Soaking of loose smut infected wheat seed in garlic extract is reported to drastically reduce the infection. Extracts of onion and garlic bulbs, ginger corns, leaves of such plants as parthenium, calatropis and margosa suppress such pathogens as Macrophomina and Erysiphe. Rhizome powder of ginger gives as good control of pea powdery mildew (Erysiphe polygoni) as the fungicides Sulfex and Bavistin (Singh, U.P., 2000). Extract of margosa provides protection of sweet orange fruits against green mold caused by Penicillium digitatum (Khilare and Gangwane, 1997). Certain neem based products provide good protection against cotton bacterial blight (Hulloli, et al., 1998). Aqueous extracts of some plants have antiviral properties and when sprayed on leaves they reduce the number of infections and increase the incubation period (Jayashree et al., 1999; Pun et al., 1999).

Phenolic compounds are widely distributed in the plant kingdom. These include simple phenols, coumarines, flavonoids and complex phenols such as tannins. Many phenols are present in the intact tissues as glycosides and are released upon enzymatic fission by β-glycosidases when tissues are damaged. A detailed account of phenols in relation to disease resistance in plants is given later in this chapter. The most commonly cited example of resistance due to pre-existing phenol is that of catechol against onion smudge mentioned earlier. Presence of pyrocatechol in roots of Eragrostis curvula provides resistance to root knot nematodes. In common scab of potato (Streptomyces scabies) resistant varieties contain chlorogenic acid in abundance. The phenol has toxicity to the bacterium. Resistance to root knot nematodes in tomato has also been ascribed to chlorogenic acid. The acid is present in roots of some potato varieties. Due to presence of this phenol some varieties of potato show resistance to Verticillium wilt (*Verticillium albo-atrum*). Young plants are resistant to the wilt and this resistance has been correlated with higher amount of chlorogenic acid in young than in old plants. Chlorogenic acid is related to stimulated cork cambium activity and rapid development of protective layers of cork around the scab or site of infection.

Saponins are glycosides widely present in plants where they exists as pre-formed substances stored in vacuoles of cell in various tissues. Sometimes saponins and alkaloids appear to be important for disease resistance. These antimicrobial compounds include avenacins, a triterpene, present in roots of oats, solanin, an alkaloid present in the green skin of potato tubers, tomatin, an alkaloid present in roots, stems, leaves and green fruits of tomato, and chaconin, an alkaloid present in leaves of potato. Although these are very active compounds, they adversely affect only those cells that contain sterols in their membrane system. Bacteria and blue green algae are insensitive to saponins. Saponins react with cell membrane sterols to form insoluble complexes. In this process, pores develop in the cell membranes through which there is leakage of electrolytes and nutrients and ultimately the cell dies. The change in the cell wall permeability is irreversible. The saponins such as tomatin and solanin are highly antifungal but their toxicity is more against nonpathogens than against pathogens which possess enzyme system for their degradation.

Roots of oats contain the saponins avenacin A and B in a ratio of 10:1 in vacuoles of the cells. Fungal invasion of oat roots triggers release of avenacins from protoplasts simultaneously with leakage of electrolytes and nutrients. A virulent fungus such as *Fusarium avenaceum* can detoxify low concentrations of avenacin A through enzymic action converting it to relatively less toxic avenacin B. However, high concentrations of avenacin A are toxic to this fungus also.

The skin and healed wound-surfaces of potato tubers contain high concentration of the fungitoxic alkaloid solanin. Greening of tubers exposed to sunlight and their bitter taste is only skin deep and is a sign of high concentration of solanin. The alkaloid prevents infection of healthy tissues by most fungi. However, infection of fresh wounds on tubers often suppresses normal alkaloid synthesis diverting it to synthesis of terpenoid phytoalexins. Nematode infection of resistant tobacco roots causes content of the alkaloid nicotine to double in leaves, thus inducing resistance to other diseases and pests. This capacity is not present in susceptible tobacco. Concentration of the alkaloid tomatin is greater in extracts of *Fusarium*-resistant tomato than in susceptible tomato cultivars. This alkaloid is toxic to *Cladosporium fulvum* also.

3) Lack of Essential Nutrients and Growth Factors for the Pathogen: The fact that may facultative saprophytes and most of the obligate parasites are host-specific and sometimes are so specialised that they can grow and reproduce only on certain varieties of a host species suggests that for these pathogens the essential nutrients and growth factors are available only in these hosts. Absence of these suitable nutrients and stimulants makes the other varieties and species unsuitable hosts. These nutritional substances must be in available form. If the pre-existing substances cannot be utilized by the pathogen or become unavailable due to reactions taking place after infection, the pathogen will fail to grow.

In the apple scab disease, the pathogen *Venturia inaequalis* has genetically controlled requirement for a growth factor in order to become pathogenic. Certain mutants of the pathogen cease to be pathogenic in absence of their ability to synthesise this factor. They can be induced to establish infection by spraying the growth factor on the leaves. The advance of infection continues only so long as the growth factor is applied externally. In seedling disease of radish and lettuce caused by *Rhozoctonia solani*, successful infection depends on formation of infection cushions from which the infection peg develop and cause penetration of the epidermis. Formation of these cushions is induced by certain essential nutrients in the host. In resistant varieties no such substances occur and no mycelial cushions are formed. Indole acetic acid and its precursor tryptophan are necessary for the reproduction of some nematodes. When these are not present in the host cells in which the nematode is feeding, its further development is checked. Reproduction does not occur and damage to the host is reduced.

4) Lack of Recognition between Host and Pathogen: The first step in the infection process is the cell (pathogen) to cell (host) communication. Plants of species or varieties may not become infected by a pathogen if their surface cells lack specific **recognition factors** (specific molecules or structures that can be recognized by the pathogen). If the pathogen does not recognise the plant as one of its hosts it may not adhere to the host surface or it may not produce infection substances such as enzymes or structures such as appressoria, infection pegs and haustoria. These recognition molecules are of various types of oligosaccharides and polysaccharides and proteins or glycoproteins (lectins). In viruses the infection fails when the host ribosomes fail to recognise the viral nucleic acid and do not produce the enzymes (replicase) necessary for synthesis of viral nucleic acid.

5) Lack of Sensitive Sites for Pathogen Toxins: In many fungal diseases the pathogen produces host-specific toxins which are responsible for the symptoms. The molecules of the toxin are supposed to attach to specific sensitive sites or receptors in the cell. Only the plants that have such sensitive sites become diseased.

ACTIVE DEFENCE MECHANISMS

Plants have induced cellular defences that prevent further colonization of the tissue once the pre-existing or passive defence has been overcome by the pathogen. These induced defences are described as active defence mechanisms because they are a response to the invading pathogen and require host metabolism to function (cf. Hutcheson, 1998). Active defence responses can be induced by all classes of pathogens including viruses, bacteria, fungi and nematodes. The response of the host can be primary, secondary or systemically acquired (Hutcheson, 1998). The primary response is at the level of the cell that comes in contact with the pathogen. It generally results in hypersensitive reaction or programmed cell death. The secondary responses occur in the adjacent cells surrounding the site of initial infection. The systemically acquired response is associated with systemic acquired resistance induced throughout the plant.

Post-infection structural defence

Resistance in many plant-pathogen interactions is accompanied by the rapid deployment of a multicomponent defence response (Dixon et al, 1994). The individual components of this include:

1) Structural: lignins and hydroxyproline rich cell wall proteins; hypersensitive response (HR)
2) Biochemical: antimicrobial phytoalexins and hydrolytic enzymes.

Signals for activation of these various defences are thought to be initiated in response to recognition of pathogen avirulence determinants (elicitors) by plant receptors. The defence response may be induced specifically (determined by the avirulent genotype of the pathogen race and the resistant genotype of the host cultivar) or nonspecifically by a range of biotic and abiotic elicitors.

The structural barriers develop in the host tissues as a result of reaction of the host to infection. Such defence mechanisms are also called **autonomous antiparasitic defence reactions or dynamic defence mechanisms**. They prevent further spread of the pathogen within the host after infection. The cellular reaction in this type of defence aims at weakening or destruction of the pathogen. The post-infection structural defence mechanisms may be of four types: histological defence structures, cellular defence structures, cytoplasmic defence reactions and necrotic or hypersensitive reaction (HR). However, many of these mechanisms are more of biochemical nature rather than structural.

1) Post-infection Histological Defence Structures: This type of structural defence consists of limiting the spread of the pathogen in tissues by creating barriers through tissue differentiation and accumulation of chemical substances. These structures consist of lignified cells and cork layers, development of abscission layers and formation of tyloses, gum deposition, etc.

Lignification of cell wall upon infection is one of the characteristic changes associated with histological defence reaction. There is strong evidence that lignification is an important mechanisms for plant disease resistance (Carver et al, 1995, Sticher et al, 1997). Lignified cell walls provide effective barrier to hyphal penetration as induced (active) resistance mechanism. They also constitute a barrier for free movement of nutrients, causing starvation of the

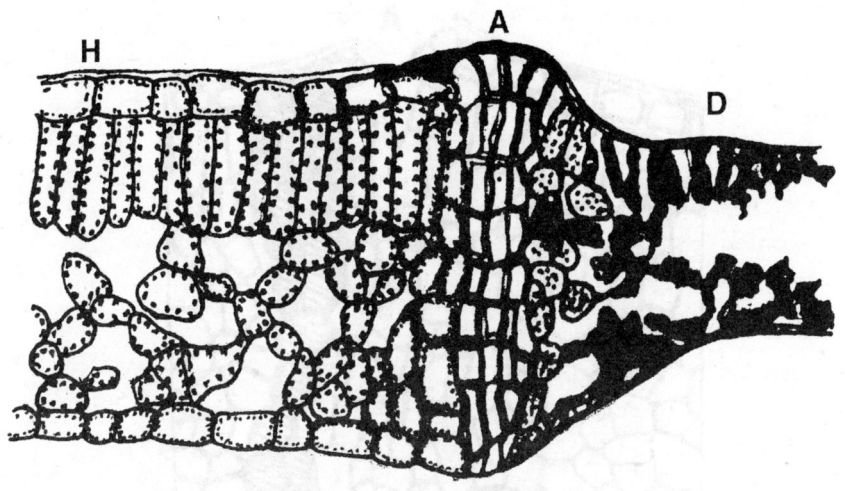

Figure 14. Formation of Cork Layer (A) between Diseased (D) and Healthy (H) Tissue.

pathogen. In *Peronospora parasitica-Raphanus sativus* system (downy mildew of radish) lignification soon after penetration of the resistant host has been shown. Other examples of induced lignification in fungal diseases are *Alternaria japonica*-radish, *Phytophthora infestans*-potato, *Septoria nodorum*-wheat, *Cladosporium cucumerinum*-cucumber, *Colletotrichum lagenarium*-cucumber, *Botrytis cinerea*-carrot systems.

Suberisation of tissues in response to wounds (injury due to penetration) is another way to isolating the pathogen from the healthy tissues. In plants infected by many fungi, bacteria, viruses and nematodes the diseased cells become surrounded by **cork layers**. These layers are mostly formed in stems, roots and unripe fruits. Usually the cork layers are thick and due to suberisation the cell walls are tough. The pathogen is therefore unable to cross them. The separation of diseased tissues from healthy tissues by cork layers prevents further growth of the pathogen. The dead tissues along with the pathogen thus remain confined, forming necrotic spots. Such tissues often are pushed upward by the underlying healthy tissue thus forming scab. Cork layers not only prevent spread of the pathogens but also prevent movement of water and nutrients from the healthy area into the diseased area and translocation of enzymes and toxins from the diseased area to the healthy tissues. As post-infection mechanism of defence, quick reaction of the host cells to infection and ability of the plant to synthesise materials for cork formation are essential pre-requisites in the formation of cork layers. All plant species do not induce cork layers. Some examples of defence by formation of cork layers are Rhizopus rot of sweet potato, common scab of potato and necrotic lesions of tobacco mosaic.

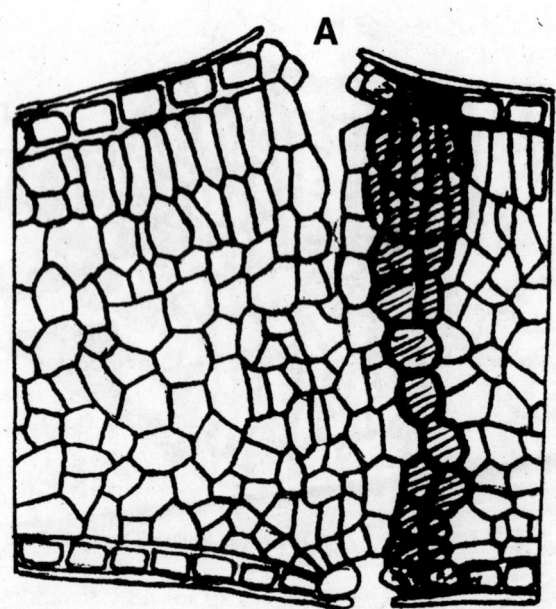

Figure 15. Formation of Abscission Layer (A) between Diseased and Healthy Tissue.

The usual role of **abscission layers** is to separate the ripe fruits and old leaves from the plant. They develop at a particular stage of plant growth or maturity of plant organs. However, the plant may form these layers in response to activities of pathogens (fungi, bacteria and viruses). These layers are not cellular. They represent the empty space between two circular layers of the host cells. The empty space is created by disintegration of middle lamella of parenchymatous tissues and thinning of the cell walls. As a result the diseased portion along with the pathogen separates and falls off. Appearance of shot holes in most leaf spot diseases is due to these layers. Abscission layer has been termed as **amputative resistance.**

Tyloses were mentioned earlier with reference to their role in pathogenesis of wilt inducing fungi. Although they obstruct movement of fluid in xylem vessels, they can also obstruct the pathogen in the vascular tissues. Tyloses are formed by protrusion of cell walls of paravascular cells through pits into the xylem vessel. They have cellulosic walls and by their size and number clog the vessels completely. Once a tylose is formed it may undergo secondary thickening. The formation of tyloses in vascular wilt diseases is triggered fast and by the time the infection is established and pathogen starts growing at one end of the vessel tyloses are formed far ahead of the site. An example of this function is that of wilt of sweet potato caused by *Fusarium oxysporum* f.sp. *batatas*. In some varieties of sweet potato tyloses develop quickly and in abundance in the xylem vessels of the upper part of the stem while the pathogen is still in the

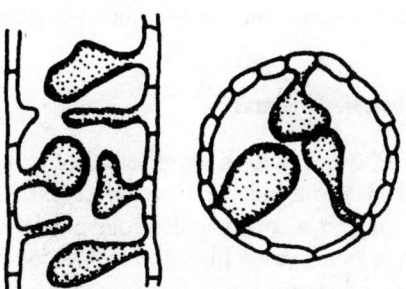

Figure 16. Tyloses formed in Vascular Elements.

roots. This suggests biochemical elicitation of tylose inducing factors by metabolites of the fungus transported up the stem. This prevents spread of the fungus to the upper parts. The fungus cannot penetrate the tyloses. Such varieties possess resistance to the disease.

Gum deposition or gels in cells along the borders of diseased tissues often serves as a protective mechanical barrier. Vascular gels that coat the walls and fill the lumen of infected vessels have been found in numerous plant species infected with vascular wilt fungi. These gels generally form within 48 hours in both resistant and susceptible cultivars of the host, but are often dissolved subsequently, presumably by fungal enzymes, in susceptible cultivars. The origin of these gels is not clear. Some believe that the gels arise from the perforation plates, end walls and pit membranes by a process of distention of pectinaceous materials, hemicellulose and other carbohydrates in the primary wall and middle lamella (Van der Molen et al, 1977). Others believe that new carbohydrates and other materials are synthesised and secreted into the vessels to form gels. The defensive role of gums or gels is that they are formed quickly and deposited in the intercellular spaces and within the cell surrounding the

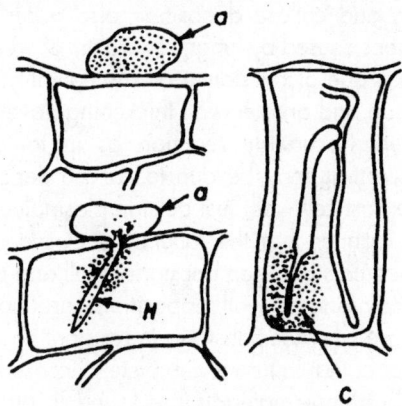

Figure 17. Gum Deposition Around Infection Thread
a = appressorium, h = haustorium, c = callose.

infection threads and haustoria. Thus, they completely isolate the fungus which starves and dies.

Post-Infection Cellular Defence Structures

Cellular defence structures involve morphological changes in the cell wall or changes derived from the cell wall of the cells being invaded by the pathogen. However, they have only a limited role in defence. Three main types of such structure have been seen: (i) carbohydrate apposition and callose deposition, (ii) cell wall thickening and (iii) structural proteins.

Carbohydrate appositions include changes in host cell walls such as callose deposition on the cell wall, in the vicinity of the penetrating hypha, appearing as swellings and sheathing of the infection thread. As fungi begin to penetrate the cell wall, either with infection hypha or with haustorium the resistant host responds by synthesising new carbohydrates particularly callose and cellulose, which are added to the inside of the cell wall just outside the plasmalemma without affecting the cuticle. These appositions may continue even after the fungus penetrates the original wall, until they become dome-shaped or elongated and are called **papillae**. Cells adjoining or near those invaded by the pathogen also may deposit new carbohydrates onto thickening secondary walls. In susceptible hosts, papillae and secondary wall thickening are often poorly developed or missing (Bell, 1981). Unless lignification or suberisation of these swellings or thickenings does not occur these may not be very efficient means of defence. But they do delay invasion of the underlying tissues. Soon the penetration hyphae may pass the swellings and the barrier is crossed. This has been observed in entry of Botrytis cinerea into pea leaves. In artificial inoculation of tomato with Pyricularia grisea and infection of oats by Helminthosporium avenae and flax by Fusarium oxysporum f.sp. lini deposition of phenols (lignin, tannin, etc.) on the swellings or thickenings has been reported.

Cell wall thickenings and callose deposition also occur in healthy cells surrounding necrotic lesions caused by fungi, regardless of whether the necrotic reaction is considered as one of resistance or susceptibility. However, more massive callose deposition and greater wall thickening have been associated with resistant than with susceptible reaction as in tomato invaded by Cladosporium. **Hyphal sheathing** may be due to callose deposition or due to stretching inward of the host cell wall, just outside plasmalemma, around the haustorium, in such a manner that the latter is enclosed in an additional membrane. This delays contact between haustorial wall and plasmalemma. In potato susceptible to P. infestans, wall appositions are poorly developed, consisting mostly of callose whereas haustoria in resistant host are surrounded by an electron dense matrix that in turn is completely encased in a callose-like material. Similar electron-dense materials are found in other carbohydrate appositions suggesting that phenolics like lignin, melanin and tannin may have to be combined with carbohydrates to form an effective barrier.

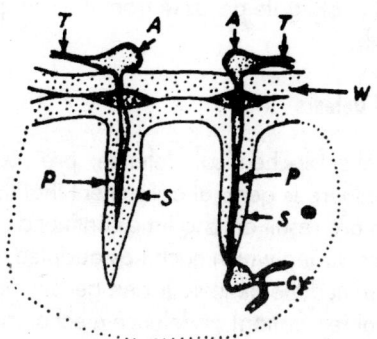

Figure 18. Cellular Defence Structure.

The **structural proteins** in cell walls are extremely important in such processes as gel and tylose formation and binding of apposition material to cell walls. Complexes of lignin with structural proteins are much more resistant to acid hydrolysis than comparable complexes with carbohydrates. (Lin and Kolattukudy, 1978). Many cell wall bound proteins have enzymatic activities also and probably are important for chemical reactions in the wall. These structural proteins, however, have been shown to be involved in resistance in only few diseases studied so far.

Post-infection Cytoplasmic Defence Reactions

Structural obstructions may also develop in the cytoplasm as a result of reaction to infection. These are the last line of defence and are effective against only slow growing, weak parasites usually causing chronic diseases or some type of symbiotic relationship such as root nodules of legumes and mycorrhizas. The cytoplasmic defence reactions weaken, localise and eliminate the endophytic nodule bacteria (*Rhizobium* spp.) or the endomycorrhizal fungi.

The Necrotic or Hypersensitive Reactions

Post-infection necrotic reactions are both structural and biochemical defence mechanisms. The topic is discussed later under post-infection biochemical defence. In may host-pathogen combinations when the pathogen has penetrated the cell wall and comes in contact with cell protoplasts there is violent reaction by the host. The nucleus moves toward the invading pathogen and is soon disintegrated. The cell shows brown granules which first develop around the pathogen and then throughout the cytoplasm. The cell walls swell and ultimately the cell dies. The pathogen mostly remains confined to the cell, starves and dies. This necrobiotic reaction is a sign of resistance in the plant. It is quite common in diseases caused by obligate fungal parasites, viruses and nematodes. The necrotic area isolates the obligate parasite from the living cells which leads to its starvation and death.

Post-infection Biochemical Defence

Regardles of how the biochemical defence, pre- or post-infectional, is expressed by the plant there is general agreement that the interactions of the host and the pathogen are result of long time continued struggle for existence between the two entities during which each has adapted to a certain situation. This struggle has determined the behaviour and genetic potentialites of the two entities. The inducement (elicitation) of defence mechanisms in the protoplams is one aspect of the struggle. Since biochemical post-infection defence is the last barrier in the process of establishment of successful infection, some of the aspects of the disease expression and their use in resistance studies should be explained.

The reaction of the host to pathogenic activities is manifested between two extremities, immunity and complete susceptibility. The immune host is not affected by the pathogen in any way. Moving down from this condition there is loss of immunity and increase of susceptibility. In between these two extremes the reactions can be grouped in several stages such as high resistance, mild resistance, mild susceptibility and high susceptibility. These categories actually indicate the degree of reaction between the pathogen and the host.

The capacity in the plant for showing resistant reaction is gene directed chemical reaction when the host cell wall or cytoplasm comes in conact with the pathogen or its metabolites. The chemical susbtances produced in these reactions to show a resistant reaction must possess following four attributes for warding off establishment of infection and match the four Koch's postulates for pathogenicity:

(i) The substance is associated with the protection against the disease at the site where protection occurs.

(ii) The substance can be isolated from the host showing protection against the disease.

(iii) Introduction of the isolated substance to the appropriate susceptible host confers protection.

(iv) The nature of the protection so induced resembles that of the natural agents of a resistant plant.

i) Toxic Materials Produced in the Plant in Response to Infection: The defensive strategy of plants exists in two stages. The first is assumed to involve the rapid accumulation of phenols at the infection site, which functions to slow or even halt the growth of the pathogen and to allow for the activation of secondary strategies that would more thoroughly restict the pathogen. The second is the activation of specific defences such as *de nove* synthesis of phytoalexins or other stress-related substances (cf. Nicholson and Hammerschmidt, 1992). Thus, the sequence of events in a defence response can be thought to include (i) the host cell death and necrosis, (ii) accumulation of toxic phenols, (iii) modification of cell walls by phenolic substituents or physical barriers such as appositions or papilae and finally (iv) synthesis of specific antibiotics such as phytoalexins.

Synthesis of inhibitory substances in response to injury caused by a pathogen or any mechanical or chemical agent is one of the most common post-infection reaction of the host. In the injured tissues a series of reactions start to isolate the irritant (pathogen) and heal the wound. The reactions are determined by the nature of the tissues. Mostly, these reactions or responses of the plant form fungicidal or fungistatic compounds around the site of infection or in the cells, such as lignin, accumulation of cell wall appositions or papillae and the early accumulation of phenols within the host cell walls. These induce formation of structural protective layers (cork layer). Depending on the intensity and duration of reactions the quantity of these compounds may be enough to inhibit or kill most microorganisms. The enzymes required for synthesis of these compounds are present in the host or may be produced in response to infection. This depends on genotype of the host plant.

a) Phenolic Compounds and Their Role in Defence: The main substances that are formed in plant in response to infection or injury are phenolic compounds such as chlorogenic acid, caffeic acid and oxidation products of floretin, hydroquinone and hydroxytyromine and phytoalexins. Low molecular weight phenols such as the benzoic acids and the phenylpropanoids are formed in the initial response to infection. Antibiotic phenols have been found in all plants investigated to date (Nicholson and Hammerschmidt, 1992). Some occur constitutively and are thought to function as preformed inhibitors associated with nonhost resistance. Others are formed in response to the entry of a pathogen or other injuries. Continuous irritation by the pathogen is essential for production of effective amounts of the above mentioned phenolic compounds. They provide both structural and biochemical defence.

The simple phenol has a single hydroxyl group on a benzene ring. When the compound has more than one hydroxyl groups on a benzene ring it is known as polyphenol. The most complex polyphenols are lignins. Lignans, phenyl glycosides, flavonoids, anthocyanin, leucoanthocyanin, anthoxanthins, etc.

In plants the synthesis of phenolic compounds takes place through shikimic acid and acetic acid pathways. Shikimic acid pathway is more important. Reaction of phosphoenol pyruvate formed in glycolysis with erythrose phosphate formed in pentose pathway results in formation of dehydroquinic acid. The pentose pathway is stimulated in a diseased plant and, therefore, more phenol synthesis can be expected. The dehydroquinic acid forms shikimic acid through several intermediate steps. Shikimic acid forms prephenic acid which is converted into phenylalanine or tyrosine. These two serve as precursors for synthesis of a wide variety of phenolic compounds such as cinnamic acid, coumarin, caffeic acid, chlorogenic acid, ferulic acid, phloretin, umbelliferon, scopoletin, isocoumarin and various other phytoalexins. In acetic acid pathway the phenolics are produced by condensation of acetates formed during breakdown of sugars in respiratory process.

The enzymes participating in these two pathways are pre-existing in the plant. These enzymes found in healthy and diseased plants are called phenol oxidising enzymes such as phenolases, phenol oxidases and polyphenol oxidases. The two most important phenol oxidases are laccase and tyrosinase. They occur in plants as well as in fungi and contain copper. In presence of oxygen these enzymes oxidise different phenolic compounds either by adding oxygen or by displacing hydrogen. Addition of oxygen to monophenols results in formation of complex polyphenols such as tannins, lignins, etc. Displacement of hydrogen forms quinones which are usually coloured and give the specific brown colour to diseased tissues.

The pre-existing phenols in the plant are called common phenols and those which do not exist in the plant but are formed as a result of interaction between the host and the pathogen are called **phytoalexins**. The common phenols are more prevalent and are usually present in the plant before infection but in concentrations not sufficient to prevent infection. However their synthesis and accumulation rapidly increases after infection. The synthesis and accumulation occurs in and around the cells invaded by the pathogen. This increase in synthesis is more rapid in resistant than in susceptible plants. A common phenol, chlorogenic acid, is found in many plants infected with various pathogens, such as in sweet potato attacked by *Ceratocystic fimbriata* and in tomato attacked by root knot nematode, *Meloidogye incognita*. Potatoes affected by late blight fungus show presence of orthodiphenol and scopoletin. Caffeic acid and umbelliferon are found in sweet potato affected with *Ceratocystis fimbriata*.

Although the rate of synthesis of common phenols increases after infection has occurred there is also movement of phenols from the healthy tissues toward the infected tissues. Both these conditions are determined by the genetic character of the plant. Several phenols may be produced at the same site and probably their combined action is more effective than their individual action. The physiological activities of the plant are also indirectly affected by phenols which may suppress the activity of indole acetic acid oxidising enzymes thus increasing the amount of this auxin or they may stimulate the activity of these enzymes.

Certain fungi hydrolyse non-toxic glycosides through the enzyme β-glycosidase produced by them or by the host cells in response to infection. Hydrolysis of phenolic glycosides yields anti-infection phenols. This has been found in leaf spot of rice (*Dreschlera oryzae*) and in some diseases caused by *Fusarium* spp.

SOME PHENOLIC COMPOUNDS FORMED IN RESPONSE TO INJURY OR INFECTION

PHENOL	STRUCTURE	HOST
CAFFEIC ACID	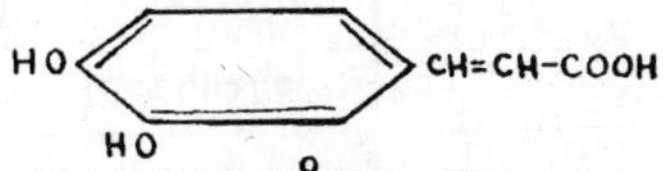 $HO-\bigcirc-CH=CH-COOH$ (with HO)	SWEET POTATO
CHLOROGENIC ACID	(structure with $CH=CH-C-O$, COOH, OH groups)	SWEET POTATO POTATO (WHITE) CARROT
PHLORETIN	(structure with CH_2CH_2-C, OH groups, rings B and A)	APPLE
HYDROQUINONE	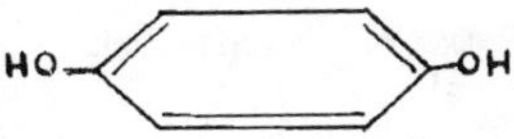 $HO-\bigcirc-OH$	PEAR
UMBELLIFERON	(coumarin structure with HO)	SWEET POTATO
SCOPOLETIN	(coumarin structure with CH_2O, HO)	SWEET POTATO

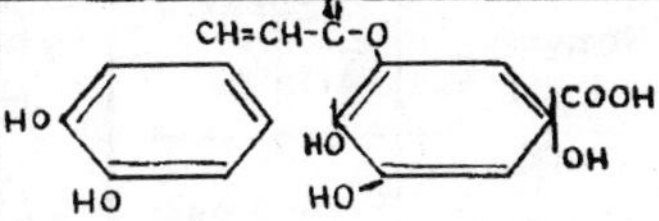

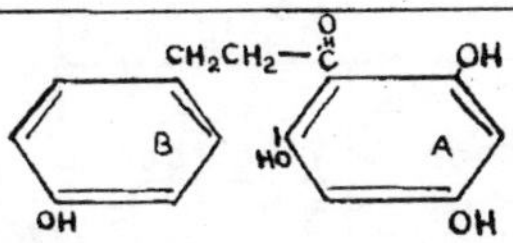

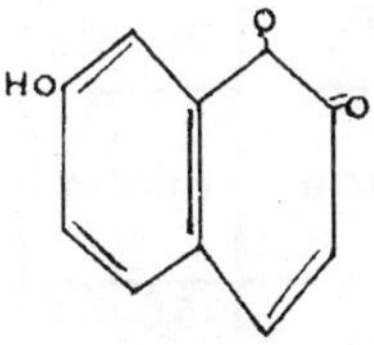

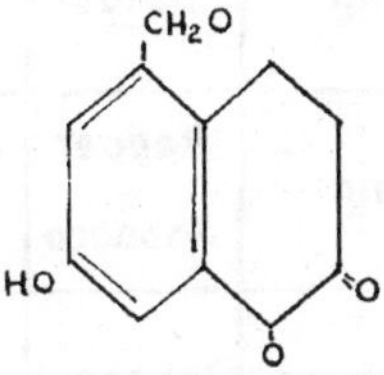

Figure 19. Some Phenolic Compounds formed in Response to Infection or Injury.

b) Phytoalexins: In the preceding paragraphs phenolic substances which may be present in the host cells but whose synthesis is accelerated after infection (post-inhibitins) as a defence mechanism were discussed. There are large number of antibiotic phenolic and other compounds which do not exist in the plant or the pathogen but are formed as a result of host-parasite interaction or any chemical or mechanical injury. Such substances have been termed **phytoalexins**. In the literal sense the word alexin means "to ward off". The

Biosynthetic pathway	Chemical group		Phytoalexin	Host	Chemical structure
	Main	Sub			
Shikimate—Polymalonate Route	Isoflavanoides	Isoflavans	Phaseollin-isoflavan	French bean	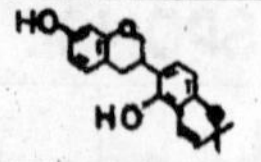
			Vestitol	Red clover Lucerne Trigonella spp.	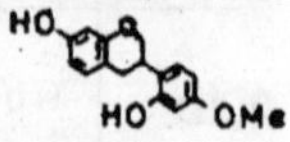
			Sativan	Lucerne Alfalfa	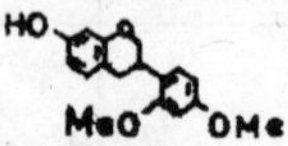
		Isoflava-none	Kievitone	French bean Cow pea	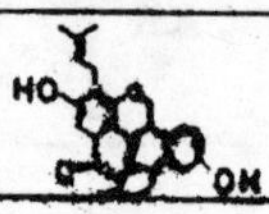
Acetate — Polymalonate Pathway (Fatty acid route)	Acetyl-ene	—	Wyrone	Vicia faba	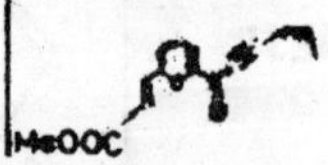

Figure 20a. Phytoalexins from Legume Plants.

Mevalonoid Pathway	Terpenoides	Sesquiterpenes	Rishitin	Potato Tomato	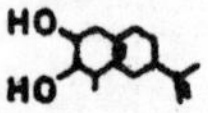
			Phytuberin	Potato	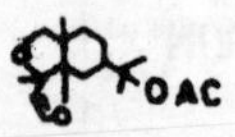
			Lubimin	Potato	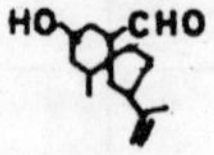
			Capsidiol	Pepper Tobacco	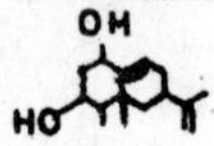
			Glutinosone	Tobacco	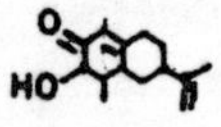

Figure 20b. Phytoalexins from Solanaceous Plants.

phytoalexin synthesis parallels the mechanism of defence through production of antibodies in the animal system except that it is not as effective as the antibodies in disease prevention. Functionally both appear to be similar. The basic concept of these antibiotic substances, proposed by K.O. Muller and H. Borger in 1941, after study of the blight of potato (*Phytophthora infestans*), was that a chemical compound, a phytoalexin, was produced during necrobiosis

Biosynthetic pathway	Chemical group	Phytoalexins	Host	Chemical structure
Shikimate-polymalonate pathway	Phenanthrene derivatives	Orchinol	Orchids	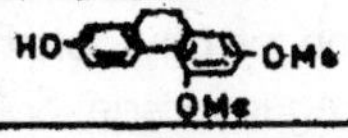
		Hircinol	Orchids	
Mevalonoid pathway	Sesqui-terpenes	Ipomeamarone	Sweet potato	
Shikimate Pathway	Isocoumarines	6-methoxy mellein	Carrot	
	Phenylalkenyl benzoxaninone derivatives	Avenalu-min-I	Oat	
		Avenalu-min-II	Oat	
		Avenalu-min-III	Oat	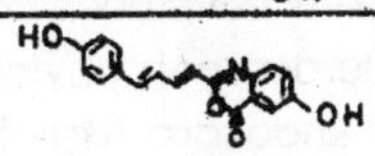

Figure 20c. Some other Phytoalexins.

by the living plant cells as a result of the interactions between the metabolites of the host and the pathogen. This substance, not detected in the plants before infection, was considered to inhibit further development of most attacking fungi although the pathogen itself was not affected. However, these compounds have now been detected even in healthy unstimulated tissues. Their production can be induced by any physical or chemical injury or by fungi, bacteria, nematodes and viruses. Even non-pathogens can induce their synthesis on the particular plant. They are more toxic to nonpathogens than to pathogens.

In the evolutionary development of host-parasite relationship, from the necrobiotic relations in which there is too much damage to host cells, through the semibiotrophism to biotrophism and to the symbiotic relations in which host and parasite become inseparable partners of the system, phytoalexins have been mostly detected in the lower categories, i.e., where there is necrotic (hypersensitive) reaction. Thus, although the basic concept of phytoalexin that plants respond to infection by accumulating chemicals inhibitory to development of the infectious agent is still valid, the original definition and conditions of their production have undergone change. Kuc (1972) had defined phytoalexins as *antibiotics produced in plant-pathogen interactions or as a response to injury or other physiological stimuli*. Occasionally, it had been proposed that any antimicrobial, secondary plant metabolite which increases to inhibitory levels following infection should be regarded a phytoalexin. However, this broader concept is not widely accepted and common phenols such as chlorogenic acid, scopoletin, caffeic acid, etc., are not included among phytolexins which are synthesised *de novo*.

Phytoalexins are low molecular weight, antimicrobial compounds produced *de novo* in plants as a result of infection or abiotic stress. The synthesis and accumulation of phytoalexins has been demonstrated in such diverse plant families as Leguminosae, Solanaceae, Malvaceae, Chenopodiacea Umbelliferae, Convolvularrae Compositae, Orchidaceae and more recently in Graminae. Their synthesis does not appear to be a characteristic of monocotyledonous plant species and to date phytoalexins or phytoalexin-like compounds have been found only in oats, rice, sugarcane and sorghum. Although many phytoalexins are phenols, including those from oat, sorghum and sugarcane, these compounds are chemically diverse and grouped together as a "family" based on their antimicrobial properties. In Leguminosae, three groups of phytoalexin compounds have been demonstarated and characterised. These are pterocarpans, isoflavans and isoflavanones. In Solanaceae only terpenoid phytoalexins have been charcterised.

The pterocarpan phytoalexin **pisatin** ($C_{17}H_{14}O_6$, mol. wt 314) was isolated from the endocarp tissues of detached pea pods inoculated with Monilinia *fructicola*, a non-pathogen of pea. The compound is a weak antibiotic with broad spectrum. Mycelial growth of *M. fructicola* is three times more sensitive to pisitin than is spore germination. Fungal pathogens of pea are generally insensitive to the amount of pisatin accumulating after infection whereas non-pathogens are sensitive. Pisatin accumulation in pods and leaves of pea is stimulated by many fungi, metabolic inhibitors, fungicides including heavy metal chlorides, spore-free germination fluids and ethylene. It is easily degraded by most pathogens of pea. Synthesis of pisatin is jointly by shikimic acid pathway and acetone-melonate pathway. Its production is accompanied by an increase in protein synthesis suggesting increased synthesis of enzymes for pisatin synthesis.

Another pterocarpan isoflavanoid phytoalexin **phaseollin** was demonstrated in spore suspensions of *Sclerotinia fructigena* or *Phytophthora infestans* when these were incubated in pod cavities of french bean (*Phaseolus vulgaris*). Against *S. fructigena* the compound is fungistatic at low concentration and fungicidal at high concentration. It shows a low ED_{50} value for a non-pathogenic fungus *Monilinia fructicola* and relatively high ED_{50} value for pathogens of french bean such as *Colletotrichum lindemuthianum* and *Rhizoctonia solani*. In resistant reactions, races of pathogens induce phaseollin production early and in greater quantities than in susceptible reactions. Monilicolin A, a sulphur containing polypeptide in the mycelium of *M. fructicola* stimulates phaseollin production in french bean pods but fails to stimulate pisatin production in pea pods. Phaseollin has also been detected in bean leaves inoculated with the halo blight bacterium *Pseudomonas savastonoi* pv. *phaseolicola*. Tobacco necrosis virus also induces phaseollin synthesis in some varieties of french bean through metabolic changes associated with cellular necrosis. Other plant species in which phaseollin has been found are *Phaseolus lunatus*, *P. radiatus*, *P. leucanthus* and

Vigna sinensis. Investigations with beans have revealed the existence of three additional chemically related phytoalexins, *viz.*, **phaseollidin, phaseollinisoflavan** and **kievitone** (an isoflavanone). Phaseollidin and kievitone are produced by *Vigna sinensis* and *Phaseolus vulgaris.*

Medicarpin is produced by leaves of alfalfa (*Medicago lupulina*) inoculated with many fungi such as *Helminthosporium turcicum, Colletotrichum phomoides, Stemphyllium loti,* etc. It is a pterocarpan phystoalexin and is also produced by *Phaseolus vulgaris, Cicer arietinum, Medicago sativa, Trifolium pratense* and *T. repens.*

Many other phytoalexins have been isolate from Leguminosae. **Triflorrhizin** was reported from leaves and root extract of clover (*Trifolium pratense*) inoculated with *Monilinia fructicola.* From the same plant species **maackiain** was reported. This phytoalexin has been detected as a minor product of infected pea in addition to pisatin and its structure is also similar to the later. Maackiain is formed also by *Cicer arietinum* inoculated with *Fusarium solani* f.sp. *phaseoli.* *Vicia faba* produces two phytoalexins- **wyerone** and **wyeronic acid**, Wyerone was first characerised as an antifungal component of extracts of broadbean seedlings inoculated with *Botrytis fabae.* These phytoalexins accumulate in necrotic cells at the infection site as well as within vacuoles of adjacent healthy cells. Alfalfa produces **sativan** (an isoflavan) which is also formed by *Lotus corniculatus* together with a related compound **vestitol.** The incompatible combination of soybean with *Phytophthora megasperma* var *sojae* yields the phytoalexin **glyceollin.** Compatible (susceptible) combinations produce very little glyceollin.

The terpenoid phytoalexins of the Solanaceae are **rishitin, phytuberin, capsidol** and **glutinosone.** The concept of phytoalexin started with the observation of Muller and Borger on potato cells expressing hypersensitivity to avirulent races of *P. infestans.* It was the work of Tomiyama *et al,* (1968) and Katsui *et al,* (1968) which characterised the chemical as rishitin, a terpenoid phytoalexin which is also found in inoculated tomato tissues. These workers isolated the compound from potato tubers of variety Rishiri of Japan, inoculated with zoospores of an incompatible race of *P. infestans.* Rishitin is a non-sesquiterpene alcohol. It is not present in intact healthy tubers or in mycelium of the fungus. The compound is also produced by inoculation of tubers with *Fusarium solani* f.sp. *phaseoli.* It has plant growth retarding properties and has been isolated from leaves also. Lubimin and phytuberin were also isolated from potato. Pepper (*Capsicum*) produces the terpenoid phytoalexin capsidol in response to fungal infection. The same compound is formed by leaves of *Nicotiana tabacum* and *N. clevelandii* following infection with tobacco necrosis virus. Another related antifungal compound glutinosone has been isolated from leaves of *N. glutinosa* bearing small necrotic lesions caused by tobacco mosaic virus. Thus, some hosts produce more than one phytoalexin and there seems to be close relationship between necrosis and phytoalexin production.

Among phytoalexins characterised in other plant families, substituted isoflavones occur in *Beta vulgaris* (Chenopodiaceae) and substituted isocoumarins in *Dacus carota* and *Pastinaca sativa*, both Umbelliferae. **Isocoumarin** derivative 6-methoxy-mellein ($C_{11}H_{12}O_4$, mol. wt. 208) is formed in tap root of carrot inoculated with *Ceratocystis fimbriata*, a nonpathogen of carrot. Some quantity of this phytoaxlexin is also produced by inoculation with *Helminthosporium carbonum*, *F. oxysporum* f.sp. *lycopersici*, *Rhizopus stolonifer* and *Thiellaviopsis basicola*. Synthesis of isocoumarin is influenced by the ability of the attacking organism to produce ethylene or cause its production by the host tissue. The cotton species in Malvaceae produce atleast two naphthaldehyde phytoalexins, **vergosin** and **hemigossypol**. In safflower (*Carthamus tinctorius*, Compositae) two aliphatic acetylenic alcohols, **safynol** and **dehydrosafynol** have been detected.

Ipomeamarone ($C_{15}H_{32}O_2$; mol. wt. 250) is produced by roots of sweet potato infected by the black root rot fungus *C. fimbriata*. It is a furanosesquiterpene ketone. The metabolic changes that occur in the diseased areas and in the surrounding uninfected tissues lead to a marked accumulation of ipomeamarone. The compound is absent or present only in traces in healthy roots and is produced in greater quantities by varieties resistant to the fungus. It inhibits mycelial growth, sporulation and protein synthesis of *C. fimbriata* by functioning as an inhibitor of electron transport and energy transfer reactions. Roots of sweet potato also produce ipomeamarone in response to infection by the violet root fungus *Helicobasidium momment* and in the presence of toxic chemicals.

Two phenanthrene phytoalexins, **orchinol** and **hircinol** are produced by the orchid species *Orchis militaris* and *O. hircinum*, respectively. Orchinol ($C_{16}H_{16}O_3$; mol. wt. 256) is one of the three compounds formed as a result of interaction between diffusible metabolites of the mycorrhizal fungus *Rhozoctonia repens* and tuber of *Orchis mario* and *O. militaris*. Orchinol is strongly fungistatic to growth of *R. repens* but *R. solani* can degrade this phytoalexin. Unlike other phytoalexins whose site of accumulation is the tissue under invasion, orchinol is not restricted to the infected area only and is extended over the whole tuber where it persists for months. Hircinol also inhibits the mycorrhizal fungus *R. repens*.

Many phenolic compounds have been claimed as phytoalexins in the Graminae. An inhibitory factor was reported in maize in response to *Puccinia graminis*. Another phytoalexin was isolated from maize with monogenic resistance to *H. carbonum*. In powdery mildew affected barley also phytoalexin activity was reported. Trivedi and Sinha (1978) and Zuber and Manibhushanrao (1979) reported such compounds in rice. However, Stoessel (1980) doubted that these compounds were phytoalexins. They seem to be more properly classified as inhibitins (Tani and Mayama, 1982). The better chemically defined phytoalexins from Graminae seem to be those reported by Mayama *et al*, (1981) and Tani and Mayama (1982). They demonstrated three phytoalexins named **avenalumin I, II** and **III** in oat leaves inoculated only with incompatible races of the crown rust fungus *Puccinia coronata* f.sp. *avenae*. These compounds are also produced by treatment of leaves with 10^3 M mercuric chloride or with cell wall components of germinating urediospores. Structurally, the three compounds are analogs of each other and are derivatives of anthranilic acid conjugated to three phenyl propanoid derivatives *p*-coumaric acid, ferrulic acid and 5-(*p*-hydroxyphenyl)-2,4-pentadienoic acid, respectively. These phytoalexins are unique in that unlike others they contain nitrogen and are highly hydrophillic. Sorghum phytoalexins identified to-date are unusual flavonoids of the 3-deoxyanthocyanidin class and include **luteolinidin, apigeninidin** and a **caffeic acid ester of arabinosyl-5-*O*-apigeninidin**. Each compound is highly inhibitory to the anthracnose fungus *C. gramnicola* at concentrations less than 9 µM. Because these phytoalexins are pigmented their accumulation can be located within 1-3 cells in living sorghum leaves. The compounds accumulate in a site-specific manner. The site restricted synthesis suggests that the initial response is by the first cells infected (Snyder and Nicholson, 1990). The mode of action has been studied in this sorghum-*C. graminicola* combination. Shortly after appressorium formation, colourless vesicle-like inclusions appear in the cytoplasm of the host cell directly underlying an appressorium. These inclusions, which are initially less than 1 µm in diameter, enlarge by coalescence, often attaining diameter of 15–20 µm, and move within the cell toward the site of appressorium attachment. During this process the inclusions become intensely pigmented and appear red due to mixtures of phytoalexins. The inclusions then burst, depositing their contents within the cytoplasm of the host cell and this results in hypersensitive death of the cell. The water soluble phytoalexins leach out of the cell and enter the overlying appressorium killing the fungus. All these events synchronously occur within 5–8 hours after formation of a mature appressorium. The observations have suggested that synthesis of the phytoalexins occurs within the vesicle and so long as the vesicle is not burst the host cell does not come in contact with the toxic compounds.

Accumulation of aromatic sesquiterpenoid phytoalexins is reported to occur in the interaction of *Xanthomonas axonopodis* pv. *malvacearum* with resistant cotton. In this interaction the flourescent cells within the infection site are isolated from healthy, non-flourescent cells surrounding the infection sites. Thus, as in sorghum, the phytoalexins occur in a concentrated fashion in cells immediately at the infection site and produce bacteriostatic effect.

Accumulation of glyceollin I occurs in soybeans resistant to the cyst nematode *Heterodera glycines*. The phytoalexin accumulates specifically within the cells adjacent to the head region of the nematode and is detected within 8 hours from the time of nematode penetration into the tissue. Phytoalexin synthesis has also been implicated with resistance of bean to *Pratylenchus* and soybean and cotton to *Meloidogyne incognita*.

Phytoalexin synthesis and accumulation is always associated with resistant reactions particularly with hypersensitive necrotic responses of the host. In spite of substantial evidence for the role of phytoalexins in resistance, it is often difficult to conclusively demonstrate their direct involvement in the *in vivo* containment of a pathogen. All plants do not produce phytoalexins. Production of antibiotic molecules which meet some of the criteria for phytoalexins has been demonstrated in more than 100 plant species representing 21 families. However such substances have not been demonstrated in interaction of host with a compatible or virulent strain of a pathogen. Most of the examples are of fungi, either weak parasites incompatible with host system, or only saprophytes. The greatest phytoalexin production is stimulated by organisms which are nonpathogens or avirulent on the particular host. This is perhaps because of violent reaction from the host which wants to reject the foreign, incompatible system. Compatible or virulent organisms elicit (induce) a lower rate of phytoalexin synthesis with a reduced yield so that the pathogen is well established in the host before appreciable amounts of phytoalexins are accumulated. Mechanical damage caused by penetration of a pathogen alone is not important in phytoalexin synthesis. Metabolites of the organism play a major role. Cell-free filtrates of the organism can induce phytoalexin synthesis. Molecules of the metabolites function as elicitors to induce the cells to synthesise phytoalexin. Quite a wide range of other compounds such as ions of copper, silver and mercury can also stimulate phytoalexin production. Pathogens which are able to infect a host but produce less phytoalexins are less susceptible to their toxic effects. The ED_{50} for *Sclerotinia fructicola* (non-pathogen of bean) is about 3 µg/ml of phaseollin whereas ED_{50} for *Colletotrichum lindemuthianum*, a pathogen of bean, is over 50 µg/ml. Pathogen strains can degrade or detoxify the phytoalexin. The high accumulation of phytoalexins in an incompatible combination of the host and pathogen is not only because of greater synthesis of the phytoalexin but also because there is suppression of biodegradation of the synthesised compound. In the compatible systems there is quantitative difference in gene activation and inhibition degrading activity. The former is slow and latter is rapid. Two pea pathogens, *F. solani* f.sp. *pisi* and *Ascochyta pisi*, completely degrade pisatin in 3 days when it is added to liquid cultures of these fungi. On the other hand *S. fructicola*, a nonpathogen of pea, which induces pisatin synthesis in pea, can degrade only half the pisatin in that period. Thus, it seems that phytoalexin synthesis is but one of a vast array of phenomena that, taken together, contribute to expression of resistance.

The biochemical or inorganic molecules functioning as elicitors are supposed to induce phytoalexin synthesis. These molecules may probably activate specific genes involved in synthesis. The biochemical mechanism of induction of synthesis of the phytoalexin glyceollin in soybean-*Phyophthora megasperma* interaction has thrown light on the steps involved (Yoshikawa and Masago, 1982). The first step after contact between the fungus cell wall and the host cell is **recognition** of the fungus associated molecules (**elicitors**) by **receptros** on the plant cell within minutes of contact. Hydrolytic enzymes of the host release the soluble glyceollin elicitor from the hyphal wall. This is followed by induction of de novo gene activation, i.e., some latent genes are activated. These genes then synthesise new types of messenger RNAs (mRNA) and proteins which are possibly required for synthesis of glyceollin.

2) Defence through induced synthesis of proteins and enzymes: Presence of synthesis of phenols is an important aspect of physiological disease resistance. The phenol oxidising enzymes present in plants induce resistance by becoming more active in response to infection. In addition to these biosynthetic compounds, synthesis of proteins and enzymes in large quantities and in modified forms also contributes to post-infection resistant reaction of plants. In many host-pathogen interactions if an avirulent strain of the pathogen or a nonpathogen is inoculated, protein synthesis and enzyme actions in cells near the point of entry are changed. The entry of a pathogen in a resistant host also results in similar responses. Due to these changes, the local tissues develop immunity or resistance.

Such enzymatic changes have been observed in black root rot of sweet potato (*Ceratocystis fimbriata*). One of the factors inducing these alternations in enzyme system is production of ethylene which itself is not antifungal. It is produced by the host cells in response to infection and moves out to adjoining healthy cells where it alters protein synthesis and enzyme activities. Extracts of *Meloidogyne* on tobacco show peroxidase activity, but only in contact with plant cells suggesting that preoxidase is a plant defence mechanism against nematode invasion (*Rev. Nematol.* 14:335.1991). The phenol oxidase enzyme not only oxidises phenolics but also increases the rate of polymerization of such compounds into lignin-like substances which are deposited in cell walls and papillae and interfere with further activity of the pathogen. Increased protein synthesis and enzymatic activities are found in many plants resistant to bacterial and virus diseases. The enzyme phenyalanine ammonia lyase (PAL) shows increased activity and greater new synthesis in diseased tissues. PAL is a key enzyme in the production of the basic molecules used for synthesis of most phenols, phytoalexins and lignin. The degree of resistance depends on the speed and limit of the protein synthesis and the speed with which the synthesised products move out to neighbouring healthy tissues to form protective layers.

3) Formation of substrates resistant to enzymes of the pathogen: The tissue disintegration by many pathogens is brought about degradation of pectic substances in the middle lamella and disorganisation of the tissue framework. The pectic substances are broken down by the action of parasite-produced pectinolytic enzymes such as pectin methyl esterase, pectin glycosidase, polygalacturonas and polymethyl galacturonase. Post-infection resistance in such cases may develop by formation of substances in the middle lamella which are not affected by these enzymes, for example, polyvalent cations of pectin-protein. Due to response of the cells to infection such substances are formed in the middle lamella and further tissue disintegration is halted. *Rhizoctonia solani* causes necrosis in bean plants. In resistant varieties, the entry of the pathogen and activity of its pectin methyl esterase (PME) separates the methyl group from methylated pectic substances and forms polyvalent cations of pectin salts in the vicinity of the pathogen and much ahead of it. These polyvalent cations contain calcium. Due to accumulation of calcium ions in diseased and neighbouring healthy tissues the pathogen fails to disintegrate the middle lamella by its polygalacturonase enzymes.

Growth regulators also aid in this process. Auxins produced by the host or the pathogen demethylate the pectic substances and, in presence of appropriate amount of calcium, pectic salts of calcium (calcium pectate) are formed. These salts are not affected by hydrolysing enzymes. The late blight of potato has been controlled by application of auxins. This is effective only when the plant contains excess of calcium.

4) Defence through inactivation of pathogen enzymes: Most necrotophic and hemibiotrophic fungi and most bacteria secrete an array of hydrolytic enzymes that often diffuse into the host tissues in advance of the pathogen, securing a nutritional base for the pathogen by breaking down complex molecules into simpler ones. If these hydrolases are inhibited or inactivated pathogenesis can be checked. In resistant reactions of a host this occurs through enhanced activity of phenols, tannins and proteins as enzyme inhibitors. The first two are most common. Several phenolic compounds are not antifungal but can inactivate the enzymes produced by fungi. Tannins have been demonstrated as inhibitors of pathogen enzymes of *Botrytis cinerea* in the skin of grape berries and also leaves. The compound is a catechol-tannin. Resistance of many young immature fruits to fungal infection is due to high content of such enzyme inhibiting compounds in the skin.

5) Defence through detoxification of pathogen toxins: In many plant diseases toxins produced by the pathogen play the major role in pathogenesis. There is a correlation between host sensitivity to the toxin and susceptibility, toxin production and pathogenicity and between host response to the toxin and to the attack of the pathogen. Thus, resistance to a toxin-induced disease is resistance to the toxin itself.

Resistance to toxin has been attributed to the ability of the plant metabolic processes to destroy or detoxify these substances. This hypothesis was proposed in studies with victorin, the toxin produced by *Helminthosporium victoriae* in oats. Resistance to the action of victorin is due to absence of receptor sites in resistant varieties and their presence in susceptible varieties. Detoxification of HC-toxin is the biochemical basis of resistance of maize to *Cochliobolus carbonum*. Rate of metabolism of toxins leading to formation of non-toxic substances has been claimed as a basis of resistance in rice to the blast fungus

(*Pyricularia grisea*). This fungus produces two toxins-picolinic acid and pyricularin. Within 3 days of inoculation about half of picolinic acid is metabolised to picolinic acid ester and N-methyl picolinic acid which are not toxic to rice tissues. In some resistant varieites the plant metabolises this acid very rapidly. The phenolics (chlorogenic acid and ferulic acid) present in resistant rice plants convert pyricularin into a non-toxic substance. In *Fusarium oxysporum* f.sp. *vasinfactum* (cotton wilt) and *F. oxysporum* f.sp. *lycopersici* (tomato wilt) resistant plants rapidly metabolise fusaric acid to form non-toxic N-methyl fusaric acid amide. The young tomato plants can degrade fusaric acid faster than older plants. Other examples of detoxification of toxins have been cited in chapter 11.

6) Defence through altered biosynthetic pathways: Respiration in diseased plants or their organs is increased with activation of dehydrogenase, peroxidase, phenol oxidase and deaminase. New enzymes, proteins and other compounds are also synthesised. Accumulation of these compounds may be in quantities sufficient to harm the pathogen. Most of these compounds are formed through shikimic acid pathway and modified acetate pathway. It is not known which mechanism activates these pathways. In diseased plants a part of the glycolysis is replaced by pentose pathway which forms the 4-carbon compound erythrose phosphate essential for initiation of shikimic acid pathway. Although pentose pathway predominates in diseased plants, glycolysis also continues and forms the 3-carbon compound phosphoenol pyruvic acid that combines with erythrose phosphate to initiate shikimic acid pathway leading to synthesis of common phenols and phytoalexins. These aspects were discussed in the light of their role in resistance to disease development.

Infected cells of plants have been found to contain larger amounts of nucleic acid (RNA), acid proteins and total proteins. The amount of nucleic acid has been related to resistance and susceptibility. It is possble that in early stages of infection the biochemical reactions affect the RNA and proteins, without affecting the DNA and thus activate certain specific genes which alter the metoblic processes. These alterations in metabolism finally lead to synthesis of different compounds involved in resistant or susceptible reactions.

7) Defence triggered by biotic and abiotic environments on leaf and root surface and by induced resistance: The surface of leaves and roots is covered with innumerable microorganisms majority of which are non-pathogens and subsist on the exudates. Cross protection, induced resistance or induced systemic resistance (ISR), systemic acquired resistance (SAR) and localised acquired resistance (LAR) are terms used to denote the condition in the life time of the plant, in which biological or chemical treatment or stress trigger mechanisms of resistance (Kuc, 1981, 1982, 1995) which may become systemic or remian localised In the latter case the signal that propagates the enchanced resistance capacity throughout the plant is lacking (Van Loon *et al*, 1998). Induced resistance is non-specific and is not creation of resistance where there is none but activation of the "turned off" resistance genes. It can be triggered by non-pathogens (common rhizosphere and phyllosphere microflora), avirulent forms of pathogens, incompatible races of the pathogens, virulent pathogens under circumstances where infection is stalled owing to environmental conditions and by substances that promote the growth and activity of rhizobacteria. It can also be triggered by exogenous application of water extracts of compost and inorganic and organic chemicals. The resistance is induced in a manner comparable to immunization in mammals but the underlying mechanisms differ (Sticher et al, 1997). The induction of resistance is due to one or more activities such as accumulation of pathogenesis related proteins (PR), activation of lignin synthesis, enchanced peroxidase activity (oxygen burst), a defined change in plant metabolism, etc. The relationship of PR proteins with systemic acquired resistance has been extensively studied in the recent past (Ryals et al, 1994; Kessmann et al, 1994, Sticher et al, 1997). Most of the PR proteins accumulate in extracellular spaces and in the vacuoles. Among the PR proteins, PR-1-β-1,3-glucanases (PR-2), chitinases (PR-3), PR-4 and osmotin (PR-5) are antimicrobial proteins. They accumulate in greater quantities at the site of infection and in lesser amounts in non-infected parts of the infected plant.

In cross-immunity or cross protection the pre-inoculation with a weak or avirulent strain of the pathogen provides protection against a virulent strain of the pathogen. Cross protection against citrus tristeza virus is an example. Cross protection against bacterial wilt caused by Ralstonia solanacearum in many plant species by prior inoculation with non-pathogenic, avirulent, bacteriocin producing or incompatible strain of the bacterium or by application of heat killed bacterial cells has been reported (Chen and Echandi, 1984; McLaughlin and Sequeira, 1988). The efficacy of heat killed cells suggests non-antagonistic nature of the interaction. Similarly cross protection against tomato wilt (F. oxysporum f.sp. lycopersici) by pre-inoculation with a non-pathogenic Fusarium oxysporum, against cotton wilt (F. oxysporum f.sp. vasinfectum) by pre-inoculation with a mild or avirulent strain of the pathogen and many other diseases has also been reported.

However, pre-inoculation with a pathogen-related inducer is not always possible under natural conditions. Thus, induction of resistance under natural conditions is more common by the action of non-pathogenic phyllosphere and rhizosphere microflora. Rhizobacteria, particularly the pseudomonads, are common inducers of resistance even against foliar pathogens. The plant growth promoting rhizobacteria (PGPR) applied to the seed or soil remain localised at the root surface but induce resistance in the leaves and the stems. The PGPR induce systemic resistance in cucumber against cucumber mosaic virus, Pythium species causing crown rot, and many other diseases (Kloepper, et al, 1993; Raupach et al, 1996; Wei et al, 1996). Among the rhizobacteria, strain 89B-27 of Pseudomonas putida induces resistance in cucumber against anthracnose

(*Colletotrichum orbiculare*), angular leaf spots, Fusarium wilt (*F. oxysporum* f.sp. *cucumerinum*), bacterial wilt (*Erwinia tracheiphila*), cucumber mosaic virus and the insects *Diabrotica undecimpunctata* and *Acalymna vitatum*. In the same host a strain of *Serratia marcescens* induces resistance against *F. oxysporum* f.sp. *cucumerinum* (vascular wilt), *Colletotrichum orbiculare* (anthracnose), cucumber mosaic virus, *Pseudomonas syringae* pv. *lachrymans* (angular leaf spot), bacterial wilt (*Erwinia tracheiphila*) and the insects *Acalymna vittatum* and *Diabrotica undecimpunctata* (*cf.* Van Loon *et al*, 1998).

Salicylic acid in plants is known to trigger the resistance genes or prime them to become active when challenged by the pathogen (Kessmann *et al*, 1994). Rhizobacteria produce salicylic acid. Exogenous application of salicylic acid induces resistance in grapes against downy mildew and powdery mildew and in several plant species against many diseases (*cf.* Sticher *et al*, 1997). Treatment with acetyl salicylic acid (aspirin) decreases the symptoms of tobacco mosaic virus in tobacco and could lead to accumulation of pathogenesis related proteins (White, 1979). Salicylic acid acts as a signal for systemic acquired resistance and activates the resistance genes.

Phosphate and carbonates as inorganic inducers of resistance are reported in many plant-disease systems. Foliar spays of phosphates and carbonates reduce the intensity of grapevine powdery mildew by inducing resistance (Reuvini and Reuvini, 1998; Fuhr *et al*, 1998).

8) Hypersensitivity as defence mechanism: The hypersensitive response or necrogenous reaction of plant tissues to invading agents includes rapid death of host cells, tissue browning and accumulation of antimicrobial components. It is generally characteristic of resistant rather than susceptible tissue. These necrotic reactions provide both structural and physiological defence. Hypersensitive response occurs only in incompatible combinations of the host and the pathogen. In such combinations no difference is observed in the manner of penetration of the epidermis in susceptible and resistant plants. In resistant plants, the invaded cells rapidly lose turgor, turn brown and die. This does not occur in susceptible plants. In the resistant plants when the pathogen comes in contact with the host cytoplasm, the host nucleus moves towards the pathogen. Soon it is disorganised and brown granules are formed in the cytoplasm. The changes that occur in such cells include loss of permeability of cell membranes, increased respiration, accumulation and oxidation of phenolic compounds and production of phytoalexins. This necrotic or abortive defence reaction causes the organs of the pathogen to degenerate, its nuclei are disorganised and the cytoplasm becomes dense. Fungal, bacterial and nematode pathogens are isolated in the area of hypersensitivity and cannot move out. In virus diseases, hypersensitivity always results in local lesion in which the virus may survive for considerable time but generally in low concentrations and its spread to beyond the lesion is checked.

The incompatibility, on a biochemical basis, can be explained as ability of the host to recognise the pathogen, coupled with failure of the pathogen to suppress the subsequent biochemical changes, leading to rapid death of the host cells. This could, thus, be due to one or more of the following conditions: (i) the invading pathogen contains the components necessary for its recognition by the host cell, (ii) resistant host cells have the ability to recognise these components of the invading pathogen, (iii) the invading pathogen may lack factors which inhibit the host recognition system, (iv) even if recognition of the invading pathogen takes place, the pathogen may not be capable of inhibiting few or more of the subsequent steps leading to resistant reaction, and (v) even if all processes of the host resistant reaction are activated the invading pathogen may not be tolerant of them. Programmed cell death (PCD) had been advocated as an inducer of necrotic response. The cells are programmed to commit self destruction or suicide when they find that they are of no more use or can not by themselves face the onslaught of the invader. They die to save the others (Gilchrist, 1998).

Necrogenous reactions have been noticed in wart disease of potato (*Synchytrium endobioticum*), late blight of potato (*P. infestans*) and blast of rice (*Pyricularia grisea*). However, the best examples of defence through hypersensivity are the diseases caused by obligate parasites or biotrophs such as rusts, powdery mildews, nematodes and viruses. The pathogenic activity of these agents depends on the living condition of the host cells, i.e., the host and the pathogen must accomodate each other as long as possible. If necrosis occurs, invaded and surrounding cells are dead and the pathogen is either sealed off from healthy tissues or is killed.

The rust fungus is a biotroph. In varieties showing immunity of very high degree of resistance it has been found that as soon as contact is established between the germ tube of the urediospore and host cell membrane after penetration, the cytoplasm reacts so violently that tissues at the site of infection are killed. Toxic compounds (bound phenols, etc.) or hydrolytic enzymes (lysozomes) released from dead cells may either cause death or inhibit further advance of the fungal hyphae. At the same time, as a consequence of host cell necrosis, the pathogen is starved as it cannot obtain nutrition from dead cells. A necrotic lesion is formed at the site of infection in which rust pustules do not develop.

The fungus causing potato wart is also a biotroph although it does cause some amount of tissue disorganization even in successful infection. The immune varieties show necrogenous reaction to infection by the pathogen. The infected epidermal cells die quickly and their protoplasm is changed into a brown gum-like mass. The rice variety Kan Non Sen is resistant to blast disease. Infection takes place in this variety also quite early but the necrogenous reactions are still faster. As in the wart disease of potato, in this disease also the protoplasmic changes occur and obstruct growth of the pathogen. Only small necrotic lesions are formed on the leaf.

In many nematode diseases resistance of the plant is based on tissue hypersensitivity to infection. The violent reaction of the resistant host to enzymes and toxins released by the nematode during penetration and feeding leads to host cell necrosis. The necroses form around the nematode head, walling it off and delaying its development and establishment of nurse cell system and causing it to die of starvation.

The strong correlation of hypersensitive response of higher plants with disease resistance or with "defence" was discovered by H.M. Ward in 1902 and was later stressed by E.C. Stakman in 1915 who introduced the term "hypersensitivity". The phenomenon has been considered as the most common and widely distributed defence mechanism. Since the early investigations were made with biotrophs and viruses, emphasis had been placed on the role of hypersensitive necrosis in arresting the growth, development and multiplication of the infecting biotrophs. Later, it was recognised that plants can exhibit a hypersensitive response not only to biotrophs but also to such pathogens as bacteria and to saprophytes, non-pathogens or weak pathogens. Phytoalexin synthesis is usally associated with appearance of necrotic response and the synthesis is induced mostly by non-pathogens or weak pathogens.

9) Summary of biochemical defence reactions: In a plant showing resistant reaction one or more of the following condtions can be found:

1. On entry of the pathogen, a temporary increase in cellular metabolic activities occurs in the host. The metabollic porcesses deviate from the normal. Due to stress caused by increased metabolic activity the cells die rapidly showing hypersensitive reaction. Rapid death of cells is correlated with increased degree of resistance in most disease systems.

2. When the infected tissues are reaching the necrotic stage metabolism of neighbouring tissues is also increased and phenolics and other compounds are accumulated. In this process, the synthesised compounds move from healthy to diseased tissues. The neighbouring metabolically active tissues become ready to obstruct the pathogen.

3. The reactions expressed by hypersensitivity form common phenols, phytoalexins, and other abnormal substances. These may include toxins by the pathogen. The oxidised products of phenolics may detoxify the toxins or inactivate other mechanisms of the pathogen.

4. When spread of the pathogen is checked, the neighbouring healthy tissues with accelerated metabolic activities try to isolate the damaged parts by forming new tissues and eliminate the disease. In this way lignification, cork layer formation and similar protective structures are formed.

5. The isolated part may separate from the plant.

13

Genetic Variability in Plant Pathogens

Structure of the pathogen, its biochemical activities and response of the plant to these activities are not the only requirements for a complete understanding of pathogenesis. Knowledge of aggressiveness of the pathogen not on an individual in the population but on the whole population of the plant in a given area under different environmental conditions is equally important for clear insight of pathogenesis. Since both participants in the disease syndrome act biologically and resistance or susceptibility of the plant and virulence or avirulence of the pathogen are hereditary the disease incidence may not be uniform in the entire plant population (crop in the field) consisting of thousands of individuals. For instance, if stem rust of wheat attacks a field it is possible that all the plants will not be infected with same speed and intensity. Even if the inoculum has landed uniformly on all plants some may show higher degree of rust intensity than others. On a genetic basis, this difference in disease incidence on individuals could be attributed to two reasons:

1) In the plant population, some individuals have developed resistance.
2) The spore population of the pathogen has units having no affinity for the plant. Thus, while some spores are successful in invasion of plant tissues others fail to do so.

The two reasons are supplementary to each other in the same manner as the genetics of the host and the pathogen.

No plant population can remain genetically pure for a long time. Genetically different individuals may develop due to methods of pollination and due to mutation. These self-generated variations can make the plant resistant or susceptible. Similar causes induce variations in plant pathogens also. Same species of the pathogen produces numerous spores which may possess different characters including aggressiveness and adaptability for the host. This chapter gives an account of this aspect of genetic variability in plant pathogens. Viruses, bacteria, fungi and nematodes undergo genetic changes by one or more methods. Viruses and bacteria do not reproduce sexually. Therefore, most of the knowledge of variability has been obtained from the study of fungi in which sexual reproduction involving union of gametes from two unlike sources and some sexual-like processes (heterokaryosis, parasexualism, heteroploidy, etc.) are now known to cause variability. Mutation and hybridization are general methods common in all category of microorganisms while heterokaryosis, parasexuality and heteroploidy, etc. are mostly found in fungi.

VARIABILITY IN VIRUSES

Viruses are not cellular organisms. However, molecular structure of viruses has the capacity of transferring its characters to its duplicates formed during replication in the host cell. The principle of breeding true or like tends to beget like applies to viruses as much as to living organisms. Since the laws of heredity apply to viruses, they can exhibit variability. Virus strains can be produced artificially by treatment of virus suspension *in vitro* with mutagens such as nitrous acid or hydroxyl amine which disturb the nucleotide sequence in the genome. In nature, mechanisms of variability in viruses can be grouped as follows:

1) Mutation: When a small quantity of virus particles is inoculated in a suitable host the particles show immutability and the newly synthesised particles produce the same effect on the host as the original particles. This continues so long as the virus is maintained continuously on the same susceptible host. However, sometimes, in addition to the immutable particles, virus particles having characters different from the original can be synthesised. These are known as **virus mutants.**

Passage through resistant cultivars of the host induces mutation. In areas where breeding for resistance to bean common mosaic is active, 10 strains of the virus were distinguished that differed in some 7 pathogenicity genes, 4 of which were postulated to interact with their host in a gene-for-gene relationship (Bos, 1983).

In this manner the virus can develop several races, strains or variants. Tobacco mosaic virus (TMV), cucumber mosaic virus, (CMV), sugarcane mosaic virus (SMV), potato virus Y and sugar beet curly top virus have such strains.

2) Mixed inoculations: Natural inoculation often consists of mixture of virus strains which can be demonstrated by inoculating a local lesion host with dilute sap and then sub-culturing the virus from individual lesions. Presence of many strains or genetic material in the same environment may lead to any of the following consequences resembling **hybridization.**

a) Pseudo-recombination: In mixed infection of viruses having split genome, fragments of the genome may aggregate and re-assemble to form a new type of progeny. This mechanism differs from true genetic recombination found in cellular organisms where the recombination occurs from the exchange of genetic information between parental genomes at crossing over of chromosomes during meiosis.

b) Heterologous encapsidation or genome masking: This can result from direct interaction between different viruses in mixed inoculations. The genome of one virus is encapsidated in the protein of another strain of virus. Since the effect of a virus on the host or the vector is mediated through its protein, this may assist the virus in infecting a host not normally susceptible to it or being transmitted by vector or vector species that normally does not do so.

If two strains of a virus are inoculated into the same host plant or if a vector introduces two strains of a particular virus in the host, one or more new virus strains may be recovered with properties (aggressiveness, symptoms, etc.) different from either of the two original strains used for inoculation. These new stains can be called hybrids and develop by recombination of their genetic material (RNA or DNA) in the same manner as the DNA in the cellular organisms.

The antigenic properties of different strains of the same virus are identical. Antigens are proteins. Synthesis of proteins and enzymes takes place through line up of amino acids in specific sequence dictated by genes (in viruses, nucleotide triplet in RNA). Therefore, if antigens are similar among different strains it proves that genetically they are closely related. This proof of affinity among virus strains is obtained by serological tests. In a suitable medium each antigen produces specific antibodies. Even serologically distinct viruses may be distantly related and may have overlapping host range. Thus, there is no clear cut difference between viruses and strains. This makes defining a species in viruses impossible.

VARIABILITY IN BACTERIA

Strains differing in pathogenicity, colony appearance, toxin production, etc. occur in bacterial plant pathogens. In the cotton blight bacterium, *Xanthomonas axonopodis* pv. *malvacearum*, 32 races varying in their virulence on different species of *Gossypium* are known. Similar variants in citrus canker organism (*X. axonopodis* pv. *citri*) and mango blight bacterium (*X. campestris* pv. *mangiferae-indicae*) are also reported. These races can evolve by mutation or gene recombination. Inheritance of genetic characters is on the same principles as in other organisms. The genetic material is the DNA. In bacteria the DNA is present in chromosomes of the unorganised nucleus (nucleoid) and also in plasmids. The transfer of the genetic material or characters coded in this material to daughter cells occurs at the time of binary fission, the method by which bacteria including mycoplasmas multiply.

The genetic variability among bacteria may be caused by following methods resembling sexual reproduction.

1) Conjugation: This plasmid-motivated process occurs in two steps and both are governed by the non-chromosomal plasmid DNA. In the first step, a series of reactions between the surfaces of two conjugating cells, known as *donor* and *receptor* cells, results in the formation of a conjugation bridge. In the second step, passage of a molecule of the plasmid DNA occurs through the conjugation tube.

Surface to surface interactions between the donor and receptor cell are the consequence of the specific properties of the surface conferred by the plasmid to the donor cell. Every cell does not conjugate. Only cells with transmissible plasmid become donor and conjugate. Such cells carry on their surface a number of special appendages called *sex pili*. A specific receptor site for conjugation occurs on the receptor cell. This cell gets attached to the tip of a sex pilus. This pilus is then retracted by the donor cell and within a few minutes the two cells move into a position of contact. The chromosomal DNA can also pass through the sex pili into the recepient cell at the receptor site. During transfer, only the DNA is moved and no other cellular material is transferred.

Prior to conjugation, the plasmid exists as a double stranded circular DNA molecule. The molecule is not transferred as such. Actually, one strand of the plasmid DNA is broken at the replication origin and enters the recipient cell. Complementary strands are synthesised by DNA polymerase in both the donor and the receptor cell, thus both retain replica of the original DNA molecule.

2) Transformation: In this process, the bacterial cells absorb the genetic material exuded by a compatible cell or freed by lysis of the cell wall. The compatibility is supposed to be due to the presence of a specific protein on the surface of the recipient cell. The recipient cell then contains altered genetic material since new genes are added to it. This cell now reproduces to develop a new race.

3) Transduction: In this phenomenon, a small piece of the bacterial chromosome is incorporated into a maturing bacterial virus (phage) particle that has infected the donor cell. The bacteriophage particle is released after the death of the infected cell. It carries the genetic material to a new bacterial cell which it infects. If this cell survives, recombination of genetic material takes place.

GENETICS AND VARIABILITY OF FUNGI

Like other living organisms, the nucleus with its genetic material (DNA) helps in determining the heredity of fungi. The fungi reproduce sexually as well as asexually. A large number of them reproduce only asexually but variability is common in them also. Whatever the method of reproduction, due to presence of the genetic material the characters of the fungus are transferred from generation to generation. Mutation and gene recombination are the two common methods by which variability is induced in fungi. In addition parasexuality and other methods are also reported in many species.

Hybridization: Hybridization as a mechanism of variability is common among fungi. Mating of different strains, varieties or, sometimes, species results in different kinds of recombination of genes. Thus, new biotypes and races are produced. In hybridization, two haploid (1N) nuclei with some difference in their genetic material combine to form a diploid (2N) nucleus or the **zygote**. In higher plants and in nematodes, the zygote divides mitotically and forms dikaryotic vegetative cells. But in most fungi the zygote divides by meiosis and haploid (monokaryotic) cells are formed. Since in the process of hybridization genetically different nuclei participate, the diploid nucleus contains characters of both parents. The combined characters may be harmful or advantageous for the progeny, which, in any case, will be different from the parents. The transfer of genetic characters to the progeny actually takes place when the diploid nucleus undergoes meiosis (reduction division). Not only the number of chromosomes are halved but the genetic material also gets distributed in different daughter nuclei. In meiotic division, the homologous chromosomes of the two parents align themselves parallel to each other in pairs. The paired chromosomes break into chromatids and then the genetic material is exchanged between the chromatids. This exchange of genetic material at the time of division of the diploid nucleus results in formation of haploid nuclei which differ from parents and from each other.

In fungi the hybridization is accomplished by haploid nuclei coming from genetically different mycelia, hyphae or spores. These fungal organs may be different varieties or strains or even species of a genus. In *Puccinia graminis* (a heterothallic fungus) different varieties, races, etc., have a common platform for sexual reproduction. They all have their spermatial and aecial stage on common barberry. Basidiospores (the haploid stage) from different races or varieties may infect the same leaf or plant. Thus, chances of dikaryotization of haploid mycelium or spores developing from this source are high. The dikaryotic stage that develops in the form of aecial stage may initiate the life cycle of a new form of the fungus. In smuts, such as in *Ustilago maydis* (maize smut), there are thousands of haploid biotypes within the species. Since in smuts mostly the dikaryotic phase causes infection, the haploid mycelium or spore is always dikaryotised by other spores or hyphae which may be genetically different. In this way biotypes of the fungus may appear even during the crop season.

Heterokaryosis: Heterokaryosis, the capacity of haploid nuclei to form various association with vegetative cells, is regarded as a prominent feature of fungi. The term precisely describes the condition of a cell containing two or more genetically different nuclei. These dissimilar nuclei occur in a single cell even if it is only one cell in a thallus. The term heterokaryosis is used to designate a vegetative condition of the thallus to differentiate it from a heterokaryotic condition present in sexually produced organs such as ascogonia and ascogenous hyphae of heterothallic fungi. The definition of the term also implies that heterokaryosis is possible in thalli with bi- or multinucleate cells unless uninucleate cells have been made bi- or multinucleate by anastomosis of compatible cells and hyphae.

When these genetically different nuclei in a cell fuse, a dikaryon with different genetic combination is formed. In many Basidiomycetes, the normal vegetative mycelium is dikaryotic throughout and, therefore, heterokaryosis exists throughout in all heterothallic forms. In *Rhizoctonia solani* (*Thanatephorus cucumeris*) the vegetative cells are multinucleate and in heterokaryotic thalli most of the cells appear to be heterokaryotic. The mechanism of conjugate division conserves heterokaryosis in such fungi and the condition is fairly stable.

Heterokaryons can arise in several ways. A mutation in any bi- or multinucleate cell produces a heterokaryotic cell. The formation of binucleate sexual spores also may lead to heterokaryosis. Such spores are formed regularly in some species of Ascomycetes and Basidiomycetes. *Thanatephorus cucumeris* occasionally produces binucleate basidiospores which might give rise to heterkaryons. Bi- or multinucleate asexual spores from heterokaryotic thalli may also produce heterokaryotic thalli.

Anastomosis is undoubtedly an important cause of heterokaryosis. It involves fusion of compatible vegetative hyphae of same species, movement of one or more nuclei into one or other of the fused cells, and the establishment of a compatible heterokaryotic state. Incompatible anastomosis results in necrosis of cells. Vegetative compatibility groups (VCG) are known in *Rhizoctonia solani* (*Thanatephorus cucumeris*) in which several anastomosis groups (AG) are reported and fusion occurs between hyphal cells of only same group. In *Fusarium oxysporum* f.sp. *cubense* (Panama disease of banana) also 13 vegetative compatibility groups are reported (*cf.* Singh, 2000). A similar nuclear displacement may occur in adjacent cells of the same hypha by formation of clamp connections through which nucleus or nuclei from one cell move into another.

Heterokaryosis has been demonstrated in many plant pathogenic fungi. Apart from rusts, smuts and hymenomycetes, it has been reported in *Colletotrichum lagenarium, Helminthosporium sativum, Alternaria solani, Fusarium solani, Fusarium oxysporum, Rhizoctonia solani, Verticillum albo-atrum, Phytophthora infestans* and *Botrytis cinerea.*

Parasexuality: Fungi-imperfecti have no sexual cycle of reproduction but variability is common in this group also. Mutation, heterokaryosis, and somatic recombinations are usually advocated as nuclear mechanisms conferring this variability in the Fungi-imperfecti. Together with sexual reproduction these mechanisms confer variability in fungi of other classes also.

Genetic recombination, without sexual cycle, in which there is no fine coordination between recombination, segregation and reduction as there is no meiosis, has been termed parasexual. The essential steps of the parasexual cycle are (i) heterokaryosis, (ii) fusion of unlike nuclei to form heterozygous diploids in the vegetative cells, and (iii) segregation and recombination at mitosis. Heterokaryosis ensures presence of genetically different nuclei in the same cytoplasm. Heterozygous diploid is formed by fusion of the heterokaryons. These diploids have the characters of different nuclei of the heterokaryotic cell. Mitotic crossing over and segregation of genes takes place when these diploids divide by mitosis. As a result three types of nuclei may be formed: (i) replicates of the diploid, (ii) nuclei with recombination of genes and (iii) segregates. Thus, new biotypes and races can develop from the daughter nuclei possessing characters different from the heterokaryons. In *Helminthosporium sativum* definite evidence of presence of parasexual cycle including the three steps mentioned above has been obtained. Other plants pathogenic fungi in which parasexuality, in addition to sexual cycle, has been observed are *Leptosphaeria maculans* (black leg of crucifers), *Ustilago maydis, Ustilago violacea* and *U. hordei*. Role of parasexuality in rust fungi has also been indicated. In *Puccinia graminis tritici, P. recondita tritici* and *Melampsora lini* new strains arise on the host plant from urediospore mixtures of two races.

Heteroploidy is the existence of cells, tissues or whole organisms with numbers of chromosomes per nucleus that are different from the normal 1N or 2N for the particular organism. Heteroploidy is often associated with cellular differentiation and represents normal situation in the development of most eukaryotes. This situation in many fungi results in variability. Growth rate, spore size, rate of spore production, hyphal colour, enzyme activities and pathogenicity are affected. **Sectoring** is the appearance of morphologically distinct sectors in fungus colonies on nutrient media. Sectors sometimes show difference in pathogenicity.

Mutation is more or less an abrupt change in the genetic material due to some physical or chemical shock. Usually, mutations occur due to unavoidable but very infrequent accidents at the time of cell division. Due to the accident there is change in sequence of bases in the DNA. The change is almost permanent and transmitted in a hereditary fashion to the progeny. Mutations occur spontaneously in all living organisms and viruses and result in morphological as well as physiological differences in the progeny. The tendency for mutation varies in species and strains. Normally the mutants show change in only one specific character but in successive steps multiple mutations for a variety of characters may also occur. Variability through mutation has been observed in *Phytophthora infestans, Puccinia graminis* and *Venturia inaequalis*. These pathogens develop the capacity to infect those hosts which they could not infect earlier. Strains of pathogens resistant to fungicides and antibiotics also develop through mutation.

Cytoplasmic adaptation: Pathogens often develop the capacity to perform biochemical reactions which were not present in them earlier. In this way they can utilize protoplasm of hitherto unfavourable host. This is known as adaptation to new type of cytoplasm. Three types of adaptability have been observed in microorganisms: (i) the pathogen may acquire tolerance to toxic materials, (ii) utilization of new type of cytoplasm, and (iii) change in virulence.

Although cytoplasmic adaptability can be due to mutation of the nuclear DNA, mutation of the extranuclear DNA in the cytoplasm also plays a role. Whatever characters are controlled by this DNA are likely to change by mutation. This type of cytoplasmic inheritance is common in all types of pathogens except viruses and viroids which do not have cytoplasm.

LEVELS OF VARIABILITY IN PATHOGENS

When the progeny of a pathogen shows variation in characters from the parents it is called a **variant**. The dissimilarities of characters in these progenies are hereditary. Progeny developed by a variant having similar heredity is called a **biotype**. Biotypes differ from the parent race only in a few minor characters, mostly in the type of symptoms produced on specific hosts. A group of biotypes with identical characters forms a **race** or **strain**. In plant pathogens these races and biotypes are distinguished with the help of a set of differential hosts. When a group of races of identical morphology attack and infect only a specific host genus or species the group is called a **variety** or **forma specialis**. Several varieties of the pathogen (fungi) constitute the **botanical species**. All these variations in species at different levels are the result of genetic mutation or recombination. Further explanation of variability in fungi is given below with few examples of pathogenic fungi.

The cereal rust fungus: Two categories of variants are recognized in *Puccinia graminis*- varieties and physiologic forms. The varieties differ from each other somewhat in shape and size of urediospores but the principal difference between them is their preference for groups of hosts in different members of Graminae. Each variety can attack several species of one or more genera of the grass family but it can not attack the members of other genera which may be susceptible to other varieties of the same species (*P. graminis*). The six varieties in *P. graminis* are *tritici, secalis, avenae, phleipratensis, agrostic* and *poae*. A variety, in turn, may contain several physiologic forms or races which differ from each other principally in their ability to infect varieties within one or more species of a genus. Varieties are given latin names while physiologic forms (races or strains) are designated by Arabic numerals. Thus, *Puccinia graminis tritici* race 15B means biotype B of race 15 of the variety *tritici* of the botanical species *Puccinia graminis*. There are more than 200 such physiologic forms or races of *Puccinia graminis tritici*. Each variety of *P. graminis* has a large number of physiologic races.

From the above example it is clear that within the botanical species *P. graminis* there was sequential variation leading to development of such biological units which gradually lost morphological differences and attained physiological differences such as choice of specific food or host. This type of variation from the original species is known as **physiologic specialization** which has been defined as *"presence of entities within morphologic species, not readily distinguishable by structure, but differing from each other physiologically"* including pathogenicity, biochemical properties, cultural variability, spore germination and ecological relationships. The entities have been called physiologic races or strains.

The smut and other fungi: As in the rust fungi, variability is quite common in smut fungi. As a matter of fact the smut fungi are more variable than any other group of plant pathogenic fungi. Variations abound in morphology as well as in physiology. Morphologically, the variations within a species differ in spore markings, cultural characters and gross symptom expression on the host. Physiologically, they differ in nutritional requirements, spore germination, compatibility relationships, nuclear behaviour and pathogenicity. Similar physiologic specialization is known to occur in fungi causing late blight of potato, powdery mildew of cereals, wilt of pea, tomato, etc.

LOSS OF VIRULENCE IN PLANT PATHOGENS

It is common observation that many pathogenic organisms lose their virulence to one or more of their hosts when they are kept in culture for relatively long periods of time or when they are subcultured repeatedly. If the culturing of the pathogen is prolonged sufficiently, the pathogen may lose virulence completely. The pathogens that have lost virulence partially or completely in culture or in other hosts are often capable of regaining virulence if they are returned to their host under the proper conditions. *Pseudoperonospora cubensis* (downy mildew of cucurbits) originally isolated from a particular host but grown for several generations on a different cucurbit species loses its affinity for the original host and sporulates better on the host on which it is being currently grown (Thomas and Jordain, 1992). The isolated cultures of *Ralstonia solanacearum* (bacterial wilt of potato) easily lose virulence upon cultivation on laboratory media. This is correlated with colony variation (Buddenhagen and Kelman, 1964). On tetrazolium chloride medium, virulent wild types produce an irregular to round, fluidal, white colony with pink centre while the avirulent mutants form round, butyrous, deep red colonies with narrow bluish borders. Spontaneous reversion of afluidal mutants to wild type in some strains of the bacterium is also reported (Chakrabati *et al*, 1995).

Loss of virulence in cultures or in other hosts seems to be the result of selection of individuals of less virulent or avirulent pathogen strains that happen to grow and multiply more rapidly than the virulent strain. On repeated subculturing the avirulent types finally predominate. If the virulence is not completely lost it means that some virulent individuals are present and if this inoculum is transferred to a suitable susceptible host, the reversion to virulence form may occur.

14

Genetics and Molecular Basis of Host-Parasite Interaction

Host-parasite interactions as treated in the preceding chapters seem more of a war where both host and pathogen try to eliminate each other. This is not always true. Since plants do not have a centralised immune system as in animals, the attacking pathogen is never eliminated. In plant systems, host-parasite interaction involves management of the pathogen by restricting its damaging effect. Adjustment is one of the natural laws. In order to understand the host-parasite interaction properly we should start right from the evolution of parasitism.

EVOLUTION OF PARASITISM

In the hierarchy of parasitism it is natural to consider the saprophytes as the lowest members. Saprophytes colonise only dead organic matter. However, many of them do possess the enzyme systems necessary for maceration of living tissues but they remain saprophytes, restricting themselves to dead organic matter where there is no resistance to their activity. The reason for this inability to become a parasite is their lack of capacity to breach the general defence system of living plants and/or the enzymes produced by them do not have access to the living cell wall polymers due to presence of structural barriers. In living organisms competition is one of the inducers of evolutionary developments. The intense competition among saprophytes for same source of organic matter forced some of them to develop parasitic abilities. *Pythium* is an example. It appears to have tendencies of a true saprophyte but developed the faculty to become a parasite although the parasitism in most species is restricted to attacking the juvenile and succulent plant tissues under adverse conditions of environment (anaerobic, high moisture, etc.) where resistance is minimum. Even then it likes to prefer saprophytic living by first killing tissues through enzymes and then colonising them. Thus, there is much more tissue damage than required for meeting the nutritional needs of the fungus. Gradually, more hardy pathogens like *Rhizoctonia* and *Sclerotium* evolved. These parasites, besides attacking tender tissues, could also exploit comparatively mature tissues.

But at the same time they retained their ability to lead a saprophytic life in soil independent of the host. All these fungi (*Pythium*, *Rhizoctonia*, *Sclerotium*) are necrotrophs and are non-specific, having a very wide host range. They cause extensive tissue damage by employing cell wall degrading enzymes and secondary toxic metabolites as their major weapon. As we go further up on the hypothetical heirarchial ladder of parasitism (Singh, 1995, Vol. 2) we find pathogens like *Alternaria*, *Cochliobolus*, *Verticillium*, *Nectria*, etc. which are necrotrophs but show growing dependence on living host tissues, less tissue damage and more loss of saprophytic survival ability. They rely more on toxins than on wall degrading enzymes. Although tissue damage is there but maceration of tissues does not occur. Further up are the semibiotrophs (*Colletotrichum*, *Cladosporium fulvum*, *Claviceps*, smuts and *Phytophthora* and biotrophs like rusts, downy mildews and powder mildews. They are guided by (i) lesser dependence of the pathogen on toxins and wall degrading enzymes, (ii) selective and site-specific use of enzymes to facilitate penetration and colonisation rather than tissue maceration, (iii) more involvement of phytohormones, (iv) decreased deleterious effect of the parasite and its increased dependence on living host cells, (v) restricted host range and (vi) more and more synchronization of the physiological processes of the host and the parasite. The physiological synchronization is a direct outcome of genetic synchronization between host and parasite. This genetic synchronization is the basic compatibility. Increased dependence of the host and the parasite on each other and decreased damaging effect of the parasite on the host tissue ultimately caused the development of symbiotic relationships such as vesicular arbuscular endomycorrhizae in which both partners derive benefit from each other.

In any disease with which a comparatively specialised parasite is associated there are three components-host, parasite and the interface of the two. This interface behaves like a third type of organism and its degree is dependent on the degree of genetic and subsequent physiological synchronization between the host and the parasite. A perfect (100%) synchronization between the host and the parasite leads to the development of an entirely new type of organism like the lichen where individual symbionts have no independent existence. They (alga and fungus) lose their identity and develop a new organism, lichen, which could be considered most advanced form of parasitism (Singh, US. 1995).

GENETICS OF HOST-PARASITE INTERACTION

Genetics of host-parasite interaction is considered as genetics of the interface in specialised parasites. It is also termed as **inter-organismal genetics** (Loegering, 1978). A pathogen is said to have developed basic compatibility with the host only when it is able to synchronise its own physiological processes with those of the host after breaching the general defence mechanisms of the host. For example, *Phytophthora infestans* has basic compatibility with potato but not

Figure 21. Relative Positive of Plant Parasites with Reference to their Ability to Establish Genetic and Physiological Synchrony with their Host.

with mango or wheat. However, all the races of this fungus are not equally pathogenic to all the cultivars of potato. This implies that establishment of basic compatibility does not gaurantee that all individuals in the pathogen would be able to infect all the individuals of the host to the same extent under similar environmental conditions. This second level of differentiation between host-parasite relations is termed **compatibility** or **specificity**. It is due to different degrees of specific and active defence offered by the different individuals of the host to different individuals of the pathogen through slight variation in their genetic composition. Such a variation in the genetic background of one partner is associated with genomic variation in the other partners. If the host develops resistance to a particular race of the pathogen, the latter also changes to overcome the resistance offered by the host. This interaction leads to evolution of gene-for-gene type of association between different races of the pathogen and different cultivars of the host.

Lack of basic compatibility shows that the parasite is unable to exploit the host or establish nutritional relationship. These are non-pathogenic responses where the host does not offer any type of specific and active defence against the parasite. In other words, unless there is basic compatibility the plant is not required to defend itself. Thus, genetics of host-parasite relationship is actually the genetics of specificity or compatibility, not of basic compatibility.

Gene-for-Gene relationship

Linseed (flax) rust is caused by *Melampsora lini* which is an autoecious rust with all spore stages occurring on the linseed plant. Sexual reproduction in the species is brought about by selfing between pycnia and pycnospores of the same race or by hybridization between different races. In addition, mutation, heterokaryosis and parasexual cycle are also known to induce variability in this fungus. Nearly 400 races have been identified in USA and Canada by inoculation of numerous rust collections on 18 differential varieties of flax. Flor (1942–1956) had conducted series of experiments on inheritance of resistance in the host and virulence in the parasite in relation to each other. He concluded his results in the gene-for-gene hypothesis. This hypothesis states that during their evolution the host and the pathogen have developed complimentary genetic system so that *for each gene conditioning rust reaction in the host there is a complimentary and specific gene in the pathogen that determines its virulence and avirulence*. The gene for pathogenicity in the parasite can be identified only by the specific varieties of the host. Conversely, the gene for rust reaction in the host can be identified only by the selective pathogenicity of the races of the parasite. Thus, resistance to rust in linseed is inherited as a dominant character although in some genes the dominance is incomplete. Similarly, avirulence in the flax rust fungus is inherited as a dominant character.

In *M. lini*- flax system genes conditioning resistance occur as multiple alleles in 7 loci designated K, L, M, N, P, D and Q. There are atleast 36 genes conferring resistance. Of 27 resistance gene 1 lies in K, 12 in L, 7 in M, 3 in N and 4 in P locus. The resistance genes in each locus are identified by numerical subscript. To show the specificity between the host and parasite, the symbol for the resistant gene is used as subscript to the symbol A (avirulence) and a (virulence) in the parasite. The gene for virulence of the pathogen at locus L would be AL, where A is the dominant gene conditioning avirulence of the pathogen at locus L and aL the recessive gene conditioning virulence at locus L. Generally, but not always, in the host, genes for resistance are dominant (R) while genes for susceptibility, that is, lack of resistance, are recessive (r). In the pathogen, on the other hand, the genes for avirulence (inability to infect) are usually dominant (A) while genes for virulence are recessive (a). The type of host reaction to the different combinations of the above set of host-pathogen system would be as given in the following table.

TABLE 8: Host Reaction to Gene-for-Gene Combination in *M. lini*- Flax System.

Host resistance genes at locus L	Pathogen virulence genes conditioned by host genes at locus L	Type of rust reaction
LL	AL AL or AL aL	Resistant
LL	aL aL	Susceptible
l l	AL Al or AL al	Susceptible
l l	aL aL	Susceptible

This indicates that a plant is resistant only when it is homozygous for resistance (LL) and is attacked by a pathogen that is homozygous (AL AL) or heterozygous (AL aL) for avirulence. When, however, the plant lacks gene for resistance (l l) at locus L, or the pathogen is homozygous for virulence (aL aL) for that locus, then the plant is susceptible. This interaction has been proved by inoculation of crosses of two varieties with two races of the pathogen. One variety was resistant to one race and the other to the second race. All host pathogen combinations giving dominant complimentary genes in both the host and the pathogen resulted in resistant reaction while the rest resulted in susceptibility.

The gene-for-gene hypothesis proved for flax-*M. Lini* system can explain and predict all host-pathogen systems. It has been found applicable to a large number of host-parasite relationships including the following:

Apple- *Venturia inaequalis* (scab)
Barley- *Erysiphe graminis hordei* (powdery mildew)
Barley- *Ustilago hordei* (covered smut)
Barley- *Rhynchosporium secalis*
Coffee- *Hemileia vastatrix* (coffee rust)

214

Cotton- *Xanthomonas axonopodis* pv. *malvacearum* (cotton blight)
Legumes- *Rhizobium* (root nodule bacteria)
Lettuce- *Bremia lactuceae* (downy mildew)
Linum- *Melampsora lini* (flax rust)
Maize- *Puccinia sorghi* (rust of maize)
Oats- *Puccinia graminis avenae* (stem rust)
Oats- *Ustilago avenae* (oat smut)
Oats- *Helminthosporium victoriae* (victoria blight)
Potato- *Phytophthora infestans* (late blight)
Potato- *Synchytrium endobioticum* (potato wart)
Potato- *Globodera rostochiensis* (cyst nematode)
Potato- Potato virus X
Rice- *Magnaporthe grisea* (blast of rice)
Sunflower- *Puccinia helianthi* (rust of sunflower)
Sunflower- *Orobanche* (flowering plant parasite)
Tomato- *Cladosporium fulvum* (leaf mold)
Tomato- spotted wilt virus
Tomato- Tobacco mosaic virus
Wheat- *Puccinia graminis tritici* (stem rust)
Wheat- *P. striiformis* (stripe or yellow rust)
Wheat- *Puccinia recondita* (brown or leaf rust)
Wheat- *Ustilago segatum tritici* (loose smut of wheat)
Wheat- *Tilletia tritici* (stinking smut of wheat)
Wheat- *Tilletia controversa* (dwarf bunt of wheat)
Wheat- *Erysiphe graminis tritici* (powdery mildew)
Wheat- *Mayetiola destructor* (insect)

Late Blight of Potato: Potato and the late blight fungus coevolved in highlands of central Mexico. Prior to 1950s it was believed that *Phytophthora infestans* is sexually sterile in nature although in cultures it did produce oospores which were difficult to germinate. Variability in the pathogen due to other mechanisms is known in the fungus. In the 1950s two mating types (A_1 and A_2) were discovered in Mexico. Sexual reproduction involves the association of these two sexually different mating types. While the mating type A_1 was supposed to be universally present, mating type A_2 was discovered outside Mexico in 1984 in USA, Canada, Europe and even in Asia suggesting that variability due to sexual reproduction is not uncommon. In addition, selfing in mating type A_1 in presence of other species such as *Phytophthora dreschleri* is also reported. The presence of races or existence of variability in the fungus was discovered when it was found that the isolate from potato produced only small necrotic spots on tomato leaves but the isolate from tomato was equally virulent on both hosts. Thus, the fungus was known to have two races- the *potato race* and the *tomato race*.

In search for resistant varieties, inter-specific crosses were made between *Solanum tuberosum* (potato) and *Solanum demissum* (a wild relative having resistance). From these crosses a number of late blight resistant varieties of potato were obtained. When these varieties were put to large scale commercial cultivation for some time it was noticed that certain isolates of the potato race could infect some varieties but not others which were infected by other isolates. Some varieties were infected by more than one isolate. However, all these isolates could infect the old potato varieties in which resistance had not been introduced. The observations suggested existence of physiologic races within the potato race and that these races could develop in the resistant population of the host.

The study of genetics of disease resistance obtained from *Solanum tuberosum* x *S. demissum* revealed the presence of one or more major genes for controlling resistance through hypersensitivity. The four genes for resistance originally identified in the breeding material from *S. demissum* were designated R_1, R_2, R_3 and R_4. Each gene is inherited independently according to Mendel's law of 3 : 1 in a monogenic dominant manner. The original race from potato (*S. tuberosum*) could infect all potato varieties lacking all these four genes (the recessive plants) for resistance but could not infect the hybrids containing one or more R-genes for resistance from *S. demissum*. This race was designated race O. The isolates of the fungus which could infect recessive plants and plants with only R_1 gene were designated as race 1. Similarly, those attacking the recessive plants and plants with gene R_2, R_3, R_4 were called race 2, race 3 and race 4. With these four R-gene types only 16 combinations were possible and in due course of time 16 races of the fungus were identified. Some isolates were found to infect not only recessive and those with single R-gene, but also plants with R-genes in combination, e.g. a new race 1.2 was pathogenic on recessive plants and plants with resistance genes R_1, R_2 and R_1R_2 and race 1, 2, 3, 4 on all 16 differential host genotypes. The system becomes increasingly complex as new resistance genes and pathogenicity genes become known and incorporated. Each addition of R-gene doubles the number of possible host-pathogen genotypes. Larger the number of resistance gene added to the host cultivated on large areas larger is the number of races of the pathogen. Fundamental to this host-pathogen relationship is the assumption that for each host-resistance gene there is a corresponding pathogenicity gene in the fungus that enables it to overcome resistance in the plant. On this basis, the late blight host-parasite relationship is used as an example of gene-for-gene system. It is considered similar to the flax-rust system.

Within the original "tomato race" pathogenic races on tomato were identified as soon as the monogenic dominant resistance was found. The tomato- *P. infestans* system is similar to that of potato. However, only one resistance gene has been identified and labeled Le1. The gene confers resistance to race O of the tomato isolates but plants with this gene are susceptible to tomato race 1.

Mutation is generally considered as the mechanism giving rise to new races in *P. infestans*. Studies of the mutants of the fungus affecting the R-genes have shown that single step mutations could extend the pathogenic range of the fungus enabling it to overcome the resistance in the cultivars. Race O (the potato race) has undergone mutation to produce race 4, race 1 to race 1.4, race 1.3 to race 1.3.4, etc. New races of the pathogen develop due to selection pressure against new resistance gene in the host. Adaptive parasitism in *Phytophthora infestans* has been suggested as a means of inducing this variability through mutation. In India, the cultivar Kufri Jyoti was developed for late blight resistance. It contains resistance genes 3.4.7. Large scale monoculture of this cultivar in the northern hills of the country enabled the pathogen to develop a race 3.4.7. Population of this race gradually increased and soon the pathogen could attack the crop from the beginning of the season. The race flora in India till 1965 was simple. Only races O, 1 and 4 were present. With the intensification of breeding for late blight resistance, introduction of resistant parental lines, and cultivation of certain resistant varieties in different parts of the country the physiologic race flora changed rapidly and many complex races developed. By 1992, the number of races and race complexes identified in India had increased to 82.

Powdery Mildews: The powdery mildew diseases result from interaction between hosts and obligately parasitic fungi in the family Erysiphaceae of Ascomycetes. The family has six genera: *Erysiphe, Podosphaera, Sphaerotheca, Uncinula, Microsphaera* and *Phyllactinia*. The hosts of these fungi number over 1500 and include several hundred species and many genera of angiosperms.

The genetics of this group of diseases has been studied best in the powdery mildew of barley and wheat caused by *Erysiphe graminis*. The fungus has two varieties, one that attacks barley is known as *E. graminis* f.sp. *hordei* and the other that attacks wheat is known as *E. graminis* f.sp. *tritici*. The inheritance of genes conditioning the reaction of barley (*Hordeum vulgare*) to *E. graminis hordei* is one of the most thoroughly studied characters in higher plants. The number of genes conditioning resistance in barley are estimated to be 12-17. These are distributed among different chromosomes. The genes are mostly dominant or incompletely dominant. Only a few are recessive. They are designated Mla, Mla1, Mla 2 ...Mla 6, Mlg, Mlh, etc. In wheat atleast 6 genes are known to control resistance at different stages of plant growth. These genes are designated Mlt, Nlu, Mls, Mla, Mlb, Mlc. The four genes Mlt, Mlu, Mla and Mlc condition resistance in the seedling stage whereas the remaining two become effective when the plant is in 3-leaf stage. Most of these genes are independently inherited.

The gene-for-gene hypothesis was extended to powdery mildews by studying the genes in barley and *E. graminis hordei* and the genes in wheat and *E. graminis tritici*. More than 50 races of *E. graminis hordei* and 38 races of *E. graminis tritici* have been identified. In India (Himachal Pradesh), Sharma and Singh (1990) recorded 14 races from 37 conidial isolates and 5 races from 12 ascosporic isolates. Most common races were 3, 12 and 13. Races from conidial isolates were more virulent than those from ascospores. In most variety-race combinations it has been shown that the pathogen race carries the gene for virulence (V) or avirulence (A), corresponding to the genes for resistance (Ml) or susceptibility (ml), respectively, carried by the host plant at a particular locus. The pathogenic phase of the fungus is haploid and all cultures are either virulent (V) or avirulent (A). A very high infection type results when the pathogen gene corresponding to the resistant host gene is virulent and when the host gene is for non-resistance or susceptibility, e.g., the combination mlml. Lower infection types result only when the pathogen gene corresponding to the resistance gene (MlMl) or (Mlml) is avirulent. The expression of a dominant host gene can be observed by the infection type produced only when the corresponding pathogen gene is avirulent.

Stem rust of wheat: Like the flax rust, gene-for-gene relationship in stem rust of wheat is based on genetic studies of both the host and the dikaryotic phase of the pathogen (*Puccinia graminis tritici*). There are hundreds of physiologic races of the stem rust fungus which vary in their ability to infect different varieties of wheat. Even the races have biotypes differing in the degree of virulence and ability to infect additional differential hosts. Twelve varieties from 5 species of *Triticum* are used as differential hosts. Through selfing and crossing of various races it has been shown that most of the races are heterozygous for pathogenicity on one or more of the differential hosts and that avirulence is dominant on some varieties, but on others virulence rather than avirulence is dominant. Whenever both parent races of the pathogen are virulent on a particular host cultivar the progeny is always virulent. The F_2 and F_3 generations of crosses between races include some individuals exhibiting virulence identical to that of the one or the other parents but they also include individuals belonging to numerous other races as a result of gene recombinations for virulence by which the parent races differed.

The type of resistance varies with wheat cultivars. In some cultivars resistance is conditioned by a single, completely dominant gene, a single incompletely dominant gene, or a single recessive gene. In other cultivars resistance is due to two or more genes which are independently inherited and which are dominant or recessive or a combination of both on the same plant. Resistance may also be due to interaction of several genes and for expression of resistance all these genes must be present in the plant.

The introduction of resistant cultivars exerts selection pressure and impels the pathogen to counteract resistance. Thus, within the races biotypes originate. In India, races 117 and 22 of *P. graminis tritici* have been more dominant than about 22 others. Race 117 was detected in 1954, its biotype 117-A in 1960, 117-A-1 in 1977, 117-1 in 1987 and 117-2 in 1991. The physiologic race flora of *Puccinia recondita* (leaf rust) is also quite large while that of

P. striiformis (stripe rust) it is not so large.

Loose smut of wheat: More than 30 races of *Ustilago segatum tritici* are known. The gene-for-gene relationship has been demonstrated in this disease also. It is assumed that two sets of resistance genes and two sets of incompatibility genes (dwarfing of susceptible plants but not smutted heads) in the host interact with four sets of complementary genes for virulence in the pathogen.

Apple scab: Seven different gene pairs condition lesion (susceptible) and fleck (resistant) reactions. In lesion type of reaction there is abundant sporulation of the fungus. When a fungus isolate causing typical lesion reaction is crossed with an isolate causing fleck reaction, the pathogenicity of the progeny to a given common apple variety is originally conditioned by a single lesion/fleck gene pair. Each such gene pair, however, conditions pathogenicity to one group of apple varieties but not to another.

Rice blast: *Magnaporthe grisea (Pyricularia grisea)*, an important pathogen of rice and many grasses, has received much attention in recent years as a model fungal pathogen in molecular plant pathology (*cf.* de Wit, 1992). Techniques of classical genetics, molecular biology, cytology and cell biology have been applied in the study of this fungus. Although sexual reproduction is not common in the pathogen, variability is immense through parasexuality and mitotic recombination. Thus, physiologic races are fairly common and stable (Latterell and Rossi, 1986). Cultivars of rice have been developed that carry single dominant resistance genes effective against certain races of the pathogen. Earlier the genetic study of the pathogen was not possible due to lack of fertile female isolates of *M. grisea* that infect rice. Such studies have now been performed by crossing a sterile field isolate pathogenic on rice and weeping lovegrass with a highly hermaphrodite strains that is pathogenic only on weeping lovegrass (*cf.* de Wit, 1992). The progeny of such crosses, that were still pathogenic on rice, were backcrossed several times with the sterile rice/weeping lovegrass isolates as recurrent parents. By checking the progeny of the last backcross on differentials of rice several avirulence genes were identified. The result suggest that avirulence genes specific for rice cultivars are common among nonpathogens of rice.

CRITERIA FOR GENE-FOR-GENE TYPE OF RELATIONSHIP

Although all the weapons of the pathogen (enzymes, toxins, etc.) and the preformed and induced defence barriers of the host are basically controlled by genes in the respective organism, gene-for-gene type of relationship has not been found in all host-parasite systems. It is most common in specialized or obligate parasites and totally absent in non-specialised (facultative) pathogens. According to Van der Plank (1973) gene-for-gene type of relatonship can exist only in those host-parasite systems where both the components are living for a certain critical period of time in order to let their nuclei transcribe, translate and subsequently exchange genetic information. This criterion is, however, not applicable to some host-pathogen systems which are necrotrophs producing host-specific toxin such as oat-*Helminthosporium victoriae* and necrosis-inducing peptides such as barley- *Rhynchosporium secalis*. A more important criterion seems to be the specificity of gene products of the pathogen (such as host specific toxins) in its interaction with gene product of the host such as the toxin binding sites. If product of one or few genes of the parasite react specifically with product of one or few genes of the host, any alteration in gene(s) of one component would not only drastically affect the overall disease reaction but also force the complimentary and related alteration in gene(s) of the other partner in order to maintain the disease equilibrium under natural conditions. Gene-for-gene type of relationship is bound to exist in these situations.

In the case of non-specialised parasites such as *Pythium, Rhizoctonia* and *Sclerotium* which are at the bottom of the evolution of parasitism and which produce hydrolytic enzymes and host non-specific toxins causing extensive tissue damage, chances are extremely remote that alteration in one or few gene(s) in any of the two partners would be able to affect drastically the overall disease development without endangering the very existence of the organism. Moreover, these parasites obtain their nutrition from damaged, dead tissues. These necrotrophs are thus neither required to nor they establish any sort of synchronous relationship with their host. Hence there is no gene-for-gene type of relationship in such combinations. This condition may be only temporary and limited only to the extent the parasites remain unspecialised. If during further evolution of parasitism in these fungi they acquire specialisation, perhaps gene-for-gene relationship will apply to them also.

MOLECULAR BASIS OF HOST-PARASITE INTERACTION

In any organism-organism interaction the earliest event is the cell to cell communication to establish compatibility. This communication is between cell surfaces of the two organisms and is a very fast process. It is this communication that decides compatibility, triggers host responses and determines the fate of the two partners. The communication is through chemical molecules on the two surface and is basically decided by the genes in the two partners.

Recognition and Specificity (Compatibility) Phenomenon

Recognition plays a central role in the interaction between plants and their pathogens. Pathogens must be able to recognise the presence of their host tissue in their environment and often must recognise specific surface features of the tissue in order to effect successful penetration and infection (Hoch and Staples, 1991). Successful pathogens must also be able to recognise and overcome plant defence responses. Studies on nature of binding of animal hormones like insulin and glucagens and bacterial toxins like cholera toxin with plasmalemma of animal cells, *Rhizobium*-legume interaction, pistil-pollen interaction in angiosperms in relation to inter- and intraspecific compatibility, and work on certain host-parasite interactions such as potato- *Phytophthora infestans*, soybean- *Phytophthora megasperma* var. *sojae*, tomato- *Ralstonia solanacearum*, rice- *Magnaporthe griseas* etc. have brought following points to light:

1) Outcome of the host-parasite interaction is determined as soon as the cell wall of the pathogen contacts the host cell wall or plasmalemma, probably due to interaction of constitutive factors. Most of the subsequently induced changes in the host, such as lignification, phytoalexin synthesis, etc., play only contributory or no role in determining the outcome of the interaction.

2) Constitutive components are the complimentary binding sites present on the surface of the host and parasite cell walls. Interaction of these binding sites helps in the recognition of the host by the parasite and *vice versa* as compatible or incompatible partners.

3) Binding sites determining basic compatibility are different from those determining specificity, i.e., susceptibility or resistance.

4) Binding of host with parasite at site(s) determining basic compatibility will lead to an array of metabolic changes in the host resulting in increased synthesis of biomolecules and other nutrients required for growth of the pathogen.

5) Binding at the site of specificity along with that of basic compatibility will result in resistant response due to one of the following effects:

   i) Shutting off of the metabolic changes induced by the binding at the basic compatibility sites.

   ii) Induction of metabolic reactions leading to synthesis of compounds toxic to the pathogen such as phytoalexins.

   iii) Binding of the host with the parasite at a specific site may cause death of the pathogen and host cells both. This may include the activation of programmed cell death in the host. Whether the death of the pathogen precedes that of the host cell or *vice versa* would depend on the particular host-parasite system. Both situations may exist within the same host-parasite system depending upon genetic background of the host cultivar and the pathogen race, such as in potato-*Phytophthora infestans* system.

   iv) Binding at a specific site may inhibit or slow down further advance of the parasite with or without causing death of the host cell.

As shown in Fig. 22, based on the potato-*P. infestans* system, the pathogen establishes basic compatibility with the host by exploiting the complimentary binding sites present on the latter's surface. Subsequently, in order to defend itself, the host develops alternative sites of binding which are termed as **sites of specificity**. Binding at the site of specificity is detrimental to the pathogen. To

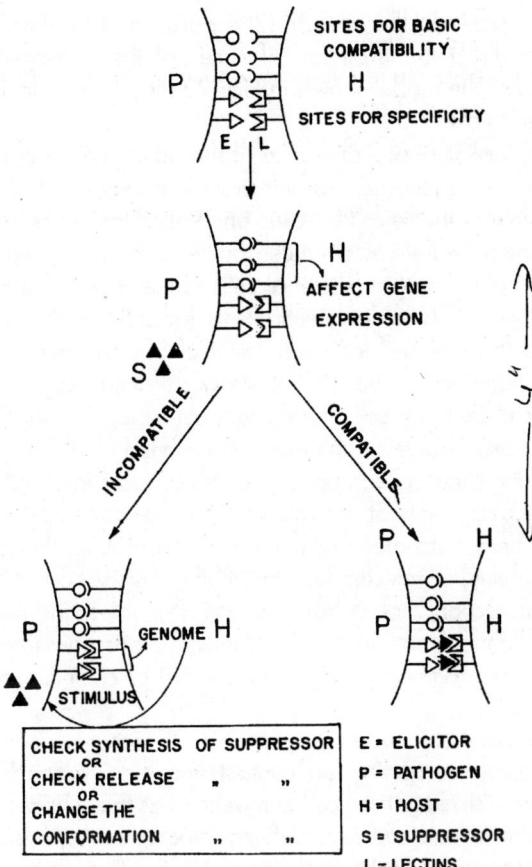

Figure 22. A Hypothetical model explaining recognition in a specialised host (H)-parasite (P) system. Recognition determined by complimentary binding sites at the surface of the host and parasite operate at two levels-basic compatibility and specificity. Binding at the site of basic compatibility leads to the genetic and subsequent physiological synchronisation of the host and parasite resulting in compatible response. However, simultaneous binding at the site of specificity induces host defence systems leading to incompatible reaction. Lectins (L), at the host surface and elicitors (E) on the surface of the parasite constitute the site of specificity. Suppresors (S) produced by the parasite bind with host lectins constituting the site of specificity and convert the incompatible reaction into a compatible one. In order to counteract this situation the host, through its specificity binding sites, tries to make it unsuitable for binding with suppressors but not with elicitors present on parasite's surface or it blocks the synthesis or release of suppressors in the parasite by producing some sort of inducer which ultimately leads to the incompatible response.

avoid this the pathogen either modifies its own specificity binding sites or it develops ability to block the complimentary binding sites on the host cell surface by producing the blocking substances (**suppressors**) well in advance of coming in contact with the host cell wall or plasmalemma. Again, in order to overcome this advantage in the pathogen, of avoiding binding at specificity sites, the host tries to modify its own specificity binding sites or block the synthesis and/or release of suppressors from the parasite into the interface.

These interactions between the host and the parasite at the level of surface contact probably direct the evolution of races of the pathogen and new varieties of the host. The model shown in Fig. 22 may be applicable to other host-parasite systems also.

The nature of the binding sites: Adherence of the pathogen propagules to the host surface is the first requirement for infection. In pathogens that are directly placed by their vectors in the cell (viruses and vessel-limited bacteria) this is not required although in the vessel-limited bacteria mechanisms of adhesion exist to prevent displacement of bacterial cells in the flow of the sap. Fungal propagules have varied types of mechanisms for adherence. Some fungal spores adhere to the leaf surface as a result of opposite electrical charge carried by them. Some spores have spiny surface that can easily adhere to a rough host surface. Bacterial cells have chemical molecules on their surface that help in adherence. However, the major contribution in this process at the cellular level is of the chemical components of the surfaces involved. This initial contact between components of the two biological surfaces determines to a large extent the final outcome of the relationship. The pathogen may be accepted (compatibility) or rejected (incompatibility) by the host at this level. In the compatible interaction, violent host reaction is avoided and the pathogen moves farther. In the incompatible reaction, violent host response results in host cell death (the hypersensitive necrotic response-HR) or chemical or structural barriers are formed.

These facts indicate that in most cases the host cell recognises some common component in the microorganism on contact and then responds to it in a predictable manner. **Cell recognition** has been defined as the *"initial event of cell-cell communication which elicits a defined biochemical, physiological or morphological response"* (Clarke and Knox, 1978). This communication is limited to the short distance of cell to cell contact. Identification and characterisation of the molecules involved in recognition between the two organisms has been extensively reported during the last 20 years (Anderson, 1984; Ebel and Cosio, 1994; Hahn, 1996). Recognition is a biochemical process and depends on the informational potential contained in the surfaces that come in contact and a response follows the complimentary interaction of the contacting molecules (Sequeira, 1980). There are two concepts of how the transfer of information between the biologically active surfaces on a molecular level takes place—the **glycocalyx concept** and **lectin concept**.

The glycocalyx are surface polysaccharides in the fibrillar mucilaginous membranes or capsular material on cell walls. These surface polysaccharides contain many useful sites for the recognition phenomenon through similarities and dissimilarities in the carbohydrates of the two surfaces. Similar carbohydrates on the two surfaces will bind (attract each other) while dissimilar ones will repel each other.

Many recognition phenomena are dependent on the specific interaction between polysaccharides and proteins that have the capacity to bind with specific sugars residues. Lectins are sugar binding proteins or glycoproteins of non-immune origin which are devoid of enzymatic activity towards sugars to which they bind and do not require free glycoside hydroxy group on those sugars for binding. Lectins have been demonstrated in almost all groups of living organisms. Each molecule of a lectin has two or more sites, possibly clefts or grooves, into which complimentary molecules of sugars or other oligosaccharides fit (see Fig. 22). Lectins are specific for saccharides and have no affinity for any other compound. Host lectins have been implicated in the binding of parasites to the host surface which, in turn, is responsible for either induced compatibility (*Rhizobium*-legume interaction) or incompatibility (potato-*P. infestans*, tomato-*Cladosporium fulvum* or tomato-*Ralstonia solanacearum* systems). In the compatible *Arthrobotrys oligospora* (nematophagus fungus) and nematode relationship a lectin-carbohydrate interaction is involved.

Elicitors: Binding components complimentary to the host lectins present on surface of the pathogen are mostly termed **elicitors**. Elicitors are cell wall components of the pathogen which are capable of inducing phytoalexin synthesis and hypersensitive response when allowed to get attached with complimentary binding sites (lectins) on the host surface. The term "'elicitor" was originally used by N.T Keen in 1975 to refer to molecules and other stimuli that induce the synthesis and accumulation of antimicrobial compounds (phytoalexins) in plant cells but is now commonly used for molecules that stimulate any plant defence mechanism (*cf.* Hahn, 1996). These include synthesis and accumulation of phytoalexins, hypersensitive necrosis, production of glycosyl hydrolases capable of attacking surface polymers of pathogens, the synthesis of proteins that inhibit degradative enzymes produced by pathogens, the production of activated oxygen species (oxygen burst) and the modification of plant cell walls by deposition of callose, hydroxyproline-rich glycoproteins and/or lignin. The recognition between plant host and pathogen, mediated by receptors at some level, lies at the root of gene-for-gene mechanisms that appear to govern race-cultivar specificity.

Oligosaccharide elicitors were among the earliest elicitors to have been chracterised in detail. Oligosaccharides play the role of biological signalling molecules and elicit the phytoalexin synthesis (Ebel and Cosio, 1994). Some oligosaccharides such as chitin induce lignin synthesis. They include glucan fragments (*Phytophthora megasperma* f.sp. *glycinea*) that induces phytoalexin synthesis in soybean, chitin that induces lignification in wheat leaves and phytoalexin synthesis in rice cells, lepto-chitooligosaccharides such as nodulation factor (Nod factor) in legume-*Rhizobium* interaction.

Glycopeptides and glycoproteins often act as elicitors in plants. They have been identified as having the ability to induce plant defence responses (Anderson, 1989). Polypeptide elicitor-binding proteins are a prominent group of signal molecules in the animal system. They are biologically active in plants also and some have been purified and characterised (such as elicitins from *Phytophthora*; harpin, a bacterial polypeptide that induces hypersensitive necrosis in plants; systemin that induces systemic accumulation of proteinase inhibitors in tomato and potato).

Elicitors have been characterized from several pathogens such as *Phytophthora infestans*, *P. megasperma* f.sp. *glycinea*, *Cladosporium fulvum*, etc. The elicitor from *C. fulvum* is a glycoprotein in the cell walls and can induce synthesis of rishitin in tomato, pisatin in pea and glyceollin soybean when applied to these hosts (de Wit and Roseboom, 1980).

In principle, elicitor-active molecules could be released from the invading pathogens prior to or during entry, they could be integral component of the microbial cell surface such as β-1-,3-glucans, chitin or chitosan that require host enzymes (glucanase; chitinase) for release or they could be synthesised and released by the pathogen in the plant in response to host signals (Dixon *et al*, 1994). However, host glucanases are not necessarily required for release of active elicitor in many cases. Elicitors may act synergistically, i.e., their combined effect is greater than the sum of the individual effects (Boller, 1995).

Suppressors: Suppressors are chemical substances present in fungal mycelium which are capable of suppressing phytoalexin synthesis and hypersensitive response in the host, induced by the pathogen or elicitors isolated from cell walls of pathogens. Their main function is to delay or prevent action of elicitors. Suppressors can convert an incompatible association into a compatible one. Chemically, suppressors are also glycoproteins in nature. However, the carbohydrate moity in elicitor and suppressor are usually different. In *P. megasperma* f.sp. *glycinea* the elicitor and suppressor both may be glycoprotein but the carbohydrate moity of elicitor is rich in glucose while that of suppressor is rich in mannose. Suppressors probably act by blocking the binding of elicitor present on the pathogen cell wall surface with the lectin, a complimentary binding cite on the host cell surface. In the interaction of *Phytophthora infestans* with potato, soluble fungal glucans appear to act as suppressors against recognition of fungal cell wall components (cf. Boller, 1995). In Ascochyta blight of chickpea (*Ascochyta rabiei*), a glycoprotein fraction from the pathogen suppresses phytoalexin accumulation in its host. A nonproteinaceous suppressor from *Cladosporium fulvum* prevents the action of nonspecific elicitors in tomato.

Common Antigens: A phenomenon which received wide interest in the past to explain the chemical basis of host-parasite interaction is the presence of common antigens or sharing of antigens between the host and the parasite. Antigens are foreign proteins, and occasionally complex lipids, carbohydrates, and some nucleic acids. In animal system it is assumed that due to presence of common antigens, also termed as cross-reactive antigen (CRA), the parasite escapes recognition by the immune system of the host which, on activation, would have otherwise destroyed it.

The research on antigen sharing in plant diseases was started following the proposal of gene-for-gene hypothesis in flax-M. lini system. It was then found that susceptible cultivars of flax share common antigen with compatible race of M. lini but not with the incompatible race as shown in the following table:

TABLE 9: Relationship of Antigen to Rust Reaction in Flax.

Flax line	Incompatible with rust races	Compatible with rust races	Shared antigen with races
1.	None	ABCD	ABCD
2.	A	BCD	BCD
3.	AB	CD	CD
4.	ABC	D	D

A number of other host-parasite combinations have been examined for sharing of common antigens. Except in few, it is a general observation that compatibility of the parasite with its host increases with increasing antigen similarity whereas incompatibility of the host is characterised by an increasing antigen disparity. Among the positive findings is the discovery that *Xanthomonas axonopodis* pv. *malvacearum* is antigenically more similar to its host, cotton, than to other species of *Xanthomonas*. Also interesting is the finding that maize smut fungus, *Ustilago maydis*, not only shares antigens with its host, maize, but also with young oat seedlings which it can parasitise. A more remote antigenic relationship was found with other oat tissues which were not parasitised. Chakraborty and Purkayastha (1983) had reported serological relationship between *Macrophomina phaseolina* and soybean cultivars. The antigenic pattern can be altered and susceptibility decreased by exogenous application of chemicals (*cf.* Purkayastha, 1998).

In spite of the above examples and evidence of presence of common antigens in most of the compatible host-parasite combinations, it has been doubted that mere presence of antigens is a proof of their essential role in parasitism and compatibility. They may be an artefact of the experiments or consequence of the parasitism, with no regulatory role. In the study reported by Charudattan and De Vay (1981) conidia of *Fusarium oxysporum* f.sp. *vasinfectum*, were found to contain an antigen that cross reacted with antiserum to cotton root tissue antigen. The cross reactive antigen from the fungal conidia was isolated, purified and chemically characterised as a protein-carbohydrate complex, similar to suppressors. Purkayastha (1998) has concluded from studies involving more than 100 host-pathogen and host-nonpathogen combinations that detection of cross reactive antigens or CRA (common antigens or matching proteins) by immunodiffusion test (not only by ELISA) is a strong indication of susceptible/compatible interaction. The most important consideration is that a threshold titre of both host and pathogen antigen is essential for compatible interaction in case of fungal diseases. It is possible that antigens act as suppressors of incompatible interactions?

HOST-PARASITE INTERACTION

In the preceding sections of this chapter we have discussed how the host and the pathogen recognise each other as compatible or incompatible partners. In this section we discuss how recognition is transformed into action and how the subsequent processes leading to compatibility or incompatibility are regulated. Although every activity of a plant, including its resistance or susceptibility to a given disease, is gene controlled, the actual operation of the host response is mediated through biochemical activities. Day (1974) had proposed a model to explain the genetic regulation between host and parasite. This theoretical model was based on the model earlier proposed by Britten and Davidson (1969) to explain gene regulation in eukaryotes. As shown in Fig. 23 the slightly modified model requires five genetic elements : (i) A *sensor gene* (the gene determining resistance or susceptibility) or nucleotide sequence which binds agents that induce a specific pattern of activity in the genome. (ii) Binding results in the activation of linked gene or genes called *integrator*. (iii) The integrator synthesises an *activator RNA*. (iv) The activator RNA forms specific complexes with *receptor genes*. (v) These receptor genes are linked to *producer genes* causing them to be transcribed, probably relieving histone-mediated repression. Producer gene codes the mRNA for the synthesis of polypeptides which may act as enzymes for the synthesis of substances involved in the host-parasite interaction (phytoalexins synthesis, etc.). Redundancy may occur either because many different sensor genes have common integrator sequences that enable them to regulate same producer genes, or because many different producer genes have common receptor sequences that enable them to respond to the same activator RNA. A battery of producer genes is controlled together when a particular sensor gene activates its set of integrator genes. The degree of required physiological coordination will determine whether the redundancy occurs at the level of integrator sequence or receptor sequence or a combination of the two. However, if each resistance gene (sensor gene) regulates a large battery of producer genes, it may be very difficult to identify any one particular end product playing the major role in disease resistance.

The model shown in Fig. 23 depicts the interaction between avirulence gene of the parasite and complimentary resistance gene of the host. There is unidirectional transfer of information. The model can be extended to explain interaction of compatible genes and bi-directional transfer of information. The question is about the nature of inducer. The first experimental proof regarding this question was provided by Rohringer *et al,* (1974). They studied the interaction of an avirulent race of *P. graminis tritici* with temperature sensitive gene (Sr-6) of the host, wheat. They isolated a small RNA molecule, probably of pathogen origin, which was capable of inducing resistance response in the

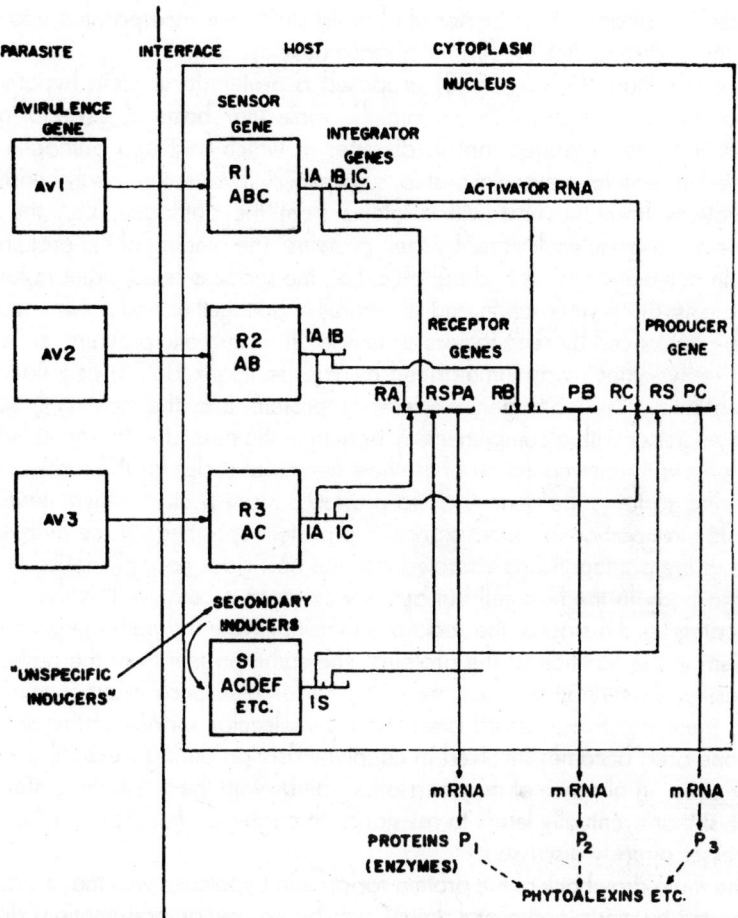

Figure 23. A Model for host-parasite interaction (Day, P.R. 1974). The sensor genes R_1, R_2, R_3 are resistance genes in the host that bind with the product of specific avirulence genes Av_1, Av_2 and Av_y respectively, from the parasite. This binding activates linked integrator genes to form activator RNA. Diffusion of activator RNA to receptor genes results in the activation of their linked producer genes to synthesize m-RNA. These m-RNA are transported to cytoplasm where they synthesize the enzymes which in turn bring the changes in metabolism associated with resistance.

host having gene for resistance. This molecule can be considered parallel to inducer molecule proposed in the above model. However, the RNA nature of inducer cannot be extended to other host-parasite systems.

Chakrabarti and Shaw (1977) also proposed a model to explain host-parasite interaction at molecular level. In their model they postulated that exchange of information between the host and the parasite is bi-directional and might take place at several levels such as (i) at cistron level, (ii) at the level of mRNA, and (iii) at the level of polypeptides. There is experimental evidence for transfer of information of cistronic (genomic) level in case of crown gall disease where part of the bacterial plasmid DNA gets incorporated into host genome and regulates the further disease development.

Van der Plank (1975, 1978) proposed a protein-for-protein hypothesis, similar to antigen parity to explain the molecular basis of gene-for-gene interaction. He suggested that in diseases in which host and pathogen are involved in gene-for-gene relationship, susceptibility involves the copolymerisation of proteins from the host with proteins from the pathogen, i.e., the two partners recognise each other by their proteins. The binding of the proteins or protein polymerization is hydrophobic, i.e., the surfaces reject water in favour of proteins. Cooplymerisation and susceptibility go together and polymerization can be influenced by such factors as temperature, osmotic pressure, solvents, etc. The hypothesis was summarized by Van der Plank (1978) as follows. In susceptibility, the pathogen excretes a protein into the host cell which copolymersises with a complimentary protein in the host. The copolymerisation interferes with autoregulation of the host gene that codes for the protein, and by doing so turns the gene "on" to produces more protein which serves as food for the pathogen. In resistance, the protein specified by the avirulence gene in the pathogen and excreted into the host does not polymerise. It is a foreign body in the host cell but actually exists as a catalyst. It is the elicitor that catalyses the start of the reactions that inactivate the pathogen. There is dualism in the function of the proteins. The same protein from the pathogen makes for susceptibility if polymerised and for resistance if unpolymerised. Thus, there are two involved areas on the molecular surface of the protein. The one area becomes involved in copolymerisation during susceptibility, the other area, in absence of polymerisation, binds with the substrate to start the catalyst that eventually leads to resistance through post-infection biochemical processes already discussed.

The main drawback of the protein-for-protein hypothesis was that it was not supported by any experimental data. It was based only on presumptions drawn from circumstantial evidence. The hypothesis could not explain the interactions in those systems where host-specific toxins are involved. The basic facts, like all other interactions at molecular level, are physico-chemical in nature. Information coded in any gene (DNA molecule) is translated in the form of protein and that same protein may perform more than one function, cannot be taken as evidence in support of the above hypothesis.

The various aspects of host-parasite interactions, discussed in the preceding pages can be summarised as below:

1) Variability is inherent character of living organisms including plants and their parasites. Genes determine the variability but through the structure and configuration of chemical molecules in or on the cell surface. Like fingerprints, these molecules retain their individuality (the characters) and can copolymerise whenever similar molecules come in contact within the distance of chemical bonds.

2) Amount of variability in the host and its pathogen is very vast. For example, if there are 20 genes for resistance in the host and 20 genes for virulence in the pathogen, there can be roughly a million pathogenic races (Van der Plank, 1978). Apart from nucleic acids, only proteins and carbohydrates (polysaccharides) are capable of storing this vast amount of information.

3) Compatibility and incompatibility are determined by a system of recognition between the host and the pathogen. The self is accepted and foreign is rejected.

4) The recognition mechanism involves proteins and glycoproteins (lectins) and/or polysaccharides. These molecules also contain the chemical information of variability. The protein or polysaccharide molecules with a particular set of information will recognise and accept only the molecules with similar set of information.

5) Matching of host and parasite genes is, therefore, at the molecular level; matching between the molecules of proteins and/or polysaccharides.

6) When there is resistance the proteins or polysaccharides on the two surfaces (host and pathogen) do not recognise each other as they have different molecular configuration. The reverse is true for susceptibility.

15

Effect of Environments on Pathogenesis

Plant disease results from the interaction of a pathogen with its host but the intensity and extent of this interaction is markedly affected by the environmental factors. Although these factors are not the causal agents of infectious diseases, they are the final determinants of almost all the events that constitute the infection chain leading to pathogenesis and also the events that follow, *viz.*, spread of the disease in the population. The fact that most plant pathogens are often present in a geographic area but the diseases caused by them become serious only occasionally, is one of the indications that the ever-changing environments influence the development of a disease. The role of environments in pathogenesis is as important as susceptibility of the host and pathogenicity of the causal agent. Any consideration of disease in the crop, therefore, involves the **disease triangle**.

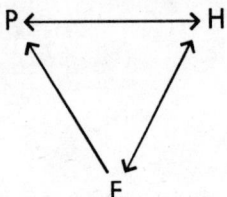

The pathogen interacts with the host and *vice versa*. Both influence each other, the host providing nutrition to the pathogen and the latter causing the disease in the former. The host also influences the environment through crop canopy, root and leaf exudates, withdrawal of nutrients and water from soil and other activities mediated through these effects. Environment affects the host through physical, chemical and biotic factors involved in plant growth and metabolism.

These interactions and their effects are not spontaneous. In epidemiology, time over which the interactions are taking place and the population, rather than the individual, is affected is the fourth component added to disease triangle which can be modified as:

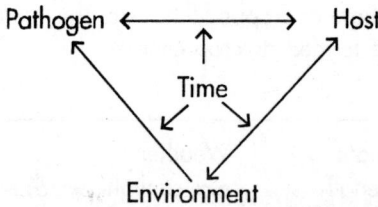

Pathogen ⟷ Host

Time

Environment

The first step in the infection chain is the survival of the pathogen. While pathogen's own characters and presence of suitable medium are basic determinants, the environment is equally important. Adverse physical, chemical and biotic environments can limit survival ability and reduce the density and capacity of the inoculum. The transport or dispersal of the inoculum in speed and distance is also dependent on environment. The germinability of spores, and the number and flight range of insect vectors are directly influenced by prevailing weather. At the time of penetration, the structural defence barriers of the host, stability and germination of spores on the host surface, and their penetration are influenced by meteorological conditions. Invasion of tissues by the developing parasite after penetration may not be directly influenced by external environments but the effect the latter produces on the host may lead to favourable or unfavourable conditions for the pathogen in the tissues. Resistant varieties may tend towards susceptibility under some temperature conditions. Also the factors such as light, temperature and humidity on the host surface definitely determine the exit of the pathogen, its sporulation, number of generations, and amount of secondary inoculum produced for dispersal. Thus, it is obvious that inspite of pathogen being virulent and the host being congenial, disease may not develop in a population unless environmental conditions are favourable for it.

The environment affecting the plant and the pathogen consists of two parts: the atmospheric environment and the soil environment. Although both are interrelated so far as physical parameters are concerned, the soil environment is chemically and biologically more complex but stable. Soil environments directly affect the soil-borne root pathogens and indirectly the pathogens of foliar parts. Atmospheric environment is mostly related to pathogens of aerial parts.

The atmospheric environments, and to some extent the soil environments too, are determined by meteorological factors. Long term generalization of weather is **climate**. The climate of a region describes the annual progression of the weather and expresses the more or lees permanent limits within which weather ranges over a period of time. The climate can be subdivided into macro-, micro- and crop climate. Macroclimate is the general weather conditions over the field while microclimate is the outcome of macroclimate acting on limited environmental units, such as slopes or valleys, light or dark soil, or on units under different environmental management, e.g. tilled or untilled, irrigated or unirrigated fields, bare or cropped land, etc. The interaction of microclimate with the crop can be termed as crop climate.

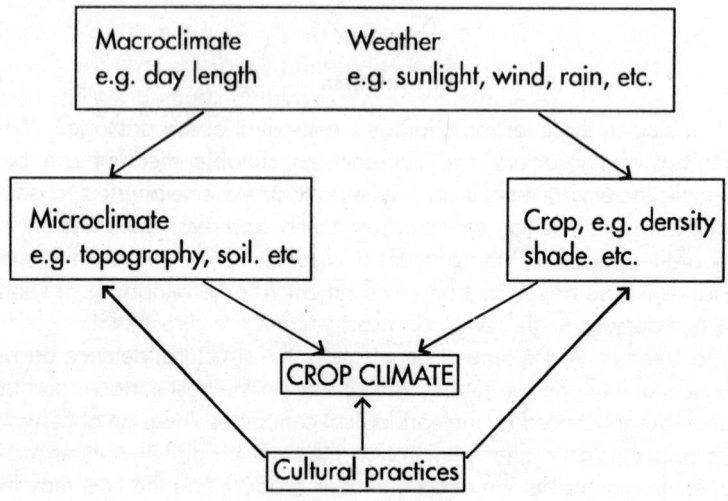

Figure 24. The Interaction of Macro-, Micro- and Crop Climate.

Although the crop climate is determined by macroclimate through microclimate and crop conditions and cultural practices, the nature of the crop growth and cultural practices may keep the conditions fairly stable for longer periods as compared to the climate outside the crop. This is very important for pathogenesis of organisms flourishing under moderate temperature and high humidity conditions.

In the same manner as the atmospheric environments, the soil environments can also be divided into two parts, the environment of soil away from the plant root system and the soil environment in the root zone under the influence of biologically active roots (the rhizosphere). The conditions in these two zones are altogether different and can influence soil-borne plant pathogens differently.

EFFECT OF TEMPERATURE ON INFECTIOUS DISEASES

Seasonal and regional occurrence of plant diseases is determined, to a great extent, by temperature. Certain plant diseases occur during winter, others during summer and for many diseases rainy season (hot and humid) is most favourable. Causes for this seasonal occurrence can be attributed to the host as well as to the pathogen. Annual crop hosts are available only during particular seasons and therefore the disease also occurs during that season.

The temperature of the season is usually favourable for growth and activity of the pathogen. It is possible that before and after the season the temperature is unfavourable for the pathogen but not necessarily.

Some pathogens achieve their optimal growth only at low temperatures. Peach leaf curl (*Taphrina deformans*) occurs only in cool (temperate) areas. In India the plant is grown in the hills (cool) as well as in the plains. The disease occurs only on the plants grown in the hills. There also the disease assumes serious proportions when the spring is cold and humid. The late blight of potato is rare in areas where the mean atmospheric temperature exceeds 25°C (77°F). It develops better in areas or seasons when the temperature is low and humidity high. Stripe rust of wheat (*Puccinia striiformis*) and stripe disease of barley (*Drechslera graminea*) are other examples where the disease development is favoured by low temperature. Number of pathogens and diseases favoured by relatively high temperature is large. The distribution of such diseases is common in the warm climate of the tropics and subtropics. Such diseases include many vascular wilts caused by *Fusarium*, bacterial wilt of eggplant, tomato and potato caused by *Ralstonia solanacearum*, soft rots caused by *Erwinia carotovora* subsp. *carotovora*, charcoal rot caused by *Macrophomina phaseolina* and Phymatotrichum root rot of many crops. There are many more examples which show that pathogens and the diseases they cause are distributed in seasons and geographic regions according to their temperature requirements. A pathogen thriving better at low temperatures can occur in tropical and sub tropical climates only at places of high altitudes where temperatures remain low.

Stages in the disease cycle of a pathogen are affected by temperature. Many rusts in the plains of India fail to survive the summer season after crop harvest due to high temperature. The viability of the chickpea blight fungus (*Ascochyta rabiei*) in crop debris is completely lost in chickpea growing areas of India due to high summer temperatures that follow the crop harvest. In regions where the crop harvest is followed by winter and snow it remains viable in crop debris and even produces its sexual stage. The process of infection also depends to some extent on the prevailing temperature but the most significant effect is on the incubation period which can be increased or decreased. Rusts, downy mildews and powdery mildews appear only on few leaves in the beginning. The secondary spread of these diseases occurs in the plant population through spores produced on these initial pustules or lesions. The incubation period determines the number of life cycles (spores-infection-spores) the pathogen can complete during the season and the number of crops of spores the pathogen will produce for spread of the disease. If the incubation period is long the number of generations will be reduced and spread of the disease will be slow. When the incubation period is short, crops of infective propagules will be produced quickly, at short intervals, and the disease will spread fast. The following data for *Puccinia graminis tritici* illustrate the point

Temperature (°C)	4.5	10.0	19.0	24.0
Incubation period (days)	22	13	9	3

It is obvious that as the temperature rises from 4.5 to 24°C the number of days required for production of a new crop of urediospores after infection is considerably reduced and quicker generations of spores are produced and dispersed for rapid spread of stem rust. This accounts for late appearance of stem rust in plains of India toward later part of the crop season and hence less losses than by other rusts. The incubation period in *Fusarium oxysporum* f.sp. *vasinfectum* (cotton wilt) is only 12 days at 27°C but 58 days at 16°C. As stated earlier the disease is serious at high temperatures. In downy mildew of grapevines (*Plasmopara viticola*) the incubation period may range from 7–8 to 20 days depending on host susceptibility, prevailing air temperature and relative humidity. As temperature rises above 5°C to 26°C the incubation period decreases. At this temperature with 100% RH the incubation period is 11–12 days. At 100% RH and temperature of 28°C the incubation period is 5–6 days. In *Uromyces ciceris-arietini* (rust of chickpea) incubation period is 15 days at 11°C –12°C, 11 days at 20°–25°C and 9 days at 25°–30°C. The disease occurs most severely in later part of the winter season.

In India, the root knot of vegetables crops (*Meloidogyne javanica* and *M. incognita*) occurs less during the winter months but is more serious during warm and wet part of the year. This is also due to effect of temperature on the time required for completion of life cycle. Females of the root knot nematode develop from the infective second stage larva to the egg laying stage in 17 days at 27.5°–30°C and in 57 days at 15.4°C. (*cf.* Singh and Sitaramaiah, 1994). Different life activities such as hatching, reproduction, movement and development have different temperature requirements.

Temperature can influence the morphological variations in spores and infection caused by them. In the rust of pea caused by *Uromyces fabae* the infection by aeciospores at relatively low temperature of 17°–22°C results in the formation of secondary aecia and the aeciospores function as the repeating spores. At 25°C the infection by aeciospores results in the formation of uredia. Since aeciospores and urediospores both function as repeating spores, the period for disease spread is long. In Phomopsis blight and fruit rot of eggplant, the pycnidia of the fungus *Phomopsis vexans* (teliomorph *Diaporthe vexans*) produce two types of spores. The alpha conidia are sub-cylindric while the beta conidia or stylospores are filiform. It was generally believed that beta conidia do not germinate and have no role to play. Kumar and Sugha (1999) have claimed that production of the two type of conidia is temperature-dependent. At low temperatures of 10°–16°C beta conidia are produced and at higher temperatures of 25°–28°C alpha conidia are formed. They get inter-converted according to temperature.

Effect of temperature on disease incidence can be seen in the blast disease of rice caused by *Pyricularia grisea* (*Magnaporthe grisea*). In this disease plant age and host nutrition may be of minor consequence if temperature and moisture conditions are favourable for infection. These factors influence the disease in many ways, viz., helping production and germination of conidia, their dissemination, and altering the host metabolism and resistance. With proper humidity the dispersal of conidia is highest at 25°–27°C. The night temperature of 20°C affects the metabolic pattern of the rice plant. At low temperatures the absorbed nitrogen tends to accumulate in the leaves as soluble nitrogen. Further, at low temperatures under conditions of high nitrogen level very little silicon is absorbed by the plant. Silicon is supposed to provide resistance to infection. Due to low silicon absorption resistance is reduced and invasion occurs easily. At very low temperature (15°C) even a resistant variety may tend to be susceptible. As temperature rises the varieties tend to show resistance. In red rot of sugarcane (*Colletotrichum falcatum*), under low temperature conditions of January (in India) reaction of cultivars shifts towards resistance while the opposite is true for higher temperatures during August–September when moderately resistant varieties may shift their reaction toward susceptibility (Singh, R.P. *et al*, 1988). In cabbage yellows (*F. oxysporum* f.sp. *conglutinans*) polygenic (horizontal) resistance is broken at high temperature while monogenic resistance is not affected. However, if temperature rises much above 24°C even plant with monogenic resistance succumb to cortical decay. In tomato, resistance to bacterial wilt (*Ralstonia solanacearum*) is lost at high temperatures (Mew and Hó, 1977). Such temperature related failures of resistance are ascribed to temperature sensitivity of the resistance genes. Genes that express at low temperature fail to do so at high temperatures. Root knot resistance gene Mi (Mi-1) derived from *Lycopersicon peruvianum* is not effective at temperatures above 30°C but the other Mi genes (Mi-2, Mi-4, Mi-5, Mi-6) express resistance at 32°C. There are some genes that express at 25°C (Williamson, 1998).

Temperature effect on infection and invasion through the effect on the host is seen in many other diseases. The optimum temperature for growth of *Armillaria mellea* (citrus root rot) is 21°–25°C but it cause infection of roots at 10°–18°C when root growth is slow. The optimum for root growth is 17°–31°C. For root growth of peach and apricot optimum temperature is 10°–17°C. Maximum damage by *A. mellea* to these roots is at 15°–25°C.

The temperature range for spore germination and for infection of the host may be different. In powdery mildews, the maximum and minimum temperatures for infection are a few degrees higher and lower than the maximum and minimum temperatures for germination of conidia. Conidia of *Erysiphe cichoracearum* on lettuce can germinate at 5° but fail to cause infection until the temperature rises to about 10°C. Similarly, the fungus can germinate at 33°C but the maximum for infection is 27°C. *E. polygoni* on pea can germinate at temperatures of 10°–30°C but the optimum infection occurs 20°–24°C. In *P. infestans* temperature affects the mode of germination of sporangia thereby determining the number of infection units. The optimum for germination of sporangia by zoospores is 12°–13°C and by germ tube 24°C. Thus, at the lower temperatures many more infection units (zoospores) are produce while at the higher temperature one sporangium produces a single germ tube.

The temperature can affect disease incidence through effect on insect vectors also. The bacterial wilt of corn (maize) caused by *Pantoea stewartii* (*Erwinia stewartii* subsp. *stewartii*) is a temperature-related disease only because of its vector (*Diabrotica undecimpunctata*). The bacteria survive during winter only through the vector. When the vector is killed by sub-zero temperatures the bacterium is also killed. Temperature controls time of flight and flight range of many aphid vectors of plant viruses. Beet yellows disease is much more serious in years when peak flights of its vectors occur relatively early and the time of flight is correlated with temperature. Temperatures between 25° and 36°C and relative humidity of 40% are congenial for multiplication of *Bemisia tabaci*, the vector of Okra Yellow Vein Mosaic Virus. Temperature of 38°C in May-June and low temperatures in January-February cause low vector population. Thus, in the crop sown in February the incidence of the disease is low and in crops sown later than June the incidence is high. The environment in the plant rhizosphere that affects many root pathogens and also their antagonists is predominantly determined by the quality and quantity of root exudates. Temperature affects root exudation pattern changing the nature of populations of microflora present in the rhizosphere.

The temperature may be favourable for the pathogen and unfavourable for the host at the same time. Under this situation the disease development is most rapid. It is also possible that the temperature may not be very favourable for the pathogen but is highly unfavourable for the host. In this situation the stress and strain caused by high temperature on the host may pre-dispose it to infection. The commonly cited example of temperatures stress as a pre-disposing factor is that of charcoal rot of potato caused by *Macrophomina phaseolina*. In northern India this disease occurs in the crop towards end of the season when irrigation is stopped and atmospheric temperature rises. As a result the soil becomes warm. Some examples of temperature stress as pre-disposing factor are given in Table 10.

In many diseases the optimum temperatures for development of host, pathogen and pathogenesis are different as shown in Table 11. In these situations also temperature stress may be one cause of higher disease incidence. The temperature at which pathogenesis occurs at its maximum is a pre-disposing factor. The temperature, at which growth of the host is adversely affected, will favour the pathogen even if the temperature is not very favourable for the pathogen itself.

Even if the temperature is favourable for both host and pathogen, the latter gets some advantage because of its other characters. The field crops are grown only in a particular season, when temperature fluctuations are normally within the limits of tolerance. In these situations, other environmental factors combine with temperature to accelerate disease incidence.

Effect of temperature on viral diseases of plants is very variable. However, it is known that in many diseases temperature not only determines ease of infection but also virus multiplication and symptom expression. Many plants get easily infected by viruses if they are first kept at 36°C for 1–2 days. At low temperatures the incubation period is increased and rate of appearance of symptoms decreased. The nature of symptoms is also changed. Masking of symptoms at high temperatures is common. Symptoms may appear as local lesions at high temperature and as systemic spots at low temperatures.

TABLE 10: Pre-disposition of Crops to Soil-borne Pathogens Due to Temperature Stress where Temperature Gives Relative Advantage to the Pathogen Over the Crop Growth (Palti, 1981).

Crop	Disease	Pathogen	Soil temperature at which plant is predisposed (°C)
Apple	Root rot	*Sclerotium rolfsii*	30-35
Barley	Stripe	*Drechslera graminea*	10
Maize	Crown and foot rot	*Fusarium graminearum*	8–16
	Seedling stunt	*Pythium* sp.	11
Potato	Charcoal rot	*M. phaseolina*	30–35
	Wart	*Synchytrium endobioticum*	13–15
Tobacco	Root rot	*Thielaviopsis basicola*	17–23
Watermelon	Damping off	*P. ultimum*	12–20
		R. solani	below 20
Wheat	Root rot	*F. culmorum*	24–28

TABLE 11: Differences in Optimum Temperatures for Development of Pathogen and Pathogenesis.

Disease	Pathogen	Optimum temp. (°C)		
		Host	Pathogen	Pathogenesis
Black root rot of tobacco	*Thielaviopsis basicola*	28–29	22–28	17–23
Root rot of wheat	*Gibberella zeae*	25	25	28
Root rot of maize	*Gibberella zeae*	25	25	16
Wilt of watermelon	*Fusarium oxysporum* f.sp. *niveum*	30	27	27

EFFECT OF MOISTURE AND HUMIDITY

Moisture is closely related to temperature and similarly affects initiation and development of infectious diseases. The geographic distribution of many diseases is determined by moisture in the same manner as temperature. In most foliage diseases the suitable combination of temperature and humidity is important. Late blight of potato is of rare occurrence in dry areas. If it occurs in dry areas the occurrence is due to heavy irrigation and fertilization which causes excessive vegetative growth and luxuriant canopy, thereby retaining the moisture conditions favourable for the pathogen even if the atmospheric humidity outside the crop is low. Leaf and stripe rusts of cereals are more serious in wet than in dry areas. The rusts are generally aggravated by heavy irrigation.

Moisture is important for sporulation and take off of spores for secondary spread in many fungal pathogens. In anthracnose and many other fungi the spores are held in a gelatinous matrix and are released only when water drops fall on the mass. The water droplets help in dispersal by air also. In downy mildew of maize caused by *Peronosclerospora philippinensis* both spores and conidiophores are turgid cells so that there is a tendency for each to round off at the flat region of contact. Change in humidity suddenly overcomes the adhesive force and the base of the conidium and tip of the sterigma round off shooting the spores to a distance of a few millimeters or so. This helps in dispersal of the spores by wind. In blue mold of tobacco (*Peronospora tabacina*) the branched conidiophore is aerial and bears single conidia at the tip of pointed sterigmata. With sudden change in humidity from high to low the main axis of the conidiophore dries and twists violently, shaking off its spores. In these fungi conidiophores usually mature during night because of favourable temperature and moisture and conidia are dispersed in the morning with fall in humidity soon after sunrise.

In making the plant surface soft, moisture pre-disposes leaves to attack of several foliar pathogens. Leaf wetness periods are important for infection of leaves in many diseases. In apple scab, along with suitable temperature, leaf wetness period of minimum 9 hours is essential to initiate the disease. In this disease the discharge of ascospores (primary inoculum) from perithecia in fallen leaves also occurs when the late autumn and early spring rains have wetted the leaves.

The downy mildew of grapevines (*Plasmopara viticola*) is very intimately associated with rains and humidity under the vines. Generally, the oospores (survival structures) of this pathogen germinate more abundantly and in shorter time early in the season when they have been subjected to frequent rains and mild temperatures. Rainy season with heavy rains during the oospore formation period hastens their maturity and there is more disease in the next crop season. Sporulation of the fungus on the leaves requires continuous relative humidity ranging between 90 and 100% or saturation. Poor maintenance of grapevine orchard encouraging growth of weeds and bushy canopy helps in accumulation of moisture on the foliage and encourages development of downy mildew. In areas where dew deposition is heavy during night, required humidity for many diseases is available from this source also.

Dependence of potato late blight on humidity is well known. Although temperature also determines sporulation, germination and infection, its role is interrelated with moisture. In dry weather, even if temperature is favourable, the size of lesions is restricted while in wet weather the blighting is so rapid that the entire foliage is destroyed within 2–3 days. A humid period is necessary for sporulation, a drier period is desirable for dissemination and a wet period is necessary for germination and penetration. In presence of film of water on leaf surface the sporangia germinate by producing zoospore which swim in the free water and increase the number of infection on the leaf. The time, temperature and humidity limits for sporulation, germination and penetration are closely interrelated. One night of heavy dew or light rain is not adequate for completion of all these processes. If the period following sporulation is dry and hot most of the spores may be killed before germination. With cool humid nights, followed by light showers, the periods between showers are likely to be cool and humid and the spores are more likely to survive till they have an opportunity to germinate. In the plains of north India late blight generally appears soon after winter rains (end of December and early January). On the hills, the pathogen has better conditions for survival and occurs throughout the growing season of the crop but greater incidence of late blight occurs in June-July when rains occur. In the valleys, due to higher temperature and low humidity the disease is not so severe.

In blast disease of rice (*Pyricularia grisea*) conidia are not produced below 88% relative humidity and at least 90% saturation is essential for their abundant production. The optimum humidity-temperature combination is available during night when the conidia are abundantly produced. In south India, peak concentration of conidia in the atmosphere over the field occurs at 4 a.m. in the main crop and at 6 a.m. in the second crop season. Dispersal is highest during night hours when the relative humidity is 86-98%, temperature in the range of 25°–27° and wind is calm. A minimum night temperature range of 20°–26°C in association with a relative humidity of 90% and above lasting for a week or more during any of the susceptible stages of the crop growth (seedling, post-transplanting tillering and neck emergence) is ideal for outbreak of the disease in the crop.

Response of some powdery mildews is an exception to the generally recognized view that foliar pathogens are dependent on moisture for their development. Many powdery mildews are reported to be favoured by dry condition of the host surface. Different species have been placed in 3 broad categories on the basis of their humidity requirement for conidial germination (Schnathorst, 1965). Some require 70–100% relative humidity, some 20–100% relative humidity, some only up to 70% RH while other are indifferent to relative humidity for conidial germination. Conidia of *Erysiphe polygoni* (pea powdery mildew) fall in the last category. The spores of these fungi have large water content and are among the shortest lived air-disseminated spores. In *Uncinula necator* (powdery mildew of grapevine) the conidial germination can occur at RH 40–100% but is erratic. Humidity has greater effect on sporulation than on spore germination. Sporulation occurs at RH above 57.4% at temperatures of 12°–30°C and does not occur at RH below 7.4% and temperature below 8.4° or above 34°C (Chavan *et al*, 1995). In *Podosphaera leucotricha* (powdery mildew of apple) conidia do not germinate below 85.5% RH. High relative humidity is a must for penetration. Conidia of *Oidium mangiferae* (powdery mildew of mango) are quickly destroyed by dry conditions.

In almost all powdery mildews rains disrupt the mycelium and conidiophores and there is less production of conidia. Free moisture (water film) is inhibitory to some powdery mildews. Not only germination of conidia is inhibited but growth of the mycelium is also abnormal when moisture film exists on the leaf surface. Immersion of conidia in water for 1–3 hours is reported to kill them in *Erysiphe graminis*, *Podosphaera leucotricha* and *E. cichoracearum*. Negative chemotropism to carbon dioxide has been given as an explanation for the growth of powdery mildews away from water.

The favourable and unfavourable effects of water on root disease have long been known. Root rots and damping off caused by species of *Pythium* and *Aphanomyces* are most severe in wet soils. The favourable effect on damping off caused by these fungi is due to formation, release and easy movement of zoospores and also due to unfavourable effect on seedling strength. In *Aphanomyces* root rot of pea (*Aphanomyces euteiches*) the damage to roots is maximum when soil is very wet but the effect on the plants is seen when soil is dry. It is because of greater demand of water by the plant under water stress conditions but roots are dead. In the soil, moisture determines the concentration of oxygen and thus affects root growth, uptake of nutrients and susceptibility of roots to attack of pathogens. It also dissolves the nutrients which may be essential for breaking dormancy of fungal structures. In presence of excess of moisture around roots, soluble salts accumulate in toxic concentrations and thus roots become prone to attack of pathogens. The most important role of moisture, however, is its effect on spore germination and subsequent penetration of the root. The citrus gummosis and root rot caused by *Phytophthora parasitica* is most severe in continuously wet soils which hinder fibrous root development and promote zoopsore formation and dispersal. In irrigated orchards and nurseries when interval between irrigation is increased to allow the soil to dry, the sporangia and zoopsores are killed as the soil dries. Common scab of potato (*Streptomyces scabies*) is favoured by dry soil during tuberisation but is checked by wet soil. In dry soils activity of antagonistic bacteria is low while it is enhanced in wet soils. In some diseases the moisture in soil affects the survival of fungal propagules. In *Phytophthora cactorum* (collar rot of apple trees) a temperature range of 20°–25°C with high soil moisture (> 75% WHC) is most suitable for production of sporangia. Mycelium of this fungus can survive only for short duration in presence of high soil moisture and at high temperatures due to enhanced lysis of hyphae. Sclerotia of many fungi survive longer in dry soil than in moist soil.

Diseases favoured by Dry Soil or Water Deficiency:
Seedling blight of cereals caused by *Fusarium roseum*
Stem rot of sweet potato (*F. solani* f.sp. *batatas*)
Pox of sweet potato (*Streptomyces ipomoeae*)
Common scab of potato (*Streptomyces scabies*)
Root and stem rot of peas (*F. solani* f.sp. *pisi*)
Charcoal rot of different crops (*Macrophomina phaseolina*)
Stalk rot of maize and wilt of sugarcane (*F. moniliforme*)
Brown spot of rice (*Drechslera oryzae*)
Safflower root rot (*Phytophthora cryptogea*)
Wheat seed decay (species of *Penicillium* and *Aspergillus*)

Diseases favoured by Wet Soil or High Relative Humidity:
Root rot of cotton (*Phymatotrichum omnivorum*)
Stripe of wheat (*Cephalosporium gramineum*)
Take all of wheat (*Gaeumannomyces graminis*)
Bare patch of wheat (*Rhizoctonia solani*)
Black root rot of tobacco (*Thielaviopsis basicola*)
Armillaria root rots (*Armillaria mellea*)
Sclerotium root rots (*Sclerotium rolfsii*)
Sclerotinia blight and wilts (*Sclerotinia sclerotiorum*)
Aphanomyces root rots (*A. euteiches* and *A. cochliodes*)
Wart disease of potato (*Synchytrium endobioticum*)
Powdery scab of potato (*Spongospora subterranea*)
Gummosis and root rot of citrus (*Phytophthora* spp.)
Late blight of potato
Damping off and stem rot of papaya
Root knot and other nematodes
Downy mildews of maize, sorghum, pearl millet, grapevines
Bacterial blight and streak of rice
Bacterial wilt of potato and tomato

Fusarium wilts and Verticillium wilts are apparently favoured by wet soils but many reports are at variance. Fusarium wilt of tomato (*F. oxysporum* f.sp. *lycopersrici*) and cabbage yellows (*F. oxysporum* f.sp. *conglutinans*) appear severe following hot dry weather while they are absent or occur in very mild form in cool moist weather. However, in greenhouse tests Fusarium wilt of tomato, pea (*F. oxysporum* f.sp. *pisi* race I), cotton (*F. oxysporum* f.sp. *vasinfectum*) and celery (*F. oxysporum* f.sp. *apii*) were found to be more severe in wet than in dry soils. The type of soil seems to be responsible for these variations. While in the USA cotton wilt is more common in sandy soils and is favoured by soil moisture of 80–90% saturation, in India it occurs in heavier soils and shows no correlation with soil moisture.

In Verticillium wilt of tomato and eggplant (*V. albo-atrum*) soil saturation for atleast a day is essential for infection. The bacterial wilt of potato (*Ralstonia solanacearum*) is favoured by high soil moisture which helps in dispersal of bacterial cells and increases the size of lenticels. A warm and wet soil is conducive to invasion of tissues and development of the disease. Most bacterial diseases are favoured by soil moisture in the same manner. Another factor that enhances infection by *R. solanacearum* is the increased activity of nematodes in wet soil. Excess of moisture or waterlogging is always detrimental to the plant and predisposes it to pathogens. The role of water-logged condition in red rot epidemics of sugarcane was stressed by Dutta Mazumdar *et al*, (1990). Even resistant varieties succumb to red rot under water-logged conditions. It is due to pre-disposition of the host as well as changes in the behaviour of the pathogen. Floating mycelial aggregates form acervuli and produce shorter than normal conidia under water-logged conditions. Fusion of conidia occurs and the fused conidia germinate to produce secondary conidia. The floating mycelial aggregates and the conidia have better chances of getting attached to vulnerable sites of the host and also being transported to long distances.

The effect of moisture on viral diseases is mostly through its effect on the plant and the virus vectors. The maximum multiplication of viruses occurs in young, growing parts of most plants. In presence of high humidity such organs are more abundantly available for activity of viruses. The concentration of virus particles is high in such tissues. Some viruses are favoured by hot and dry weather. They do not express any symptom on the host in wet conditions. The reproduction and flight of insect vectors is also determined by humidity in the air. Therefore, incidence of insect-transmitted virus diseases is affected by relative humidity of the atmosphere.

There is close relationship between plant pathogenic nematode and soil moisture. In very dry soils, the water content of pore spaces is so low that there is absence of suitable medium for the movement of nematode larvae. In very wet soils the pore spaces get completely filled with water and there is lack of aeration. This also adversely affects movement or activity of nematode larvae. In both conditions, due to slow movement of larvae, incidence of the disease is reduced. The best soil moisture condition is in between the two extremes. The moisture content of the soil is, to a great extent, determined by the soil type. In sandy soils, pore spaces are large but are emptied quickly. The above mentioned first condition applies in this case. In heavy soils water-filled pore spaces retain water for longer time. The nematodes such as *Meloidogyne* spp. (root knot) are most destructive in sandy loam or silt loam soils.

Indirect effect of moisture or free water on incidence of plant diseases also includes the effect on stability of spores on plant surfaces, quality and quantity of root exudates, nature of antagonism in soil, etc.

EFFECT OF SHADE

The role of crop canopy in maintaining relative humidity suitable for development of *P. infestans* was mentioned earlier. The crop canopy is involved in a shading effect and alters the temperature and moisture conditions under the plants. The canopy is decided by rate of sowing or planting, application of water (irrigation) and nutrients, nature of the crop and management of the canopy in special crops like grapevines. The effect of a given density of crop cover is dependent on a variety of factors such as direction of the crop row in relation to the prevailing wind, distance of the foliage from the ground, varietal characters of the crop and topographical and other soil factors. These factors determine the rate at which the top soil dries and, in turn, it affects the atmospheric humidity in the crop canopy.

In years when macroclimatic conditions are unfavourable for incidence of late blight, close planting, heavy irrigation and high levels of fertilizers produce a crop with dense canopy which prevents rapid loss of soil moisture and maintains high relative humidity in the crop. The disease starts from few infected tubers and spreads under the canopy without being affected by external environments and without being visible from outside. Due to high humidity the sporangia fail to detach from sporangiophores and remain trapped under the canopy. Disease spreads by contact between leaves. In grapevine diseases such as downy mildew (*Plasmopara viticola*), bunch rot (*Botrytis cinerea*) and black rot (*Guignardia bidwellii*), the microclimatic conditions under the canopy are closely related to infection, inoculum production and spread of the diseases. To control these diseases in vineyards, leaf removal method is followed (English *et al*, 1989). In this method, basal portion of shoots near the clusters is defoliated shortly after bloom. This increases wind speed under the canopies causing quick drying, lowering the relative humidity. Upto 69% reduction in the incidence of bunch rot is reported when shading is reduced by this method. Canopy structure coupled with irrigation has direct influence on the development of diseases caused by *Sclerotinia sclerotiorum* (Blad *et al*, 1978) also.

In addition to crop canopy, shade is also provided by slope of the land and any vertical object in the vicinity of the crop field, such as trees, building, etc. The morning shade provided by slopes and vertical objects suffices to increase attack of potato late blight, many downy mildews and maize leaf spot (*Helminthosporium maydis*). Use of trees for shading is of particular importance in tropical crops. It reduces light intensity during summer, increases duration of stomatal closure, reduces dew deposits on leaves, provides protection from strong winds, adds organic matter through mulch and absorbs excess of water in low sites.

However, shading has predisposing effect also in many diseases. Diseases of coffee such as rust (*Hemileia vastatrix*), berry disease (*Colletotrichum coffeanum*), leaf sports (*Cercospora coffeicola*) and wilt (*Fusarium oxysporum* f.sp. *coffeae*), diseases of tea such as blister blight (*Exobasidium vexans*) and of citrus such as wither tip and die back (*Colletotrichum* spp.) are favoured by shade. Role of shade in increasing susceptibility of rice to bacterial leaf blight due to increased physiological weakness of the plant has also been reported (cf. Singh, 1998).

EFFECT OF WIND

Role of wind in dispersal of pathogens was discussed in chapter 7. In addition to this role, wind can influence disease incidence in other ways also. It accelerates the drying of wet top soil or the wet plant surfaces. This reduces incidence of many diseases (examples are cited above). Strong winds, especially when accompanied by rain or sand particles cause minute injuries to leaf and fruit surface and facilitate infection by bacteria and fungi. Rubbing of leaves and fruits with strong or thorny stems and twigs promotes infection of citrus canker bacterium.

EFFECT OF LIGHT

Although light strongly influences development of certain diseases it is not a very important determinant of seasonal and regional occurrence of plant diseases. The intensity of light and day length (time from sun rise to sunset) usually affects plant growth and may make the plants favourable or unfavourable media for disease causing fungi, bacteria and viruses. The intensity of light affects survival of infective organs of a pathogen, stages before infection, penetration, incubation period, amount of secondary inoculum produced and sometime symptom expression.

Strong light accompanied by high temperature causes sunscald of fruits in tomato. Low light intensity leading to more or less etiolation pre-disposes plants to attack of non-obligate pathogens such as *Botrytis* but decreases susceptibility to obligate parasites such as rusts. Tomatoes grown under conditions of low light intensity become highly susceptible to Fusarium wilt while plants maintained in strong light show very little disease. Thus, long daylength and clear skies suppress wilt of tomato. Similarly, leaf mold of tomato (*Cladosporium fulvum*) and powdery mildew of rye (*Erysiphe graminis*) develop best under reduced light conditions. Light controls the opening and closing of stomata and all those pathogens that enter the host exclusively through stomata are favoured by light.

Light has positive effect on sporulation, growth and infection type produced by rust fungi. In general, relatively high light intensity and long days are most favourable for rapid development of mycelium and abundant sporulation in susceptible hosts. The incubation period of *Puccinia graminis tritici* is reduced and amount of sporulation increased if the inoculated plants are kept in strong light. The incubation period of *Melampsora lini* (flax rust) can vary from 6 days in continuous light to 14 days in low light intensity.

Strong sunlight kills conidia of many fungal pathogens such as *Uncinula necator* and *Oidium mangiferae*. Light is essential for germination of oospores or *Phytophthora cactorum* (collar rot of apple). Since sunlight is involved in photosynthesis it affects plant metabolism and root exudates. Reduced light intensity generally increases the susceptibility of plants to virus infection. Number of lesions and sharpness of symptoms are affected by light. Therefore, seasonal variations in nature of symptoms according to daylength can be found. Daylength is short during winter. Thus, the plants show same symptoms as if they were kept under shade. Tobacco necrosis, tomato leaf curl and many other viral diseases are more conspicuous during winter than during summer. Due to low light intensity the number of lesions on leaves is high and whole leaf may be destroyed.

EFFECT OF SOIL REACTION (pH)

Most crops can grow within pH limits wider than those conducive to development of their root pathogens. Although most fungi, bacteria and nematodes can tolerate the pH range in which their hosts normally grow, there are some diseases in which the tolerance limits of the pathogen are very narrow while the host can flourish at pH outside these limits. This provides an opportunity for management of such diseases by manipulation of soil pH. Directly or indirectly, soil pH affects speed of disease cycle and, therefore, the disease intensity in a large number of cases.

Effect of soil pH is on the pathogen as well as the host. In general actinomycetes such as *Streptomyces scabies* grow best at soil pH close to neutral while fungi prefer acidic than neutral soils because they are more adaptable to hydrogen ion activity than bacteria including actinomycetes. Therefore, they suffer less from competition or antagonism under acidic conditions. The effect of soil pH through host is mainly through tolerance of the plant to a given pH and its better nutrition at that pH. Acid tolerant plants, when grown in acidic soil, become highly resistant to such necrotrophs as *Pythium*. This fungus attacks tissues lacking secondary thickening. The vigorous growth enables the plants to develop such barriers. Although, *Pythium* also is tolerant to acidic soils, the host gets better advantage of pH. If such plants are grown in alkaline soil, they become susceptible because of poor nutrition and poor growth.

Diseases Favoured by Alkaline Soil:

Texas root rot of cotton (*Phymatotrichum omnivorum*)
Take all of wheat (*Gaeumannomyces graminis*)
Verticillium wilts (*V. albo-atrum* and *V. dahliae*)
Wilt of pea (*F. oxysporum* f.sp. *pisi* race I)
Common scab of potato (*Streptomyces scabies*)

Diseases Favoured by Acidic Soil:

Cotton wilt (*F. oxysporum* f.sp. *vasinfectum*)

Tomato wilt (*F. oxysporum* f.sp. *lycopersici*)
Clubroot of cabbage (*Plasmodiophora brassicae*)
Powdery scab of potato (*Spongospora subterranea*)
Wart of potato (*Synchytrium endobioticum*)
White root rot of apple (*Rosellinia necatrix*)
Root rot of tobacco (*Thielaviopsis basicola*)
Sclerotium diseases (*Sclerotium rolfsii*)
Bacterial wilt of potato (*Ralstonia solanacearum*)
Bacterial canker of stone fruits (*Pseudomonas syringae* pv. *syringae*)

The common scab of potato (*S. scabies*) is severe in dry soils with pH ranging from 5.2–8.0. Below pH 5.2 intensity of the disease is drastically reduced. The effect is through effect on the life activities of the bacteria. The clubroot of cabbage (*P. brassicae*) is favoured by wet acid soils. Optimum pH for its development is 5.7. Intensity of the disease rapidly declines between pH 5.7 and pH 6.2 and at pH 7.8 the disease does not occur. The spores germinate poorly or not at all under alkaline conditions. If the pH of the infectious soil is raised to 7.2 and above by liming and the soil is kept continuously wet there is no disease and plants grow normally. In dry soil liming does not help. It is because of the fact that the soil moisture film around the rootlets may be acidic due to carbon dioxide released by roots. In moist soil there is sufficient movement of alkaline particles to neutralize the acidity around the rootlets. In dry soils this is not possible. Uptake of calcium is important for resistance of the plant to the pathogen. Calcium uptake is more at pH 6.8–7.2 than at lower pH values. The soil pH above 7.2 reduces infection and clubbing because thalli abort before producing zoospores. (*cf.* Singh, 1998).

Soil acidity is well known for its effects on soil microbiological activities. The acidity in soil is due to H-ions or mixtures of H-ions and Al-ions. Ordinary cultivated soils do not accumulate toxic levels of H-ions. However, even a very low concentration of Al-ions may be toxic for the plants. Therefore, at the same pH the effects on soil microflora may vary according to presence or absence of Al-ions.

Apart from its direct toxic effects soil acidity may influence disease development indirectly also. This is mediated through the effects on activity of antagonistic microflora. Such strong antagonists as *Trichoderma* and *Penicillium* grow rapidly in acidic soils and obstruct the activities of plant pathogens. Black root rot of tobacco (*Thielaviopsis basicola*) is strongly decreased by acid soil reaction. The same fungus causes root rot of bean also. The disease is differently affected by acidity and the effect on disease development is probably indirect. In cultures the fungus grows best at pH 4.7–5.5 and growth is markedly reduced at pH 7.0. But in non-sterile soil the disease does not occur below pH 5.5 while the root system of the host (bean and poinsettia) is completely destroyed within 2 weeks in neutral or alkaline soil. Rapid colonization of host roots by antagonistic fungi such as *Trichoderma viride* and *Penicillium* spp. in acidic soil is considered as an explanation for this.

High alkalinity in soil also has a strong influence on survival of certain fungi. Application of ammonia, sufficient to cause the soil pH to exceed 9.8 for a period of 24 hours, kills the sclerotia of *Sclerotium rolfsii* not only by ammonia toxicity but by high alkalinity also. Use of high doses of fertilizers containing ammonia nitrogen had been recommended as one of the methods of control of sugarbeet root rot caused by this fungus.

As regards nematodes, most of them grow best at pH values of 5.0–6.0. The exception is *Heterodera avenae*, cereal cyst nematode (molya disease), which prefers high pH values.

EFFECT OF OXYGEN AND CARBON DIOXIDE CONCENTRATION

The ratio of oxygen and carbon dioxides in the atmosphere above the ground is almost fixed and shows no fluctuation. Therefore, these gases normally have no marked effect on pathogenesis of foliar pathogens except some special situations such as when at the microsites of the infection court water imbalance or closure of stomata causes accumulation of carbon dioxide which repels germs tubes of pathogens with low tolerance to this gas. Low oxygen and high carbon dioxide environment stimulates growth of *Geotrichum candidum* that causes sour rot of citrus and some other fruits. Storage and transport of harvested fruits under low oxygen (5%) and high carbon dioxide levels (5–20%) have been used to suppress respiration of the fruit and the fruit decay pathogens thereby suppressing the post-harvest rots. Growth of *Sclerotinia minor* is greatly reduced in CO_2-enriched atmosphere.

Soil aeration is the process by which gases produced or consumed under the soil surface are exchanged for gases in the aerial atmosphere. In the consideration of soil aeration oxygen and carbon dioxide are the two gases of greatest importance although other volatiles are also released during decomposition of organic matter. The interaction of plant root, microorganisms and biochemicals affects the balance between oxygen and carbon dioxide. This balance between oxygen and carbon dioxides in soil is highly variable and affects host root development and the root pathogens and their pathogenesis. Oxygen present in soil is used by plant roots and microorganisms and carbon dioxide is released. This gas then diffuses from the areas of high respiration through pore spaces. Thus, aeration of soil depends on soil type, structure, moisture and vegetation present on the surface. The amount of free oxygen influences activity of saprophytes and parasites in the soil and at the same time it determines the root growth, nutrient uptake and many biochemical activities in soil.

The direct effect of aeration on plant pathogenic fungi in soil includes the requirements of fungi for free oxygen for their vegetative growth and reproduction. Many species of *Phytophthora* are known to require some oxygen for their zoospore production. *P. parasitica* and *P. citrophthora* grow vegetatively at low oxygen and absence of oxygen stops their growth. Aeration is essential for germination of sclerotia of *Sclerotium oryzae* and this is always available under field conditions even if sclerotia are present deep in soil because of presence of cracks in the field soil. The availability of oxygen (or oxygen diffusion rate) determines the vertical distribution of fungi in soil.

Amount of oxygen is not as important as the concentration of carbon dioxide in determining the effect of soil aeration on activities of pathogenic microorganisms. Carbon dioxides affects metabolism of autotrophic organisms, acts as differential inhibitor for heterotrophic microorganisms and influences the pH at the microsites where most organisms survive and are active in the soil. Increased concentration of CO_2 inhibits growth of *Rhizoctonia solani*, especially the pathogenic stages, in soil. The take all disease of wheat (*Gaeumannomyces graminis*) is more prevalent in sandy alkaline soils due to low carbon dioxide concentration in soil atmosphere. Growth of the pathogen is greatly reduced if the soil is moistened or compacted or if clay is added.

Fusarium culmorum and *Gibberella zeae* cause root rot and seedling blight of wheat. At the normal CO_2 concentration of the soil (0.03%) wheat shows maximum germination but at 1% or 1.5% concentration of the gas its germination is greatly reduced and at 2% plant growth stops. On the other hand, these fungi continue to grow up to 7% concentration of carbon dioxide. Many other fungi such as *Fusarium oxysporum* f.sp. *melonis* (muskmelon wilt) have good tolerance for carbon dioxide concentration in soil. Obviously, high carbon dioxide concentration in soil makes the plant weak and increases activity of such fungi thus causing increased incidence of the disease. In Panama disease of banana (*F. oxysporum* f.sp. *cubense*) the pathogen is very sensitive to low partial oxygen pressure in flood followed soil. The carbon. dioxide in submerged soils causes continued germination of conidia but prevents chlamydospore formation. Absence of these survival structures ultimately leads to elimination of the pathogen from the flooded soil. Harmful effect of carbon dioxide in flooded soils is also responsible for elimination of *Sclerotinia sclerotiorum*.

Aeration affects general soil microflora and these affect the plant roots or pathogens through their metabolites. The root rot of wheat and sugarcane (*Pythium arrhenomanes*) is more serious in wet, ill-drained soils. In such soils anaerobic bacteria are active and liberate toxic metabolites. These toxic substances weaken the roots thus facilitating easy infection by the pathogen. In submerged soils, one of the causes of elimination of *F. oxysporum* f.sp. *cubense* (banana wilt) is the effect of toxic substances such as acetic acid liberated during anaerobic decomposition of organic matter. Same applies to destruction of nematodes in flooded soils.

The role of available oxygen in soil appears most important in determining the population limit reached by nematodes species. Survival of nematodes in inactive state does not appear to be influenced by oxygen. Growth and motility are prominently influenced by aeration. Increased oxygen level stimulates activity of larvae while reverse is true when carbon dioxides concentration is increased. Thus, conditions which reduce aeration also decrease the activity of nematodes. Usually, under excess carbon dioxide nematodes do not die for some time but become quiescent. In this condition motility ceases, metabolic processes are stopped and the larvae do not spend their lipid reserves. This may prolong their survival ability. However, many nematodes such as *Ditylenchus dipsaci* are killed when exposed to high carbon dioxide concentration for 24 hours. In general, increase in oxygen or decrease in carbon dioxide enhances emergence of larvae from eggs or cysts and motility of the second stage larvae. The movement of larvae through soil pores is also affected by oxygen or carbon dioxide level in the pore spaces. When the pore spaces are completely filled with water oxygen is reduced. This causes a tendency among larvae to become slow in movement. Thus, inspite of presence of suitable medium (water) for movement, the optimum movement does not occur. There is no evidence to suggest that CO_2 is toxic to nematodes. It is even stimulatory at low concentrations or short exposure times. In *vitro* studies have proved that nematodes are attracted to low levels of carbon dioxide in the root zone of host plants. Larvae of *Meloidogyne* spp., wandering in soil, are attracted towards small quantities of carbon dioxide emanating from plant roots during respiration of the latter. Together with role of root exudates and electrical potentials, carbon dioxide explains the orientation of nematode larvae towards roots.

HERBICIDES AND INSECTICIDES

Insecticides and herbicides are commonly used against insect pests and weeds. Directly or indirectly these chemicals influence plant diseases also. The main relationship of insecticides with plant diseases is through their effect on insect vectors of pathogens especially the viruses. The use of herbicides is directed to control weeds in the field so that survival of inoculum and its build up are destroyed. Certain herbicides may have direct relationship with a disease (Altman and Campbell, 1977). Herbicides have been shown to increase the severity of certain diseases such as *Rhizoctonia solani* on sugar beet and cotton, Fusarium wilt of cotton and tomato and Sclerotium stem rot of various crops. In many host-pathogen combinations, herbicides have been shown to decrease disease severity. The herbicides propanil, benthiocarb, dinitroaniline and many others have been shown to reduce the activity of *Rhizoctonia, Fusarium, Verticillium, Alternaria, Peronospora* and *Xanthomonas* (cf. Singh, R.S. 2001). Herbicide applications can alter the activity of fungicides also.

PLANT NUTRITION

The host plant obtains mineral nutrition from soil and the parasite obtains mineral nutrition and source of energy from the plant. Along with temperature, light and other environmental factors, the mineral nutrition of the plant influences its suitability as a substrate for the parasite. Development of the host, rate of its growth and its various physiological activities are influenced by quality and quantity of available nutrients and in this way influence readiness of the plant for defence. The effects vary with the type of the disease, form of the nutrient, proportion of other nutrients, and season. Effect of same nutrient may vary under the influence of these factors.

The host parameters of nutrient effects include (i) vigor of growth which in turn may affect the crop climate, (ii) anatomical and histological characteristics such as thickness of cuticle and epidermis, silification, lignication, etc., and biochemical and physiological reactions, (iii) rate of growth determining escape from susceptible stages or speed of tissue thickening and (iv) water economy.

The pathogen parameters of nutrient effect include (i) rate of penetration, colonisation and reproduction of the pathogen, (ii) relative rate of development of pathogens and their competitors and antagonists on aerial host surfaces and in rhizosphere, (iii) direct toxic effect of some fertilizers such as urea on certain pathogens, (iv) change in soil pH which may be favourable for the pathogen or which may stimulate development of antagonists of pathogens.

Since the effect of nitrogen on the plant is most pronounced, it has been studied in more detail than other nutrients. In nitrogen deficiency, the plant is weak, its development is incomplete and its maturity is hastened. Pathogens favoured by slow growth of the host are, thus, favoured by low nitrogen level in soil or its reduced availability. Some examples of diseases aggravated by low nitrogen are tomato wilt (*F. oxysporum* f.sp. *lycopersicae*), early blight of potato and tomato (*Alternaria solani*), root rot of sugarbeet (*Sclerotium rolfsii*), bacterial wilt of solanaceous crops (*Ralstonia solanacerum*), downy mildew of muskmelon (*Pseudoperonospora cubensis*) and purple blotch of onion (*Alternaria porri*).

Application of nitrogen, more than required for normal development of the plant, causes new succulent vegetative growth and delays maturity. Those pathogens which attack such plant organs are favoured by high nitrogen. Rusts (*Puccinia* spp.) and powdery mildew of wheat (*Erysiphe graminis*), blast of rice, (*Pyricularia grisea*), some wilt diseases caused *Fusarium oxysporum*, sheath blight of rice (*Rhizoctonia solani*), leaf spots of groundnut (*Cercospora* spp.), black leg of potato (*Erwinia* spp.), black spots of brassicas (*Alternaria* spp.), and tobacco mosaic are known to be more severe in excess nitrogen conditions. Soft rot of potato (*Erwinia carotovora*) is more severe in tubers raised in nitrogen rich soil.

Nitrogen affects diseases through its effect on the host as stated above and/

or through the effect of excess nitrogen in the host on metabolic activities of the pathogen. Thus, while excess nitrogen in rice interferes with silicon uptake thus making the leaves prone to attack of *Drechslera oryzae* (brown spot) and *Pyricularia grisea* (blast), in bacterial leaf blight of the same host it enables the bacterium (*Xanthomonas oryzae* pv. *oryzae*) to secrete more toxic metabolites. Excess nitrogen (more than 100 kg/ha) promotes most of the disease of rice including bacterial leaf streak (*X. oryzae* pv. *oryzicola*), stem rot (*Sclerotium oryzae*), false smut (*Ustilaginoidea* virens or *Claviceps oryzae-sativae*), bunt (*Neovossia horrida*), and leaf smut (*Entyloma oryzae*).

The chemical form as well as the amount and distribution of a nutrient influence biological activity in soil. Those nutrients that are capable of existing in different oxidized states fall primarily in this category. Apart from nitrogen, other nutrients in this category are phosphorus, sulfur, iron, manganese and copper.

Nitrogen is available to the plant in the form of ammonia or nitrate. The form of nitrogen decides whether it will increase or decrease incidence of a particular disease. Some diseases are favoured by ammonia nitrogen and some decreased by this form of nitrogen. Similarly, a large number of diseases caused by soil-borne pathogens are decreased by nitrate nitrogen while many foliar diseases are increased by it (Table 12).

TABLE 12: Effect of Inorganic forms of Nitrogen on Plant Diseases.
(Ref. Huber and Watson, 1974)

Pathogen	Disease	Host(s)	Nitrogen Nitrate	form Ammonia
Rhizoctonia	Seedling disease	sugarbeet	Decrease	Increase
	Root rot	bean	Decrease	Increase
	Root/stem rot	potato	Decrease	Increase
Sclerotium	Blight	tomato	Decrease	Increase
Pythium	Root rot	pea, maize	Increase	Decrease
Phytophthora	Root rot	citrus	Decrease	Increase
Fusarium	Root rot	Bean, maize	Decrease	Increase
	Stalk rot	maize	Decrease	Increase
	Wilt	cotton, tomato	Decrease	Increase
	Yellows	cabbage	Decrease	. . . . .
Verticillium	Wilt	potato, tomato	Increase	Decrease
Pyricularia	Blast	rice	Decrease	Increase
Erysiphe	Powdery mildew	wheat	Increase	
Puccinia	Stripe rust	wheat	Increase	Decrease
	Stem rust	wheat	Increase	Decrease

Thus, root rot and wilts caused by *Fusarium* spp., clubroot of cabbage caused by *Plasmodiophora brassicae* and damping off and stem rot caused by *Sclerotium rolfsii* are increased in severity when ammonium fertilizer is applied while cotton root rot caused by *Phymatotrichum omnivorum*, take all of wheat (*Gaeumannomyces graminis*) and common scab of potato (*Streptomyces scabies*) are favoured by nitrate nitrogen.

F. solani f.sp. *phaseoli* (root rot of bean) and most other forms of pathogenic *Fusarium* spp. survive in soil through chlamydospores. New infections are caused by germination of these spores. Exogenous nitrogen and carbon are required in soil for formation and germination of chlamydospores. Ammonia nitrogen is more effective than nitrate nitrogen. In addition, exogenous nitrogen favours early penetration and pathogenesis by this pathogen. Reduced form of nitrogen is more effective than oxidised nitrogen as nitrate for this purpose also. Verticillium wilt of cotton (*V. albo-atrum*) is decreased by nitrogen application as ammonium nitrate but not by ammonia and nitrate separately. However, this selectivity for form of nitrogen is not found in all members of the same group of fungi *F. roseum* is not affected by forms of nitrogen and can utilise both ammonium and nitrate nitrogen. *Thielaviopsis basicola* (root rot of bean and tobacco) also is not affected by form of nitrogen and both forms enhance survival of its endoconidia.

Suppression and stimulation of some diseases by specific form of nitrogen has also been attributed to a shift in soil pH, especially in the root zone. Ammonium nitrogen generally lowers the pH while nitrates raise it. The effect of pH on some diseases caused by soil-borne pathogens was mentioned in the section on effect of soil reaction. Diseases favoured by ammonium nitrogen are generally more severe in acid soils while those favoured by nitrate nitrogen are severe in soils with neutral to alkaline reaction. Nitrate nitrogen is known to suppress damping off of sugar beet and table beet caused by *Pythium ultimum* but ammonium nitrogen fails to do so. According to Smiley (1975) this is due to change in rhizosphere pH toward alkalinity by nitrate nitrogen and beets being alkali-tolerant plants develop more vigor under these conditions, thereby developing resistance to *Pythium* quickly through tissue maturity.

Not all the effects of nitrogen are achieved via its effect on host tissues. Some nitrogen compounds are directly toxic to fungi and nematodes. Ammonia is known to be toxic to *Sclerotium rolfsii, Phymatotrichum omnivorum, Plasmodiophora brassicae, Gaeumannomyces graminis* and some parasitic nematodes. *Pythium, Fusarium oxysporum* and many nematodes are suppressed by urea due to direct toxicity.

Mycorrhyzas are considered as natural barrier in plant roots against several root infecting fungi and nematodes. A high level of nitrogen suppresses mycorrhizal development as a consequence of nearly complete utilization of carbohydrates by the host. Thus, this barrier to infection is lost. In addition, the absence or elimination of mycorrhyzae disturbs the phosphorus uptake by the plant. This unbalances the nutrition of the plant resulting in disturbed metabolism and consequent effect on disease.

Other major and minor nutrients also influence disease incidence. Phosphorus has been shown to reduce severity of take all of wheat (*Gaeumannomyces graminis*) and potato scab (*Streptomyces scabies*). It increases the severity of cucumber mosaic virus on spinach, glume blotch of wheat (*Septoria*) and leaf spots of groundnut (*Cercospora* spp.). Deficiency of phosphorus promotes clubroot of cabbage. Phosphorus seems to increase resistance either by improving the balance of nutrients in the plant or by accelerating the maturity of the crop and allowing it to escape infection by pathogens which prefer younger tissues.

Usually high potassium (K) application has been reported to reduce incidence of plant diseases. It seems to directly affect the various stages of the pathogen establishment and development in the host, and to indirectly affect infection by promoting wound healing, by increasing resistance to frost injury and thus preventing infection by pathogens that infect through such injuries and wounds. It delays maturity and senescence in some crops. Among diseases reduced by potassium application to soil are stem rust of wheat (*P. graminis tritici*), powdery mildew of wheat and barley (*Erysiphe graminis*), brown spot (*Drechslera oryzae*), stem rot (*Sclerotium oryzae*), narrow brown leaf spot (*Cercospora oryzae*), and bacterial leaf blight (*Xanthomonas oryzae* pv. *oryzae*) of rice, early blight of tomato (*Alternaria solani*), Fusarium and Verticillium wilts of cotton, Fusarium wilt of melon, Alternaria leaf spot of tobacco, cabbage yellows (*F. oxysporum* f.sp. *conglutinans*), downy mildew of cauliflower (*Peronospora parasitica*), wildfire of tobacco (*Pseudomonas tabaci*) and root knot (*Meloidogyne incognita*). However, excess of potassium predisposes tomato to wilt and citrus to Phytophthora root rot and gummosis.

Nutritional imbalances, primarily those of N and K, predispose rice leaves to infection of *Drechslera oryzae* (brown spot or Helminthosporiose). Severity of the disease decreases with rise in the K level of the soil during growing period of the plants. Conidia formed on plants grown in K-deficient soil are more pathogenic. About 90% of conidia penetrate the leaf through motor cells and 10% through the stomatal pore. Following the invasion, motor cells develop granular deposits which are abundant in plants grown in soil with high nitrogen and potassium and sparse in plants with deficiency of these nutrients. The extent of granular deposit indicates resistance of the tissue. Infection hyphae appear slightly narrower in cells with granular deposits (*cf.* Singh, 2001). However, excess of nitrogen alone (above 100 kg./ha) predisposes the plants to the disease.

Among micronutrients, the role of zinc in disease incidence is well established. Zinc is a part of enzymes involved in auxin synthesis and in oxidation of sugars. Deficiency of zinc in soil or its low uptake is known to cause deficiency disease in rice and guava. In zinc deficient soils the incidence of downy mildew of maize is enhanced. It is responsible for rosette of apple, mottle leaf of citrus and little leaf of grapevine. Prominent leaf veins, pale yellow colour, bronzing of lamina surface, crinkling and reduced size of leaves are other effects of zinc deficiency. Susceptibility of citrus to the nematode *Tylenchulus semipenetrans* is increased in zinc deficient soils. Zinc, boron and manganese are reported to restrict the saprophytic activity of *Fusarium udum* (pigeonpea wilt) in soil.

Manganese has a controlling effect on the disease. Manganese reduces severity of potato scab also. Boron deficiency in soil may result in disintegration of core tissue in applies and other fruits. It also promotes black rot of cauliflower caused by *Xanthomonas campestris* pv. *campestris*.

Calcium is another important nutrient, especially in diseases that show soft rot symptoms of fruits, tubers etc. The principal role of calcium in host-pathogen relationship is the formation of calcium pectate in the cell walls thus making them resistant to degradation by wall-degrading enzymes of pathogens. Potato tubers produced in calcium-rich soil and apple fruits from plants supplied with calcium are less susceptible to attack of soft rot fungi. The role of calcium in clubroot has already been mentioned. The pathogens whose effects are suppressed by calcium include *Rhizoctonia solani, Sclerotium rolfsii, Botrytis cinerea, Erwinia carotovora* subsp. *atroseptica, and Monilinia fructicola* causing brown rot of apples and pears (Conway *et al*, 1992, 1994; Biggs *et al*, 1997). The vascular wilts caused by *Fusarium oxysporum* are also reduced by calcium, especially in conjunction with nitrogen. However, calcium favours black shank of tobacco caused by *Phytophthora nicotianae* var. *parasitica* and potato scab.

It is generally believed that if balanced nutrition is given to the plants they have better tolerance to diseases. This is not always true. Balanced fertilizers can be given only within a limit. Application of nitrogen in excess of what is required by the plant is likely to favour disease development even if it is balanced with phosphorus and potassium.

THE BIOTIC ENVIRONMENT

The interaction of pathogens with the biotic environment around them is as important as, if not more, than the physical and chemical environment. Although it had been realised before the middle of the last century, it has assumed a highly significant position among various aspects of the approaches to management of plant diseases, particularly the soil-borne pathogens.

The biotic environment around the pathogen consists of the plant itself and the microbiota associated with the aerial and soil plant surfaces and its immediate vicinity. This environment is determined by the nature of the plant under the influence of physical and other environmental factors including the management of the plant and soil through different cultural practices. The microbiota, in turn, add to the chemical environment through their metabolites. The total density of these microbiota is much larger than the pathogens. The non-pathogenic microbiota form different types of associations among themselves and with the pathogen when it lands on or comes in contact with the host surface. The association may affect the pathogens or the other component of the association favourably or unfavourably. The disease may be enhanced or suppressed.

ANTAGONISTIC ASSOCIATIONS

In the biotic environment in which the pathogen operates and disease develops, antagonistic associations are most important. Antagonism implies that in any association of two or more species atleast one of the interacting species is harmed due to activity of one or more of the rest. Antagonism is a general term used to denote harmful effect of biotic environment on a living organism. It can be between two plants, two or more members of microflora and fauna and between a plant and microorganisms not necessarily producing symptoms of a disease. The mechanisms of antagonism could be briefly stated as below:

a) **Competition** is found in the indirect rivalry of two species for some feature of the environment that is in short supply. Broadly speaking, competition could involve all kinds of interplay between organisms in which one is favoured at the expense of the other. In strict sense, competition has been defined as *"more or less active demand in excess of the immediate supply of material or condition on the part of two or more organisms"* (Clark, 1965). The microorganisms mainly compete for available oxygen and nutrients or substrates. The organisms with better tolerance for low oxygen supply or high carbon dioxides concentration can compete with others advantageously. The food supply in soil is perpetually short due to kinds and numbers of microbiota living there. With each increase in supply of food there is proportionate increase in microbial activity which depletes the supply. With this depletion, the organisms having better competitive saprophytic ability predominate over those that have poor competitive ability. The latter then have to find ways and means of survival and undergo dormancy.

b) **Competitive saprophytic ability** is defined as *"summation of physiological characteristics that make for success in competitive colonization of dead organic substrates"*. The various physiological traits are high birth rate, rapid germinability of spores, good enzyme system ensuring high metabolic turnover and rapid hyphal elongation, resistance to deleterious chemicals in soil including toxic metabolites (antibiotics) produced by other microorganisms, ability to produce own antibiotics and ability to withstand starvation, parasitism, etc.

c) **Antibiosis** is defined as *"the condition in which one or more metabolites excreted by an organism have a harmful effect on one or more other organisms.* In this type of relationship the producer of the harmful metabolite(s) does not get any direct benefit from those affected by the metabolite. But, it gets the indirect benefit of being a better competitor for substrate colonisation. Usually, antibiotic production in soil take place on dead organic substrates such as pieces of straw or seed coat buried in soil. There is evidence that antibiotic production occurs in the rhizosphere also.

The antibiotic substances become a part of the chemical environment of the soil. They diffuse through soil with soil water as a chemical substance. This diffusion is not large because eventually some component of the microflora will metabolise the antibiotic. In addition to their direct harmful effect on soil microflora, the antibiotics reduce the competitive saprophytic ability of other microorganisms. The antibiotics produced in soil are absorbed by plant roots and may induce resistance to specific pathogens. Physiology of the plant is affected and, thus, quality and quantity of the root exudates is also affected. The relationship of antibiotics with soil microflora is usually specific. All microorganisms are not harmed by a particular antibiotic. As a result, there is rapid degradation of antibiotics in the soil.

d) Exploitation is a term used to denote *parasitism and predation of one organism by another*. In this type of relationship the exploiter gets direct benefit from the attacked organism. The two terms, parasitism and predation, have basically the same effect, i.e., destruction of the host or the prey. In parasitism some sort of etiological relationship between the parasite and the host organism is established and the host is not rapidly killed. Predation physically eliminates its prey by directly feeding on it without etiological relationship.

The phenomenon of hyperparasitism or mycoparasitism is quite common on plant leaves and in soil and has been demonstrated in numerous studies as a method of biological control of many pathogens. Lysis of fungal mycelia, parasitisation and destruction of fungal sclerotia, perforation and digestion of fungal spores by bacteria and giant amoebae, destruction of nematodes by other nematodes or by fungi etc., are common features in soil, especially, the one with rich organic matter content.

The Role of Metabolites: Metabolites are produced in the environment by plants, microflora and fauna and insects. These metabolites have different functions in the biotic environment.

(a) Sterols of fungal, plant and animal origin induce (i) sexual reproduction in species of *Pythium* and *Phytophthora* (ii) formation of larger sporangia in *Phytophthora nicotianae* var. *parasitica* and (iii) increased tolerance to temperature in *Pythium*. Induction of oospore formation at the cost of vegetative growth by older tissues, with secondary thickening, was proposed as the mechanism of age based resistance against damping off caused by *Pythium*.

(b) Certain metabolites of plants induce an attracting (chemotactic) response of motile organisms. This helps in infection of the host. Chemotaxis has been demonstrated in many plant-parasite interactions including fungi, bacteria, nematodes and phanerogamic plant parasites. Mixed constituents of root exudates of sugarbeet attract zoospores of the root rot fungus *Aphanomyces cochlioides*. Organic acids (especially gluconic) and neutral fractions of the root exudates (especially glucose and fructose) attract zoospores but do not influence their germination and development while amino acid fraction helps, in some cases, in germination and growth but not attraction. In the case of *Pythium*, sugars stimulate germination of zoospores, amino acids stimulate germ tube elongation, but mixtures of several sugars and amino acids are essential for overall favourable effect. On the other hand plant species resistant to *Pythium* damping off contain amino acids in their root exudates which are toxic to the fungus. Germ tubes on leaf surface are guided to stomata by chemical emanations from the host. Similar chemotactic responses of bacteria are also reported. Chemotaxis allows bacteria to find the environment which provides them with the greatest supply of energy. The attraction of nematode larvae to their host roots is through the exudates (and carbon dioxide) emanating from roots and diffusing in the soil forming a gradient. In the infection of plant roots by the phanerogamic parasite, *Striga*, seedlings of the parasite in the soil are attracted by chemical signals from the host roots (within 2–3 mm) enabling it to attach to the root and form haustorium.

(c) Bacterial metabolites in soil are reported to induce chlamydospore formation in wilt causing *Fusarium oxysporum*. The effect of metabolites is stronger on spores than on hyphal fragments. Detrimental effect of metabolites on fungi and bacteria in soil is discussed under fungistasis.

(d) Metabolites may induce resistance to disease in plants. Antibiotics produced or synthesized on root surface not only obstruct growth and infection by root pathogens but also may induce systemic resistance in the aerial parts of the plant. This aspect is discussed under rhizosphere. Wilt resistant cultivars of pigeonpea are reported to harbour a high population of antagonistic actinomycetes.

(e) In very wet, saturated or flooded soils, anaerobic bacteria are very active, especially when organic matter is available for decomposition. These bacteria produce toxic substances which can be harmful not only to plant roots but also to plant pathogens. *Clostridium* spp. is reported to produce such substances under anaerobic conditions of decomposition of organic matter which kill larvae of nematodes within few minutes. Such conditions have been observed in flooded rice fields.

Phenomenon of Fungistasis: The term "soil fungistasis" or soil mycostasis describes the phenomenon whereby (a) viable propagules not under the influence of endogenous or constitutive dormancy do not germinate in soil under conditions of temperature and moisture favourable for germination, or (b) growth of fungal hyphae is retarded or terminated by conditions of the soil environment other than temperature and moisture. The phenomenon is encountered in soils from tropical, subtropical, temperate and sub-arctic areas. Level of fungistasis in soil differs among soil types and microorganisms. It is more pronounced in surface than in sub-surface layers of any one given soil profile. Seasonal variations also occur. Fungistasis of a given soil may reach its maximum in summer and minimum during winter.

One kind of fungistasis is supposed to be of microbial origin and is called **microbial mycostasis** or heat and sugar sensitive mycostasis because it is annulled by heat or sugar treatment of the soil. Factors that reduce soil biological activities and compensate for depletion of energy sources reduce the level of fungistasis in a soil. The other kind of fungistasis, known as **residual fungistasis** or heat and sugar resistant fungistasis, is of abiotic origin and is caused by presence of some inherent harmful component in the soil.

Among structures affected by fungistasis, spores have been studies in much detail. A vast majority of pathogenic fungi show susceptibility to this phenomenon in soil and their spores remain dormant. The same phenomenon has been reported for bacteria also under the name of **bacteriosis**. The fungistasis has three stages:

i) induction by inhibitory soil factors other than temperature and moisture but including microbial metabolites and competition,

ii) maintenance, i.e., the inhibitory factors persist, and

iii) release from fungistasis when the inhibitory factors of biotic origin are removed by addition of sugars, heat or fungicidal treatment, or influence of roots exudates.

The induction of fungistasis can be through any of the following mechanisms: nutritional sink due to intense microbial competition, production or presence of toxic materials, toxic volatiles, antibiotics and presence of self inhibitors. According to Watson and Ford (1972) the general but complex phenomenon of soil fungistasis is caused by the presence in soil environments of complex inhibitors of biotic and abiotic origin effective at low concentration of the substances stimulatory for the microorganisms, such as nutrients, present in the soil. In other words, when inhibitory factors in soil dominate over the stimulatory factors there is fungistasis. The suppression of fungistasis is effected by specific balance of inhibitor and stimulator combination.

Biological significance of fungistasis for soil biota lies in the fact that it enables an organism to survive as dormant structures through a period of adverse conditions. The effect of fungistasis on plant pathogens is twofold: (i) it promotes survival by avoiding germination under wrong conditions when infection of the host is not possible and (ii) the survival of the dormant organs is eventually limited and fungistasis hinder the **germination-sporulation cycle** that halts senescence, hence decreasing the chances of long term survival.

Fungistasis can prevent formation of conservation structures such as chlamydospores through **starvation-lysis**. Macroconidia of *Fusarium solani* f.sp. *cucurbitae* are not sensitive to fungistasis and can germinate and produce chlamydospores in soil. However, when the soil is amended with chitin, germination is checked by fungistasis induced by specific microflora in the amended soil and no germ tubes or chlamydospores are formed.

Release from fungistasis can have different consequences according to germination inducer used and biotic and abiotic factors associated with the interaction of the soil, plant and the pathogen. In presence of the host roots, release from fungistasis results in better infection while in absence of host roots it leads to destruction of the germ tube by forces of antagonism (**germination-lysis**). However, if antagonism is not there, the possibility of formation of new resistant structures cannot be ruled out (**germination-sporulation**). According to Watson and Ford (1972) any procedure, chemical or cultural, that prevents the induction and maintenance of fungistasis, stimulates inopportune release or prevents opportune release of soil fungistasis with respect to a particular pathogen, will tend to reduce the inoculum density of the pathogen in soil. However, the reverse may also happen.

SYMBIOTIC ASSOCIATIONS

The plant symbionts of practical importance in crop production are mycorrhizal fungi and nitrogen fixing root nodule bacteria (*Rhizobium* spp.). Both are important in plant nutrition. The mycorrhizas help in uptake of mineral nutrients, especially phosphorus, from the soil while the root nodule bacteria of legumes fix atmospheric nitrogen and transfer it to plant.

The mycorrhizal roots are often resistant to fungal and nematode pathogens (Caron, 1989; Hussey and Roncodori, 1982; Schoenbeck, 1979). The mycorrhizal symbionts and root pathogens such as species of *Pythium*, *Phytophthora*, *Rhizoctonia* and *Fusarium*, are intimately associated with succulent, fine feeder roots. Where an ectomycohhirzal covering develops on the roots before they are invaded by the pathogens, the latter rarely succeed in penetration. In addition, the mycorrhizas change the microbial composition of the rhizosphere which may be disadvantageous for the pathogen. Most studies on ectomycorrhizal associations and root pathogens have been carried out with forest trees but there are some examples of fruit trees where presence of ectomycorrhizas inhibits several fungi such as *Phytophthora* and *Pythium*.

The endomycorrhizal fungi form haustorium-like structures, called arbuscules, in the host cells and hyphal swellings (vesicles) in or between host cells, hence they are called vesicular arbuscular mycorrhiza (VAM). The mycorrhizal fungi can increase host tolerance to the root knot nematode under field conditions. Prior establishment of *Glomus faciculatus* on roots of tomato seedlings significantly reduces the number and size of galls caused by *Meloidogyne incognita* and *M. javanica* when the seedlings are transplanted onto a nematode infested soil (Bagyraj *et al*, 1979). Inoculation of alfalfa with vesicular arbuscular endomycorrhiza improves plant growth and reduces the number of nematode and development of adults in roots of a susceptible cultivar. Wilt of alfalfa caused by *F. oxysporum* f.sp. *medicaginis* and *Verticillium albo-atrum* is low in plants with VAM. The number of propagules of the pathogens in the soil is also reduced (Hwang, 1992). Mycorrhizal citrus roots show significantly less damage from *Phytophthora parasitica* than non-mycorrhizal roots in low phosphorus soils but not in high phosphorus soils (Davis and Menge, 1980). Resistance of tomato to tomato mosaic virus and potato virus X, of beans to rust (*Uromyces phaseoli typica*) and anthracnose (*Colletotrichum lindemuthianum*) and of cucumber to powdery mildew (*Erysiphe cichoracearum*) is decreased by endomycorrhizas. Although many plant diseases are reduced by endomycorrhizas (Table 13), in some cases the plants are weakened by these associations and pre-disposed to pathogen attack. The favourable effect of mycorrhizas on the host against a disease is through general improvement in plant vigor as a result of better nutrition, mechanical barrier against penetration, and production of antibiotic substances.

TABLE 13: Effect of Endomycorrhizae on Soil-borne Pathogens of some Crop Plants (Schoenbeck, 1979).

Crop	Pathogen	Effect of mycorrhiza
Soybean	*Pythium ultimum*	None
	Phytophthora megasperma	Few plants killed
	Meloidogyne incognita	Fewer galls, increased yield
Cotton	*Thielaviopsis basicola*	Less stunting
	Meloidogyne incognita	Less stunting
	Pratylenchus brachyurus	Fewer larvae on roots
Tobacco	*Olpidium brassicae*	Reduced infection
	Thielaviopsis basicola	Less stunting, Chlamydospore formation inhibited
	Heterodera tabacum	Fewer nematodes
Tomato	*F. oxysporum* f.sp. *lycopersici*	Less stunting and infection
	Meloidogyne incognita	Fewer nematodes
Cucumber	*F. oxysporum*	Less stunting and infection
	Meloidogyne incognita	Fewer nematodes increased fresh weight

The effect of symbiotic nitrogen fixing bacteria on plant pathogens has not been studied much. At pH 7.2 and 7.6, but not at pH 5.2, ample rhizobial nodulation of soybean greatly reduces, or even eliminates, the root rot caused by *Fusarium*. Simultaneous inoculation of soybean in pots with *Rhizobium japonicum* and *Phytophthora megasperma* protects the plants from root rot (Tu, 1978). Similar results with same fungus was reported for lucerne. This protective effect of root nodulation is restricted only to prior colonization of roots by the bacterium.

SYNERGISTIC ASSOCIATIONS

Earlier in the chapter on dispersal of plant pathogens it was mentioned that many soil-borne plant viruses are transmitted by selected groups of parasitic nematodes and some chytrid fungi. Apart from transmission, synergism between plant pathogens has been reported in a large number of diseases. It is not surprising since many fungal and nematode parasites of plants are present in the same rhizosphere. There are proven examples where the damage done by a complex of two or more pathogens is much more than what the pathogens could cause independently. Such diseases have often been termed as **disease complexes.**

Pythium ultimum and *F. oxysporum* f.sp. *pisi* race 2, both pathogens of pea, cause only slight reduction in plant growth or rate of maturity in a susceptible variety when present alone but together they cause severe stunting and early death of the plant. Plants of wilt resistant variety are only slightly stunted and do not die prematurely when inoculated with both pathogens. This interaction, only in a susceptible variety, suggests that combined effect of root rot and vascular infection is more lethal than vascular infection alone.

Of the numerous references describing effect of root knot nematodes on soil-borne fungal parasites of roots, the common is the reference to vascular wilts caused by *Fusarium* spp., especially on wilt resistant plants. The incidence of wilt of cotton (*F. oxysporum* f.sp. *vasinfectum*), tomato (*F. oxysporum* f.sp. *lycopersici*), pea (*F. oxysporum* f.sp. *pisi* race 1), sweet potato (*F. oxysporum* f.sp. *batatas*) and many others has been found to increase when *Meloidogyne* spp. especially *M. incognita* have invaded the roots earlier (cf. Singh and Sitaramaiah, 1994). The increase in wilt incidence is related to the number of nematode larvae infesting the plants before invasion of the fungus, race of the nematode, and resistance of the host to the fungus, the effect being more pronounced in resistant cultivars. It would appear that wounds caused by the nematode are directly responsible for increase in wilt incidence but the interactions are much more complicated and less understood. Propagule counts of *Fusarium* in the rhizosphere of tomato infested by root knot nematode increase and actinomycete counts decrease suggesting changes in root exudates. Root galls of tobacco are much more rapidly attacked by *Fusarium* than healthy roots. Breakdown of resistance or inconsistency in the pattern of inheritance of resistance to wilt are most important effects of root knot nematode invasion. Changes in host physiology due to activities of the nematode are given as a reason for resistance breakdown.

Black shank, a destructive stem and root disease of tobacco caused by the fungus *Phytophthora nicotianae* var. *nicotianae* is greatly enhanced by earlier invasion of *M. incognita*. The fungus needs no assistance to infect susceptible tobacco cultivars. But where the interaction between the nematode and the fungus occurs the resistant cultivars suffer heavy losses. There is extensive colonisation of the hypertrophied and hyperplastic regions of the root galls and the fungus even colonises the anterior and posterior regions of the nematode females as well as the egg masses.

Bacterial wilt of potato, tomato and eggplants is enhanced by the presence of root knot nematodes. The interaction is mainly the effect of wounding of roots that permits easy entry of the bacteria. The bacterium shows a preference for the giant cells which degenerate following bacterial invasion (cf. Singh and Sitaramaiah, 1994). Cauliflower disease or fasciation disease of strawberry is a good example of how a nematode-bacterium interaction is essential for full expression of symptoms. In this disease the interaction involves the foliar nematode *Aphelenchoides ritzemabosi* and the bacterium *Rhodococcus fascians*.

The nematode alone produces localised feeding sites on the crown buds. When the bacterium is also present, proliferation of auxiliary buds on the affected crown occurs and leaf galls are formed. It is believed that the nematode acts as a vector of the bacterium to carry it to the crown buds which the bacterium could not enter by itself. In the red ring disease of coconut palm caused by the nematode *Rhadinophelenchus cocophilus* there is close relationship between the nematode and the palm weevil (*Rhynchophorus palmarum*). The nematode lives in the body of the weevils for several days. The infection of the crown is initiated by the weevils which introduce the nematode in the plant (cf. Singh, 2000).

One of the best studied nematode-bacterium interaction is the yellow slime or ear rot disease of wheat in which the nematode *Anguina tritici* and the bacterium *Rathayibacter tritici* are involved. The nematode causes ear cockle or seed gall disease. The yellow slime caused by the bacterium cannot occur unless the nematode is also present. The nematode acts as a carrier of the bacterium.

Apart from the mechanism of entry of fungal or bacterial pathogens by wounds caused by nematodes as primary pathogens, there are also physiological changes in the host that promote the effects of the interactions between nematodes and other pathogens. During pathogenesis, nematodes break down host proteins by their pectolytic enzymes into free amino acids whose concentration increases in the host cells and also in the root exudates. The amino acid tryptophan is metabolised to IAA resulting in hormonal imbalance which is not confined only to the site of action but also affects farther tissues. These hormonal imbalances and associated physiological changes in the plant attacked by nematodes as primary pathogen can influence pathogenesis by the secondary pathogens (fungus or bacterium) in one or more of the following ways:

1) Characteristics of the rhizosphere microflora can be changed due to change in quality of root exudates. The microflora antagonistic to the secondary pathogen in the rhizosphere may be suppressed. Actinomycetes in the rhizosphere of tomato attacked by root knot nematode are decreased.

2) Resistance to vascular wilt caused by fungi and bacteria mainly lies in the endodermis which is normally avoided by these pathogens. The endodermis contains large quantities of anticmicrobial compounds such as phenols, napthols and anthrols which, in later stages, give rise to benzoquinones, naphthoquinones and anthroquinones, some of which are highly antifungal and antibacterial. Invasion by root knot nematodes results in pericyclic origin of syncytia (nurse cell system) and galls and the endodermis is not properly differentiated in the affected area. The absence of endodermis removes the barrier against invasion by vascular wilt fungi and bacteria.

3) Due to change in physiology of the host, the mechanism inhibiting production of toxins by the wilt pathogens may be destroyed thus enabling the wilt

fungus to break the host resistance.
4) Tissues affected by the nematode enzymes may become preferred food for the invading secondary pathogen. Root galls in tobacco are much more readily attacked by *Fusarium* than the healthy roots. In the interaction between *Phytophthora* and *Meloidogyne* in tobacco roots the fungus extensively colonises the hypertrophied and hyperplastic regions of the root galls.

THE SUPPRESSIVE OR RESISTANT SOILS

Soils which permit the development of a disease are termed *conducive* or *sensitive*. Soils in which this property does not exist, does not reach a serious proportion, or diminishes if the susceptible host is cultivated as a monocrop, are termed *suppressive, resistant, immune* or *long life*. This property of some soils is a system of natural biological control of soil-borne pathogens.

The reduction in take-all of wheat (*Gaeumannomyces graminis*) in monoculture was one of the early example of development of suppressiveness in soil. There are many more examples (table 14). When potatoes are cultivated continuously in scab infested soil, the soil later becomes resistant and further cultivation of potato in the field is free from common scab (*Streptomyces scabies*). This suppressiveness of the soil is transferable to other soils and, therefore, is considered of biological origin. The assumption is supported by following observations.

1) Suppressiveness is lost by irradiation, steaming, or autoclaving. It is diminished by fumigants and some fungicides. It is re-established if the treated soil is inoculated with a small amount of the original soil.

2) A sensitive or conducive soil may or may not be converted into resistant or suppressive soil. If it is convertible, it can be to varying degrees. Conversion is contagious process obtained by mixing a suppressive soil with a previously disinfested convertible soil, even sometimes in as low proportion as 1 to 9 (w/w). If the sensitive soil is not disinfested, the suppressiveness of the mixture is relative to the ratio of suppressive soil to sensitive soil. Suppression proceeds as if the inoculum is multiplying in the disinfested rather than in the normal soil. The suppressiveness can be lost following addition of nutrients.

3) The acquisition of the ability of suppression is a progressive phenomenon. If a suppressive soil and a sensitive soil which has been disinfested and re-inoculated with *F. oxysporum* are mixed, the suppressiveness becomes complete only after a incubation of several months. This indicates that one or more biological factors present in the suppressive soil progressively colonise the niche created by the disinfestation of the conducive soil.

4) Ability to suppress is relatively specific in nature. In general, a soil which is suppressive to one form is suppressive to all other forms of the same

species. But it is not suppressive to other genera or species of the pathogen. There is evidence that some soils are suppressive to only some strains of the same pathogen and relatively less suppressive or not suppressive to other strains.

5) Suppressiveness is related to inoculum density of the pathogen. A conducive soil may permit infection by 5 to 500 propagules/g soil while a highly suppressive soil may give protection against inoculum densities under 3200 propagules/g soil and another soil may give protection only when the density is three times weaker.

Soil characters indirectly influence suppressiveness of the soil. In general, conducive soils are lighter in texture (with low organic matter content) than suppressive soils. The soil characters seem to operate through the biological factors. However, Ko (1985) had reported a type of suppressiveness in which no biological agency was involved. The suppressive factor was contained in the clay, silt or sand fraction of the soil and was not destroyed by autoclaving or flaming of the soil.

TABLE 14: Some Root Diseases and Pathogens Inhibited by Suppressive Soils.

Crop	Disease	Pathogen
Banana	Wilt	F. oxysporum f.sp. cubense
Bean	Wilt	F. solani f.sp. phaseoli
Cucumber	Wilt	F. oxysporum f.sp. cucumerinum
Cucurbits	Black root rot	Phomopsis sclerotioides
Flax	Wilt	F. oxysporum f.sp. lini
Lettuce	Root rot	Olpidium brassicae
Melon	Wilt	F. oxysporum f.sp. melonis
Pea	Wilt	F. oxysporum f.sp. pisi
Pigeonpea	Wilt	Fusarium udum
Potato	Common Scab	Streptomyces scabies
Radish	Wilt	F. oxysporum f.sp. raphani
Sweet potato	Wilt	F. oxysporum f.sp. batatas
Tomato	Wilt	F. oxysporum f.sp. lycopersici
Various vegetable crops	Damping off	Pythium ultimum and other spp.
	Wilt	Ralstonia solanacearum
Wheat	Root rot	Fusarium culmorum
	Bare patch	Rhizoctonia solani
	Take all	Gaeumannomyces graminis
	Stripe	Cephalosporium graminearum

In some cases, ability to suppress is independent of previous cultivation. The soil may be suppressive even if the susceptible crop has not been cultivated for 30 years in that soil. However, cultivation of the susceptible crop further increases the suppressive ability. Regular application of organic matter to the soil maintains the suppressiveness provided it is not specifically stimulatory for the pathogen. In other cases, the soil is initially conducive but develops suppressiveness with continued cultivation of the susceptible crop. In the first year the crop is severely attacked. During the following years the disease intensity declines to a very low level, even if the soil is artificially inoculated with the pathogen. Crop rotation hinders development of suppressiveness.

The mechanisms involved in suppressiveness include phenomena of fungistasis, hyperparasitism, predation as well as antiboisis (antibiotic production) and nutritional competition. However, antagonists isolated from suppressive soils are difficult to induce suppressiveness when artificially inoculated into a conducive soil unless the physical, chemical and biotic conditions of the soil are the same as in the suppressive soil. The antagonists involved in the induction of soil suppressiveness are spread over many genera and species of fungi (*Trichoderma* spp.) and bacteria (*Pseudomonas* spp.; *Bacillus subtilis*). It is generally believed that prolonged coexistence of a pathogen with soil microbiota may lead to development of organisms which may be directly antagonistic to the pathogen or indirectly shift the existing balance between the pathogen and its antagonists so that the balance becomes unfavourable for the pathogen.

ORGANIC AMENDMENTS OF SOIL

Modification of physical, chemical and biotic environment of the soil through addition of decomposable organic matter has been found to influence many plant diseases. In cultivated fields, normal cultural practices and substantial quantities of organic matter in the form of crop debris (roots, dry leaves and straw, etc.). In addition many farming practices such as green manuring, use of oil-cakes for manure, composted materials, wood sawdust, etc., also periodically add organic matter to the soil. Organic amendments may influence the soil and the incidence of a disease in one or more of the following ways:

1) Alteration of soil structure: Decomposition of leaves, roots, farmyard manure, green manure, sawdust, etc. loosens the soil and improves its aeration and water holding capacity. These improvements in soil conditions help better development of plant root system. The enhanced root system helps the plants in two ways: nutrient uptake capacity is increased and the plant becomes more vigorous thus capable of avoiding effects of many diseases; and roots damaged by parasites are quickly replaced by new roots thus harmful effect of root diseases is delayed or masked. The improved tilth, aeration and water holding capacity help activity of microorganisms also which may have favourbale effect on the plant growth and unfavourable effect on the pathogens.

2) Biotic environment is altered: Intense microbial activity in amended soil improves chances of development of antagonistic microorganisms which can reduce the population of pathogens through competition, antibiosis, parasitism and predation. Legume green manuring is known to reduce the root rot caused by *R. solani* and common scab of potato through increased activity of *Bacillus subtilis*. In the enhanced microbial activity there may be such bacteria as fluorescent psuedomonads and such antagonistic fungi as *Trichoderma* spp. and *Gliocladium virens* that act as strong antagonists of pathogens. A variety of bacterial antagonists, such as *Flavobacterium balustinum*, *Pseudomonas putida* and *Xanthomonas maltophilia*, which are rapid colonizers of organic matter, also play important role. Combinations of antagonists are more effective than single antagonists in suppression of some pathogens.

3) Biochemical effects: As stated above, decomposition of organic matter leads to different types of biological and biochemical reactions which can harm the plant pathogens. At the same time many components of the decomposition products can be absorbed by plant roots. This induces biochemical resistance to pathogen activity within the plant. Development of *Globodera rostochiensis* in potato plants grown in soil supplied with farmyard manure is slow. The metabolites of the microorganisms have the capacity to trigger induced systemic resistance in plant. Composts induce systemic acquired resistance in cucumber against *Pythium ultimum*, *P. aphanidermatum* and *Colletotrichum orbiculare* causing anthracnose (Zhang *et al*, 1996). The resistance was associated with increased peroxidase activity.

The list of diseases suppressed by organic amendments is fairly large and ever increasing. These include root rot of pea (*Aphanomyces euteiches*), root rot of beans (*F. solani* f.sp. *phaseoli* and *Thielaviopsis basicola*), wilt of banana (*F. oxysporum* f.sp. *cubense*), wilt of cotton (*F. oxysporum* f.sp. *vasinfectum*), wilt of pea, wilt of pigeonpea (*Fusarium udum*), root rot of cotton (*Phymatotrichum omnivorum*), root rot of avocado (*Phytophthora cinnamomi*), *Rhizoctonia* diseases of potato, beans, lettuce, etc., and root knot and other nematodes.

TABLE 15: Effect of Organic Amendment of Soil on Incidence of Black Scurf of Potato (*Rhizoctonia solani*). Average of Two Years Data.

Amendments	Disease index
None (check)	0.88
Castor bean cake	0.51
Margosa cake	0.40
Groundnut cake	0.27
Mustard cake	0.03
Wood sawdust	0.25
Fertilizers (120-60-60)	0.55
Sawdust + fertilizers	0.13

Singh, R.S. *et al*, 1972

Green manuring with a crucifer for two consecutive seasons before peas is reported to reduce root rot (*Aphanomyces euteiches*) by reducing the number of propagules of the fungus in the soil (*cf*. Singh, 2001). Decomposition of crucifer leaves releases toxic sulphur compounds. The pathogen is susceptible to such soil microflora as *Trichoderma* spp., *Penicillium oxalicum, Pseudomonas cepacia, Pseudomonas fluorescens* and *Bacillus subtilis* (Parke, *et al*, 1991). *Pseudomonas cepacia* is antagonistic to *Pythium* also. Some amendments such as oat straw, corn stover, lucerne hay sugarcane residue suppress the pathogens by germination-lysis process. Some immobilize nitrogen and some release ammonia for toxicity to the pathogens. Different mechanisms are discussed later in the chapter on disease management.

MICROBIAL INTERACTIONS ON THE PLANT SURFACES

A. THE PHYLLOPLANE

Existence of epiphytic microflora on plant surfaces including leaves and flowers is a natural phenomenon (Preece and Dickinson, 1971). These aerial organs of the plant act as a stage for landing of air-borne propagules of saprophytic (and also pathogenic) fungi and bacteria. The exchange of gases and guttation fluids help their survival on the surface for different durations. The interaction between the plant, the saprophytes and the parasites takes place on this surface (Dubos, 1987). The non-pathogenic microflora on leaves and flowers normally do not harm the plant. On the other hand, they play some role in resistance of the plant to foliar pathogens. There are many studies where their presence has been cited to explain the reduction of disease incidence (Blakeman and Fokkema, 1982; Narula and Mehrotra, 1987). Brown leaf sport of rice (*Drechslera oryzae*), leaf spot of rye (*Helminthosporium sativum*), fire blight of apple (*Erwinia amylovora*), Alternaria spot of tobacco and many other foliage diseases are less severe when the normal epiphytic microflora is allowed undisturbed than when the microflora has been eliminated or reduced by some treatment such as spray of broad spectrum fungicides. Possibly, the epiphytic microflora compete with the pathogenic microflora for nutrition. They also act through antibiosis and hyperparasitism and through induction of resistance in the plant. In the latter case, it is reported that the marked increase in the respiratory rate of barley leaves due to inoculation with saprophytic fungi is due to energy consuming defence reactions which can induce resistance to subsequent infection by powdery mildew. Strawberry gray mold rot caused by *Botrytis cinerea* is significantly reduced by the presence of *Gliocladium roseum*, a versatile antagonist (Sutton *et al*, 1997). Kalita *et al* (1996) isolated a stain of *Bacillus subtilis* and *Aspergillus terreus* from phylloplane of citrus which provided protection against citrus canker (*Xanthomonas axonopodis* pv. *citri*). *Erwinia herbicola*, a common bacterial epiphyte grows more rapidly than the citrus canker bacterium and eventually causes quick decline of the pathogen. Certain strains of *Erwinia herbicola* and *Pseudomonas fluorescens* have the potential to suppress the fire blight bacterium (*Erwinia amylovora*). They are easily disseminated by honey bees (*cf.* Singh, 2000). In Germany, it was observed that leaves sprayed with water extracts of composted horse manure-straw-soil mixture develop resistance to downy mildew (*Plasmopara viticola*). The extract had no direct inhibitory effect against the pathogen (Weltzien and Ketterer, 1986). Subsequently, in attempts on biological control of powdery mildew of grapevine (*Uncinula necator*) acqueous extract of compost or fermented cowdung had been used with positive results. Populations of *Bacillus, Pseudomonas, Serratia, Penicillium* and *Trichoderma* were enhanced on the host surface and they provided biological control of the pathogens (Schlosser, 1994).

B. THE RHIZOSPHERE

Rhizosphere is the zone of influence of plant roots in the soil. This influence is the result of root exudates and carbon dioxides emanating from the roots and diffusing into the soil around the roots. This influence extends from the root surface (rhizoplane) to several millimeters to centimeters. The total area of the zone of influence varies with the nature of the root growth and plant density. In a wheat field, the entire field may be considered as rhizosphere. The microbial population, its density and variety in the rhizosphere is always higher and more complex than in soil away from roots. A rich flora and fauna of soil microbiota lessens the danger of epidemic outbreaks of diseases caused by soil-borne pathogens (Palti, 1981). Biological complexity of the soil assures associative, competitive and antagonistic relations which limit population explosion and thus bring about balance. The more numerous the kinds of organisms, and the greater their number, or perhaps the shorter their generation time, the more stationary is the balance.

The quantity and quality of the rhizosphere microflora is influenced by the amount of sloughed off root tissues and the quantity and quality of root exudates or exudates from the germinating seed. These, in turn, are influenced by physical, chemical and biotic environments in which the plant is growing, the soil type, and soil or plant treatments including tillage which influence root growth, nutrient uptake and metabolism of the plant and diffusion of the exudates in soil. Thus, all factors which influence plant metabolism affect the quantity and quality of root exudates. These exudates determine the rhizosphere microflora which may not be only antagonist of the pathogen but also favourbale for their activity on the root surface including formation of infection structures and movement of the infective propagules toward the root and their accumulation on the root surface. The rhizosphere may influence disease by one or more of the following mechanisms:

1) Availability of inorganic minerals and nitrogen to the plant and the pathogen.
2) Production of beneficial or harmful microbial metabolites affecting the plant and the pathogen.
3) Induction of broad spectrum systemic resistance in the plant.
4) Interaction with symbionts and parasites. This mainly includes adverse effect

of rhizosphere microbiota and symbiosis for nitrogen fixation, non-symbiotic nitrogen fixing bacteria and associations among parasites, mycorrhiza, etc.

The role of rhizosphere bacteria in plant growth improvement, even in absence of a pathogen, and their role in suppressing root diseases has been extensively studied during the last two decades (Weller, 1988; Van Loon *et al*, 1998; Sticher *et al*, 1997). The term **rhizobacteria** was coined for bacteria having the ability to aggressively colonize the plant roots. The rhizobacteria are put in two categories: the **plant growth promoting rhizobacteria (PGPR)** and the **deleterious rhizobacteria and fungi (DRMO)**. The increased plant productivity in no-disease situations results in a large part from the suppression of deleterious microorganisms (Schippers *et al*, 1987) and often through induction of metabolic changes in the plant. The fluorescent species of *Pseudomonas* have received special attention. Strains of *Pseudomonas fluorescens* and *P. putida* applied to potato seed pieces (Burr *et al*, 1978; Coyler and Mount, 1984; Geels and Schippers, 1983; Kloepper and Schroth, 1981; Kloepper *et al*, 1980) and to sugar beet and radish increase the yield of potato by 5–33%, of sugar beet by 4–8 tons/ha, and root weight of radish by 6–144% (*cf*. Weller, 1988). The bacterization of potato seed pieces reduces the population of soft rot bacteria (*Erwinia carotovora*) on roots and daughter tubers (Kloepper, 1983, Xu and Gross, 1986). The PGPR colonize the root system and prevent infection by major root pathogens as well as suppress the DRMO which damage the root tips and root hair without producing disease symptoms but have adverse effect on root functioning and, therefore, yield of the crop (Weller, 1988).

Some of the PGPR suppress cyanide producing organisms in the root zone of potato (Bakker and Schippers, 1987). Cyanide production in root zone harms the plant root functioning. Some deprive the pathogens of iron supply through siderophores, such as pseudobactins (Leong, 1986). Siderophores are extracellular, low-molecular weight-iron (III)-transport agents produced by most aerobic and facultative anaerobic microorganisms under low-iron stress. They selectively complex iron (III) with very high affinity. The function of siderophores is to supply iron to the cell (Bakker *et al*, 1991, 1993). The antagonism between PGPR and DRMO is mediated by competition for iron (Buyer and Leong, 1986). Not only bacteria but antagonistic fungi such as *Trichoderma harzianum* are also involved in plant growth promotion (Chao *et al*, 1986). The fungus produces a growth regulating factor that increases the rate of seed germination and dry weight of shoots irrespective of pathogen control (Windham *et al*, 1986).

Often, the rhizobacteria, without being antagonistic, have given better disease control than such antagonistic as *Trichoderma* and *Pseudomonas fluorescens* (Jubina and Girija, 1998). *Plasmodiophora brassicae* (clubroot of cabbage) has been suppressed by rhizobacteria obtained from host roots from widely located field (Bhattacharya and Pramanik, 1998). Indigenous rhizobacteria (strains of *Pseudomonas fluorescens, Pseudomonas putida* and *Pseudomonas alcaligenes*) used for cotton seed treatment not only improved germination and seedling growth but also suppressed the cotton blight bacterium (*Xanthomonas axonopodis* pv. *malvacearum*) on the leaves (Mondal *et al*, 1999) suggesting induction of systemic acquired resistance. Non-pathogenic rhizobacteria can induce systemic resistance in plants (van Loon *et al*, 1998). Such resistance has been demonstrated in bean, carnation, cucumber, radish, tobacco and tomato under conditions in which the inducing bacteria and the pathogen were kept separated from each other.

FORECASTING OF PLANT DISEASES

The ultimate objective of the understanding of principles involved in host x pathogen x environment interactions is its practical application for minimising the losses from plant diseases through adoption of control measures which are economical and effective. Diseases are not regular in their occurrence and they may not be always equally severe. The losses depend on the time of appearance of the disease and progress of its incidence in the crop. While the common cultural practices for disease management integrated with agronomic crop culture practices do not involve extra resources (money, labour, time) the use of chemicals for disease management costs extra money. In the absence of any economic loss, the additional costs increase the cost of production of the crop which is a loss to the grower as well as the consumer. Therefore, there must be a system for deciding whether, when and where a particular management practice can be economically effective and unavoidable. Necessity, economy and efficacy of control measures can be determined if there is a system of prognosis or forecasting the extent and time of occurrence of plant diseases. **Disease forecasts** and **action thresholds** are the tools designed to help the farmer in enhancing the efficiency and adequacy of his disease management efforts (Fry, 1982, Sutton *et al*, 1984, Young *et al*, 1978). *Forecasting involves all the activities in ascertaining and notifying the farmers in a community that conditions are sufficiently favourable for certain disease, that application of control measures will result in economic gain, or that the amount of disease expected is unlikely to be enough to justify the expenditure of time, energy and money for control.* Action thresholds are levels of disease or pathogen population at which the management activity is warranted. This requires complete knowledge of epidemiology, that is, development of the disease in plant population under the influence of the factors associated with the host, the pathogen and the weather. Thus, forecasting is actually **applied epidemiology**.

For the farmer, the practical advantage of disease forecasting is that he can decide whether the crop should be sprayed with a fungicide or not and if to be sprayed whether immediately or he can wait for some days before starting the operation. Thus, he may avoid mandatory calender-based sprays that are recommended and this may result in saving the cost of one or more sprays. Forecasting requires adequate technology for detection and estimation of the amount of initial inoculum of the pathogen, adequate understanding of environment and host influences on the pathogen and adequate understanding of the pathogen and disease dynamics. It also requires proper equipments for measuring components of the environment within and outside the crop. Weather data are common components for all the forecasting systems and since weather is not identical everywhere the forecasting systems differ from place to place (Bourke, 1974).

Plant diseases for which forecasting can be of practical value to the farmer meet the following requirements: (i) the disease causes significant economic damage in terms of loss of quantity and quality of the produce in the area concerned, (ii) the onset, speed of spread, and destructiveness of the disease is variable, mostly due to dependence on weather which is variable, (iii) control measures (curative and preventive) are known and can be economically applied by the farmer when he is told to do so, (iv) information on weather-disease relationship is fully known. Most crucial are the availability of effective control measures and data on disease-environment interactions.

In absence of the host it is very difficult or impossible to detect small populations of plant pathogens. When moderate to large populations of the pathogen are present, inoculum propagules of soil-borne pathogens (fungi and nematodes) are estimated after extraction or trapping from soil. Airborne fungal spores and insect vectors are estimated by trapping them in various devices. In such diseases as cereal rusts and rice blast where the inoculum is not locally present the incoming spores can be trapped on early sown trap crop or on glass slides smeared with greese.

Possibility of occurrence of a monocylic diseases such as loose smut of wheat can be predicted on the basis of primary inoculum that could be detected by estimating per cent infected or contaminated seed in random samples and the extent of distribution of such seed. In such diseases one can give a rough estimate of possible extent to which primary infection of a disease can occur in an area. However, this type of forecast is of minor importance. It warrants steps on the part of the state and seed supplying agencies. Theoretically, even in a soil-borne disease long term prediction of prevalence of the disease in a monoculture system can be made if data are available about the rate of multiplication of the pathogen in the soil, likely weather during the coming years (based on long term observations of the past) and possible changes in the cropping system. Most of the diseases in which forecasting service has helped the farmer are polycyclic in nature. They have rapid cycles in the standing crop and have the potential to develop into an epidemic if suitable control measures are not followed. The examples of such diseases are late blight of potato (*Phytophthora infestans*), blast of rice (*Pyricularia grisea*), leaf rust of wheat (*Puccinia recondita*), downy mildew of cucurbits (*Pseudoperenospora cubensis*), downy mildew of lima bean (*Phytophthora phaseoli*), downy mildew of grapevine (*Plasmopara viticola*), blue mold of tobacco (*Peronospora tabacina*), and apple scab (*Venturia inaequalis*).

In developing a plant disease forecasting system, several characteristics of the pathogen, host and, most important, environment are taken into account. The host associated factor are highly variable. These include prevalence of a susceptible variety in the area, response of the plant at different stages of its growth to the activity of the pathogen and density and distribution of its cultivation. Crops on a limited scale or at scattered locations may escape serious disease incidence while a dense population of a genetically uniform susceptible variety invites epidemics. Cultivation practices such as amount of irrigation, amount and quality of fertilizers also are added to host factors. Among the factors associated with the pathogen are almost all stages of the disease cycle, *viz.*, amount of surviving initial inoculum its dispersal, spore-germination, primary infection, sporulation on the infected host, re-dispersal of spores and pattern and distance of secondary spread. Most of these stages in disease cycle are dependent on weather or meteorological factors. The initial level of inoculum from which the infection builds up is the most crucial factor but is least known in most diseases. It is difficult to forecast the amount of primary inoculum surviving at remote places (such as in cereal rusts and powdery mildews in India) and brought to the locality by wind to initiate the disease. Most prediction systems are necessarily based on the assumption of a uniform starting level of inoculum from season to season. This is reasonable assumption because when established sanitary precautions are taken to prevent initiation of a disease, many diseases fail to establish in the field or the area. The above listed factors make the number of variables in the pathogen parameters also rather large. Although they can be studied separately under a fixed set of environmental conditions, the variations in the environment make them more complex. Weather favourable for the pathogen activity has been one of the bases of developing forecasting systems but unfavourable weather after a spell of favourable weather, which may happen within 24 hours or in weeks, is also important because the former can undo the effects of the latter. Behaviour of the pathogen in the field (its natural environment) is more important than its behaviour in laboratory under uncompetitive conditions. In other words, the response of different stages of the pathogen to the natural environment should form a more realistic background for a forecasting system. Where a particular information requires the study of a simple interaction, such as germination of spores at different temperatures or in different relative humidities, laboratory studies also help.

The environmental factors (weather and climate) affect host as well as the pathogen and are the final determinants of the outcome of host-pathogen interactions. The weather records taken into consideration for developing a forecasting system are from three locations, viz., in-crop environment (actual temperature, humidity, light conditions under the crop canopy), local climate of the filed or farm outside the crop canopy, and weather conditions recorded at standard meteorological stations, local or over a wide geographic region. The difference between conditions within and outside the plant cover depend largely on the nature and density of the crop and on the level of moisture on the underlying ground. Differences tend to be greater on calm sunny days. Wind helps in establishing homogeneity. Conditions inside and outside the crop agree closely in cloudy, wet weather, which is the weather of greatest importance in most fungal diseases. Relative humidity is generally high in the crop than outside especially if the crop canopy is continuous due to thick planting and high doses of fertilizers and if soil moisture is high. Wind helps towards greater uniformity between local climate and the macroclimate recorded by meteorological stations. It also plays the role of deciding direction, speed, distance and time of fall of spores. In those diseases in which spores are disseminated by wind from remote sources of survival, wind is very important determinant of disease incidence and can be effectively used in forecasting.

The need for continuous monitoring of several components of weather (temperature, relative humidity, leaf wetness, rain, wind and cloudiness) at various locations in the crop canopy or on plant surfaces in one or more fields creates difficulties in weather monitoring. Several types of traditional and battery operated electrical instruments are used to measure the various weather factors. Temperature measurements are made with thermometers, hygrothermographs, thermocouples and thermisters. Relative humidity is measured with a hygrothermograph, ventilated psychrometer etc. Leaf wetness is monitored with string-type sensors that constrict when moistened or slackened when dry and leave an ink trace in the process or break an electrical circuit. In modern weather monitoring systems, the weather sensors are connected to data-logging devices. Computers are now used to store, analyse and reproduce the data. These equipments are costly and can not be afforded by every farmer. Thirty three per cent of land holding in world agriculture are less than 1 ha size. Small farmers till 65% of the cultivated land area of the world. This segment of the farming community cannot provide basic weather data for disease forecasting.

An effective forecasting system uses a correlated information on the above mentioned factors for the host, the pathogen and the environment. By collecting data for several years on weather conditions during the crop season and pattern of disease incidence, supported by laboratory studies, correlation between weather and disease can be determined and on that basis forecast can be made whenever meteorological conditions tend to become favourable for serious disease incidence.

SOME EXAMPLES OF DISEASE FORECASTING

Late blight of potato *(Phytophthora infestans):* Due to its devastating effects and its effective control with fungicides the late blight is one of the diseases in which forecasting has been tried since long in Britain, Europe and USA (Miller and O'Brien, 1957). The initial (primary) inoculum of late blight is not very large. In most regions of the world, the tubers (not most of the produce) are the source of primary inoculum. But the pathogen has the capacity to multiply rapidly in the crop if the weather conditions are favourable and susceptible host surface is available. In this disease, therefore, the number of cycles of secondary inoculum production are important for forecasting. The disease is initiated by a few infected plants arising from infected tubers. These plants produce sporangia as secondary inoculum. Survival and availability of this inoculum during early stages of the crop growth is crucial in deciding late blight behaviour in the season. Sporangia are formed at a relative humidity of nearly 100% or at least more than 90% at 18°–25°C for at least 6 hours or at 12°–15° for at least 12 hours. Sporangia lose viability within 1–2 hours at 20–40% relative humidity or in 3–6 hours at 50–80% RH. Thus, the favourable conditions for their germination must occur soon after their formation. These favourable conditions are the presence of moisture on the leaves and a fairly low temperature (10°–15°C) for 0.5–2 hours. Rainfall is not necessarily a source of moisture. In northern India, dew deposit during December–January provides the requisite level of moisture. Infection of potato leaves by zoospores liberated by sporangia requires at least 2–2.5 hours at 10°–25°C. Thereafter, the pathogen develops most rapidly at 18°–21°. Under these ideal conditions the incubation period is 3–7 days and if the favourable conditions persist new crop of infective propagules will be available for secondary spread every 3–7 days. The incubation period is longer at higher temperatures resulting in corresponding decrease in secondary cycles.

On the basis of these pathogen-environment relations the earliest forecasting method, the Dutch rules, included (i) a night temperature below dew point for at least 4 hours (that is, dew must stay for this period), (ii) a minimum temperature of 10°C or above, (iii) a mean cloudiness on the next day of at least 0.8, and (iv) at least 0.1 mm of rain during the next 24 hours. These rules worked well in Holland but neither weather nor the density of initial inoculum is same everywhere. Modifications suiting a particular area are essential. In southwest England these rules did not work. There, the rules were reduced to two, viz., a minimum temperature of 10°C and a relative humidity not falling below 75% for at least 2 days.

In the United States two systems of forecasting late blight were developed by Hyre (1954) and Wallin (1962). The Hyre's system is based on records of daily rainfall and maximum and minimum temperatures. The data are not collected from within the crop canopy. The initial appearance of late blight is forecast 7–14 days after the first occurrence of 10 consecutive blight favourable days. A day is blight favourable when the 5-day average temperature is below 25°C and the total rainfall for the last 10-days period is more than 3.5 cms. Days on which the minimum temperature falls below 7.2°C are considered unfavourable for blight development. In Wallin's system blight forecast is based on relative humidity (RH) and temperature. In this system, the data are required from within the crop canopy. It takes into consideration the seasonal accumulation of "severity values". Severity values are numbers arbitrarily assigned to specific relationship between duration of RH periods of more than 90% and the average temperature during those periods. The first occurrence of late blight is predicted 7–14 days after 18–20 severity values have been accumulated from the time of plant emergence. Wallin's system was superior to Hyre's system in that it took into consideration the actual humidity and temperature within the potato rows. This is particularly significant in situations where crop canopy is dense due to irrigation and fertilizers. The meteorological data outside the field (as used in Hyer's system and Dutch system) may be deceptive if recorded as unfavourable for blight.

Although both systems in the USA were used for more than 18 years they were not widely accepted and utilized by farmers because the systems were not readily available on a timely, regular and localised basis. This led to development of a computer programme which provides the farmers with a late blight control programme tailored to their local conditions. The programme is known as BLITECAST in the USA. Blitecast is a computer programme integrating both Hyre's system and Wallin's system of forecasting based on rainfall, temperature and relative humidity (Krause *et al*, 1975). The system is a service, relaying information on forecast from a central office to those who seek such forecast.

The data for the forecast are required on the following:

1) Maximum and minimum temperature for the day.
2) Number of hours of relative humidity more than 90%.
3) Maximum and minimum temperature during the period when relative humidity was more than 90%.
4) The rainfall recorded at 24-hours basis and measured to the nearest 0.1 cm.

The temperature and humidity data are recorded with sheltered hygrothermographs located between rows and within the crop canopy. Data recording must be started early in the season when emerging plant foliage starts giving the look of green rows. In the early stages there will not be much difference in data from within and outside the crop. The computer programming for Blitecast has two parts: one determining the possible occurrence of the disease which it can do even with data from outside the crop canopy, and the other determines severity and recommendation for spraying. When the farmer desires a Blitecast, he conveys the most recently recorded data to the Blitecast station. The operator feeds the data into a computer programmed to use the data which analyses them within seconds and returns the forecast and spray recommendation to operator who relays the same to the farmer.

More recent refinements in late blight forecasting include, in addition to moisture and temperature data, information on the level of resistance of the potato variety to late blight and the effectiveness of the fungicide used. Expensive equipments based on the principle of Blitecast have been developed. These are self-contained units, programmed with parameters determining epidemics of late blight, and placed in the field where they continuously monitor the weather. As soon as a critical period has arrived, warning is flashed which can be interpreted with the help of direction charts. Tomato is also a host of the late blight fungus and often these two crops are concurrently grown. Precautions recommended for potato should be taken for tomato also for better use of the forecast.

Apple scab (*venturia inaequalis*): The apple scab fungus is polycyclic pathogen with very large amount of initial inoculum and very large amount of secondary inoculum. The pathogen has two distinct stages in its life cycle: the saprophytic pseudothecial or overwintering stage on dead fallen leaves, and the conidial summer stage which is parasitic on leaves, flower buds, fruits and shoots on the tree during spring and summer. The saprophytic stage provides the primary inoculum in the form of ascospore showers (continuing for 1–2 months after bud burst) for primary infection of new leaves and flower buds in spring. In a single scab-infected overwintered leaf lesion over two million ascospores are formed. There are many secondary inoculum cycles during the period when the apple tree is susceptible. Ascospore discharge is triggered by spring rains which moisten the overwintering ascospore-producing leaves on the ground. The discharge of ascospores occurs at a time when the leaf buds on the tree are about to burst open and leaf tips are exposed to receive the inoculum. This is the critical period for initiation of the disease and if fungicidal protection is not given at this stage, secondary cycles producing conidia start and conidia-infection-conidia cycle is repeated several times throughout the season. The forewarning of initiation of apple scab can be done on the basis of the approximate quantity of ascospores present at the time when host leaves are susceptible and its intensity can be decided on the basis of the number of infection periods occurring during the season. Quantity of ascospores is determined by examination of overwintered leaves in the laboratory. Infection periods are days when weather conditions are highly favourable for discharge of ascospores, their landing on susceptible host surface, inoculation and colonization of the host tissue. It is based on leaf wetness periods and accompanying temperature prevalent in spring (Mills, 1944). For instance, if the mean temperature is 17.5°–24°C the incubation period is 9 days provided the leaves are continuously wet. Nine hours of leaf wetness results in light infection, 12 hours of leaf wetness results in moderate infection while 18 hours of leaf wetness at the above temperatures results in heavy infection. Hours of wetness periods interrupted by more than 8 hours of dry period are not included in determining the infection period. Strict adherence to these criteria has not been possible in many countries. In England, the leaf wetness duration has been substituted with hours of 90% or higher relative humidity after rains. In Germany, a mean day temperature aggregate of 105°F (over 15 days) after March 1, favourable for pseudothecial maturity is also taken into consideration along with leaf wetness period. In the Netherlands, warning of possible disease development can be issued on the basis of (i) ascospores are ready to mature, (ii) ascospore release is expected, (iii) ascospore release has taken place, and (iv) infection periods have occurred.

Predictive models and procedures have been developed in USA and Europe. These avoid the cumbersome process of microscopic examination of leaves for ascospore discharge. One such system, Vintem TM (a PC based programme for farms) developed at the East Malling Horticultural Station in UK alerts about infection periods and then forecasts the scab severity. It calculates the infection efficiency (IE) of ascospores and conidia. The IE values (0-100%) alert the grower that weather conditions favour infection. Then, it forecasts scab intensity specific to the particular orchard (Butt and Xu, 1994).

In a study by Gadoury et al (1992) it was reported that physiologically the asci become ready for discharge of mature ascospores several days after bud burst and fungicidal sprays could be delayed beyond the date predicted by estimation of ascospore showers.

Miscellaneous systems of prediction: In Stewart's wilt of corn caused by the bacterium *Pantoea stewartii* subsp. *stewartii*, the pathogen survives in the intestinal tract of its vectors *Chaetocnema pulicaria* (corn flea beetle) and *Diabrotica undecimpunctata* (12-spotted cucumber beetle). Estimate of populations of these vectors surviving through winter can give an idea about the intensity of the disease in the coming season. Corn flea beetles are killed by prolonged exposure to temperatures less than −1°C. In an area if the temperature persists during winter at this level absence of the disease can be predicted. The cucumber wilt bacterium (*Erwinia trachiephila*) survives through its vectors *Acalymna vittatum* (striped cucumber beetle) and *D. undecimpunctata*. The primary infection of the crop during spring and summer is initiated by these vectors. Estimation of prevalence of these vectors can predict the incidence of cucumber wilt.

Forecasting has been applied in case of many other diseases on the basis of (i) weather conditions during the intercrop period, (ii) weather during the crop season, (iii) amount of disease in the young crop (initial inoculum level), and (iv) amount of inoculum in air, soil, or planting material. Prediction of downy mildew of grapevine (*Plasmopara viticola*) is made on the basis of weather favourable for germination of oospores and infection of the plant. Rainy season with heavy rains during oospore formation hastens their maturity, hence, there is more disease in the next crop season. Generally, the oospores germinate more abundantly and in a shorter time early in the season if they have been subjected to frequent rains and mild temperatures. Optimum temperature for oospore germination is 20°–25°C. Downy mildew of lima bean (*Phytophthora phaseoli*) is predicted on the basis of rainfall-temperature. A day is considered favourable for the disease when the 5-day mean moving temperature (recorded graphically) is less than 26°C with the minimum 7°C or above, and the 10-day total rainfall is 3.05 cm or more. The disease is likely to appear after about 8 consecutive favourable days.

There is good amount of data for forecasting blast of rice (*Pyricularia grisea*) in India. Forecasting can be made on the basis of minimum night temperature range of 20°–26°C in association with a high relative humidity range of 90% and above lasting for a period of a week or more during any of the three susceptible stages of crop growth, *viz.*, seedling stage, post-transplanting tillering stage and neck emergence stage. Another approach is to plant a highly susceptible variety in the locality ahead of the main crop and watch for occurrence of the disease. If the disease appears in the indicator crop and above weather conditions prevail, a forecast of coming epidemic can be made.

In fire blight of apple and pear caused by *Erwinia amylovora*, the pathogen multiplies much more slowly at temperatures below 15°C than at temperatures above 17°C. The blight develops most rapidly at temperatures between 18° and 30°C. The number of days with temperatures above 18°C at the time of bloom coupled with high humidity are implicated in the rate of disease development. In California (USA) a disease outbreak can be expected to occur in the orchard if daily average temperatures exceeded a "disease prediction line" obtained by drawing a line from 16.7°C on March 1–14.4°C on May 1. When such conditions occur, application of a bactericide during bloom is recommended. The spread of groundnut leaf spot (*Cer cosporidium personatum* and *Cercospora arachidicola*) in Georgia (USA) is favoured by diurnal periods of 10 hours or longer with RH at or above 95% and with temperatures above 21°C during these periods.

16

Assessment of Disease Incidence, Severity and Loss

In measurement of disease, the three assessments that are commonly employed are (i) the **incidence**, that is the number of individuals showing the disease, (ii) the **severity** of the disease, that is the proportion of area or amount of plant tissue that is diseased, and (iii) the **loss** caused by the disease. The first two are mainly academic but help in assessment of loss. Incidence and severity are generally assessed to compare effects of treatments or practices. Accurate estimation of loss is most important measurement because it gives a practical base for comparisons and for decision making.

The fact that disease is a malfunction in the plant is in itself an evidence that whenever an individual plant is diseased it does not give as much yield as the individual which is healthy or protected from the disease. Theoretically, the difference between the yield obtained from a healthy plant and that from a diseased plant is a loss to the grower. This loss can be expressed in terms of quantity and/or quality of the produce and also in terms of money that the farmer could not get due to the difference in yield or difference in quality. When the individual is part of a population in the field many more individuals are sick and the loss is multiplied by the number of individuals sick. This fact about loss in yield due to disease had been known to the farmers ever since crop cultivation was started but anxiety to develop methods for quantifying this loss was felt only in the mid-twentieth century.

The measurement of loss due to disease directly from the field is difficult because of complexities of causes of loss in crop yields. Crop yields are determined by the combination and interaction of many and varied factors. These include losses due to natural calamities such as floods, untimely rains, drought, storms, frost, etc. which are beyond the control of the grower, and due to insects, diseases, weeds and poor management which can be controlled by the grower. In a particular field many of these causes may be simultaneously present and it becomes difficult to separate the loss from disease alone. If the loss is estimated under controlled conditions in the laboratory the data may be unreliable and may not be applicable to field conditions because of the low population studied and absene of the natural environments. Often, the difference in yield is determined from the yield of a normal crop or normal yield. Normal yield of a crop is difficult to define accurately because it varies from area to area and with quality of management. These difficulties led to the need for standardization of method. In 1971; the FAO brought out the first manual on crop loss assessment methods with the objective of introducing a uniform system of appraisal at international level (Large, 1966, FAO, 1971, James,1971, 1974).

Reliable information on crop loss due to disease and pests aims to establish the increase in yield when these enemies of the crop are controlled or eliminated at acceptable economic cost. The early descriptions of strategies had made clear the difference between quantification of disease incidence and severity and quantfication of loss from this abnormal condition of the crop. However, loss continued to be a vague term. Loss of what and to whom? Is it only the loss in quantity of yield or quality also? Loss in quantity of yield results in loss of area because, to meet the demand, more acreage will have to be put under the particular crop. To prevent the loss, money is spent and this also amounts to loss. Healthy crop would not have required this investment. Loss in yield of one crop affects the aggregate loss in yield of other crops because of withdrawal of acreage. The loss is suffered by individual grower because he gets less income due to reduced yield. But other growers, whose crop is not affected, may benefit because their produce may fetch a higher price. The businessman who distributes the produce benefits as he may charge higher prices for the short supply of the commodity. Final analysis leads to the conclusion that it is the consumer who pays for the loss in quantity and quality and for the loss to the grower through enhanced prices and taxes charged by the government to meet the expenditure on arranging suitable control measures.

The plant diseases can be disfiguring, debilitating, devastating, limiting or annihilating. Generally a loss of less than 1 per cent is considered negligible and a loss of 1 per cent over widespread area is minor one. Diseases causing 1–5 per cent loss and occasionally becoming severe in some seasons and areas constitute a third category. Diseases causing 5–10 per cent loss in most seasons are in the fourth category. Destructive diseases that cause more than 20 per cent loss in most seasons may make a crop uneconomic.

The benefits from accurate measurement of loss can be listed as follows:
1) It avoids nonjudicious use of control measures.
2) It convinces the farmer about economics of control measures and enables him to use them at resonable cost, taking into consideration the cost: benefit ratio.
3) It advises the scientists and research workers to concentratee their efforts on finding out suitable control measures including application of new fungicides.
4) It helps the industry to undertake development and production of fungicides on the basis of anticipated demand.

5) It aids the funding organizations such as the government to provide funds for research, development and use of fungecides and also undertake other measures for control of the diseases.

The strategy for loss appriasal involves two distinct phases:

1) Measurement of disease incidence and severity and its correlation with loss in yield.

2) Assessment of the disease in surveys of a number of fields using the assessment methods developed in the first phase. By knowing the quantitative relationship between amount(s) of disease and yield, the loss can be calculated from the disease data recorded in the surveys.

In the first phase relatively a small team of workers is involved while the second phase requires larger survey parties. Due to involvement of large teams in the second phase which will generate the data on amount of disease present, it becomes essential that the methods of assessment of amount of disease developed in the first phase are simple, should hold good for different observers in different areas and in different seasons and must be comparable. They should be objective rather subjective so that possible human errors are eliminated to the maximum.

Disease incidence or frequency of incidence or prevalence can be defined as the number of plant units infected in a population. The units can be whole plant, leaves, fruits, twigs, stems, etc. The term expresses the proportion (0–1) or percentage (1 to 100) of the diseased eniities. For small areas the proportion can be determined by counting the entire population and all the diseased units while for larger areas proper sampling is done and proportion can be determined from the sample. Also called **disease intensity**, disease severity is the area or area and volume of plant tissues affected by the disease and expressed as percentage of total area or volume. Disease intensity is often meant to include both disease incidence and disease severity.

MEASURING DISEASE SEVERITY

The tactics of disease measurement vary according to the nature of the disease and, therefore, no single method is applicable to all diseases. Methods for measuring diseases causing complete loss of the plant, partial loss of foliage, tubers or fruits and partial loss of the inflorescence take into cosideration the specific situation but the objective remains the same, i.e., the measurement must be as accurate and realistic as possible and could be applied with least error to estimate loss or compare treatments in evaluation of fungicides, nematicides, varietal resistance, etc. According to Large (1966) a general strategy, applicable to most diseases, should meet the following requirements:

a) A close descriptive study of the gross morphology and course of development of the healthy crop plant from sowing to harvest or from season to season. This gives a knowledge of growth stages of the plant which is

282

needed in assessment of disease incidence and severity.

b) A similar study of the course of the disease on plants in the field, over the whole range of attack.

c) With the help of preliminary drawings, sketches, notes and measurements, resulting from the study of healthy and diseased plants, as given under (a) and (b), a standard diagram or research key for the assessment of the disease is prepared. This key is later simplified to field key to be used by the survey parties.

d) Field trials are carried out over a number of years in which development of the disease is observed with the help of the field key and crop yields are recorded. Simultaneously, check plots are maintained in which the crop is kept free from the disease by chemical protection and other means and yields are recorded for comparison. These field trials are adequately replicated with sound statistical foundation. They permit more detailed analysis of disease development and crop yield loss than would be possible over large areas.

e) From these trials the methods of disease assessment which are likely to be most useful for disease survey work in the field are devised and the calibration of disease severity with likely crop loss is attempted.

In those diseases which kill plants rather quickly or which cause about the same amount of damage to all the infected individuals (vascular wilt, systemic viruses, damping off, ergot, covered smut of barley, loose smut and stinking smut of wheat) recording the percentage of diseased plants and organs is a direct measure of crop loss involved. Smuts which destroy only some of the grains in the ear can be assessed by multiplying the average number of smutted grains per ear by the percentage infected ears. Such direct counts are not applicable to diseases in which different plants show different amounts of infection. Leaf spots vary in intensity from plant to plant and from leaf to leaf on the same plant. Determining the number of infected plants can be useful only if the severity of spots is also assessed. Severity estimates are difficult and laborious. Sometimes, the percentage of infected plants, percentage of infected leaves and percentage of leaf area destroyed, all are recorded to obtain an overall figure of disease severity. This method is possible only for small experimental plots.

For such diseases where the amount of disease varies greatly on different plants in the population (rusts, mildews, blights, leaf spots) many arbitrary indices and ratings had been in practice but have been discouraged and replaced by percentage scales and standard area diagrams of disease intensity. In percentage scale methods, usually the number of plants or organs falling into known percentage disease groups are recorded. These groups are categories distinguished on the basis of per cent damage seen by human eye. In 1945, J.G. Horsfall and R.W. Barratt had proposed a 12-grade scale. They had taken into consideration the fact that the grades detected by the human eye are approximately equal divisions on a log scale and generally follow the Webber-Fechner law which states that visual acuity depends on the logarithm of the intensity of stimulus. Hence, 20 per cent increase between 30 and 50 per cent is not easier to distinguish than 4 per cent increase between and 1 and 5 per cent. In percentage disease assessment the eye actually assesses the diseased area up to 50 per cent and the healthy area above 50 per cent. In 12-grade scale, the categories were numbered as follows: 1 = 0%, 2 = 0–3%, 3 = 3–6%, 4 = 6–12%, 5 =12–25%, 6 = 25–50%, 7 = 50–75%, 8 = 75–87%, 9 = 87–94%, 10 = 94–97%, 11 = 97–100% and 12 = 100% disease. This is a logarithmic scale and is satisfactory not only for disease measurement but also for epidemiological studies, because pathogens multiply at logarithmic rate, and also for loss appraisal. Such methods should be used in conjunction with standard area diagrams which are necessary to define specific levels of disease.

A system using percentage scale, developed by British Mycological Society in 1947, for measuring late blight of potato and later used in Canada by James (1971) for loss appraisal in potato, is given below:

TABLE 16: Late Blight Key.

Blight (%)	Nature of infection
0.0	No disease observed
0.1	A few scattered plants blighted; no more than 1 or 2 spots in 10 meter radius
1.0	Up to 10 spots per plant; or general light infection
5.0	About 50 spots per plant; up to 1 in 10 leaflets infected
25.0	Nearly every leaflet infected but plants retaining normal form; plants may smell of blight; fields look green although every plant in infected.
50.0	Every plant is affected and about 50% leaf area is destroyed; fields look green, flecked with brown
75.0	About 75% of the leaf area is destroyed; fields appear neither green nor predominantly brown
95.0	Only few leaves on plants but stems are green
100.0	All leaves are dead, stems dead or drying

The percentage scale has many advantages, such as (i) the upper and lower limits of the scale are always well defined, (ii) the scale is flexible in that it can be divided and subdivided conveniently, and (iii) it is universally known and can be used to record both the number of plants infected (incidence) and area damaged (severity) by a foliage or root pathogen.

The first standard area diagram for a plant disease was that of Nathan Cobb for leaf rust of wheat (*Puccinia recodita*). It divided the rust intesity into 5 grades representing 1, 5, 10, 20, and 50% of the leaf area occupied by the visible or sporulating rust pustules. The highest grade (50%) represented about the maximum possible cover. A modified Cobb's scale was proposed by Melchers and Parker in 1922 for the estimation of stem rust of wheat. In this scale, maximum rust cover was arbitrarily taken as that occurring when 37%

284

of the leaf or stem area was occupied by the pustules and this was labelled 100%. Diagrams representing 5, 10, 25, 40, 65, and 100% stem rust on this basis were then prepared by copying Cobb's diagrams for leaf rust. In 1948, Peterson, *et al.*, proposed further modifications suggesting 1, 5, 10, 20, 30, 40, 50, 60, 70, 80, 90, and 100 per cent cover with rust pustules of various sizes. These methods are still used universally.

One of the most practical set of area diagrams was proposed by James (1971) for Canada Department of Agriculture and is being widely used. The diagrams give the actual area of leaves, stems, pods and tubers occupied by lesions in terms of per cent area covered. These diagrams are printed on loose durable plastic cards which can be used in the field. Some of the keys are reproduced on the following pages as example.

In many diseases area measurement is not possible but there is definite correlation between tissue damage or number of resting organs such as sclerotia and nematode cysts present on specific organs and general symptoms such as stunting. By observing the level of stunting, yellowing etc., disease intensity can be estimated. In root diseases caused by fungi and nematodes, this method has been used.

EXAMPLES OF ASSESSMENT KEYS

Figures 25 to 33 give keys for assessing the host surface area covered by

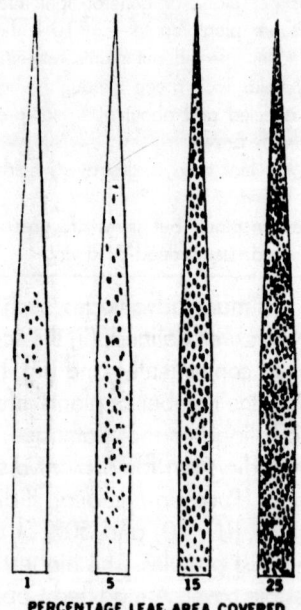

PERCENTAGE LEAF AREA COVERED

Figure 25. Disease Assessment Key to Estimate Percent Leaf Area Covered.

285

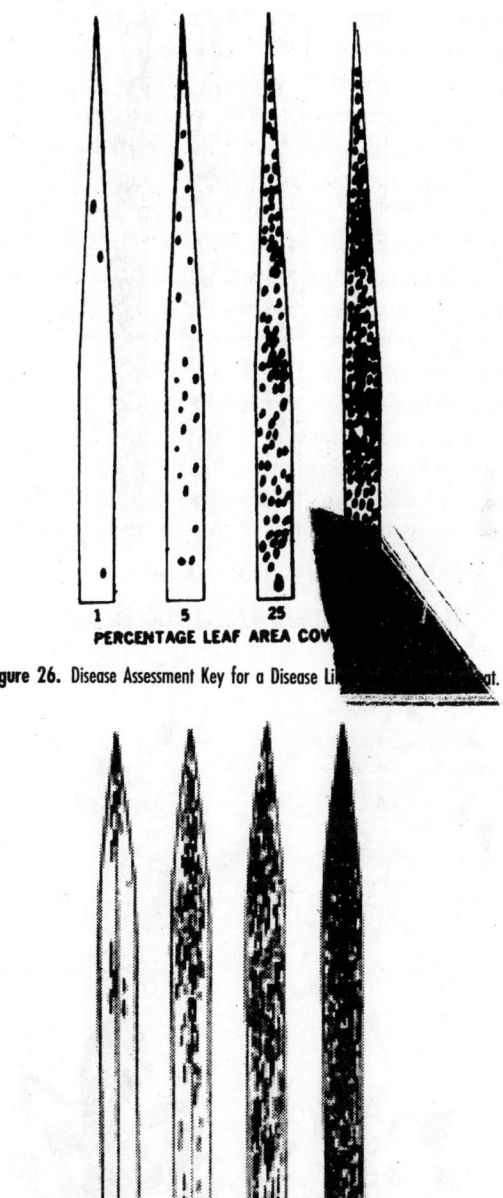

Figure 26. Disease Assessment Key for a Disease Li_____at.

Figure 27. Percent Area Covered and Index Value in Leaf Streak of Rice.

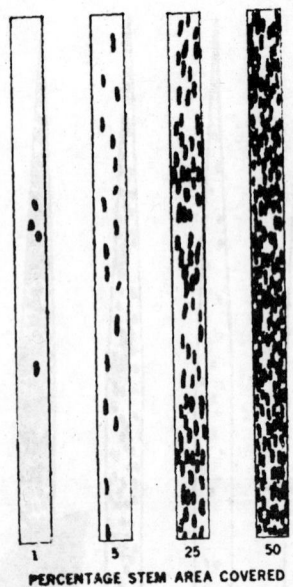

PERCENTAGE STEM AREA COVERED

Figure 28. Percent Stem Area Covered in Stem Rust of Wheat.

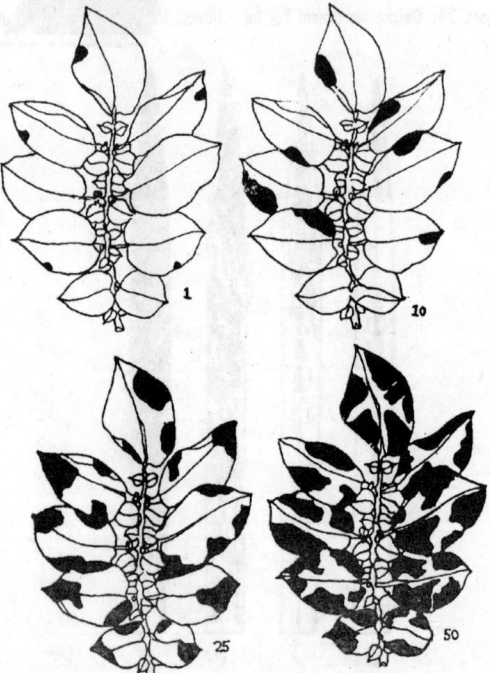

Figure 29. Percent Leaf Area Covered in Late Blight of Potato.

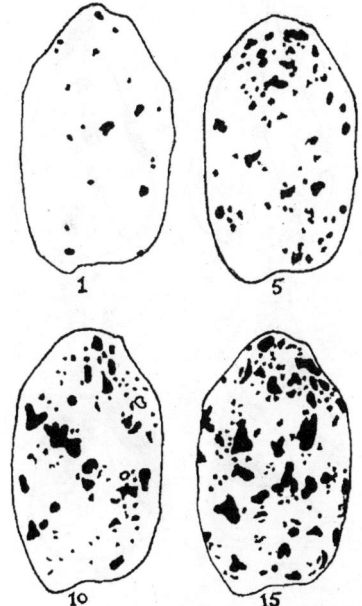

Figure 30. Assessment Key for Black Scurf of Potato Tubers.

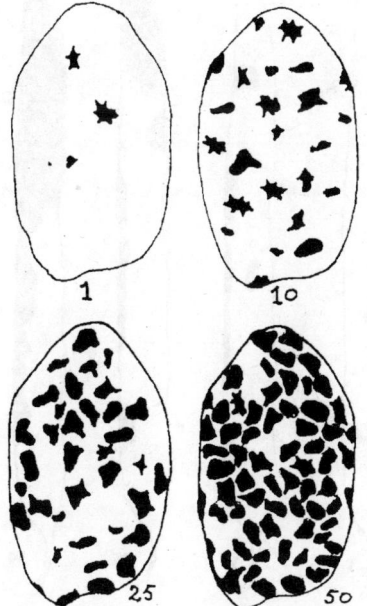

Figure 31. Disease Assessment Key for Common Scab of Potato.

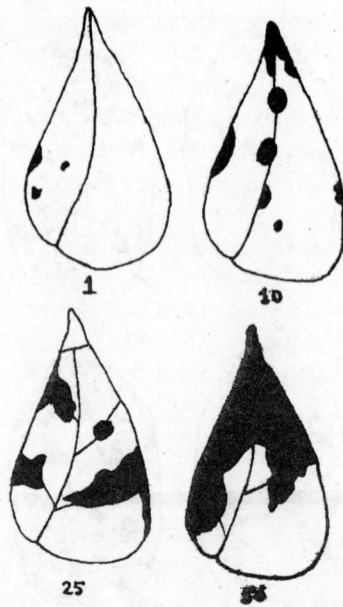

Figure 32. Assessment Key for Bacterial Blight of Bean Leaf.

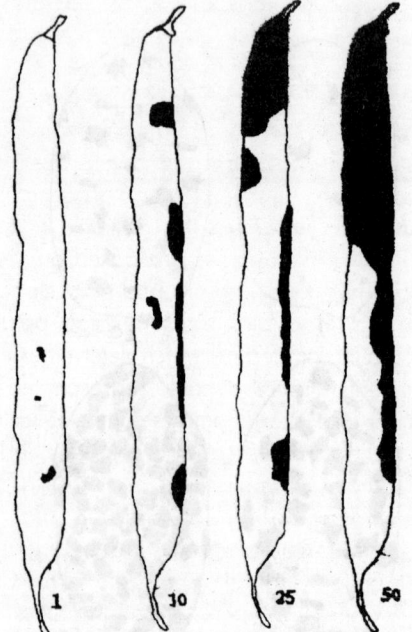

Figure 33. Assessment Key for Bacterial Blight of Bean Pod.

the disease in terms of percentages. These keys can be used for comparing the samples collected in the field and calculating the mean percentage area damaged by the disease. It can also be used for calculating infection index, disease intensity, etc. by converting the keys into grade cards, i.e. giving a number to each range of percentage area affected. Thus, in a disease of the foliage if the assessment keys give 1, 5, 25, 50% area covered, the grade numbers can be given as 0 for no infection, 1 for upto 1% area covered, 2 for up to more than 1% less than 5%, 3 for more than 5% and less than 25% area covered, and 5 to more than 50% area covered. The grade 0 can be eliminated since it does not affect the calculations provided total number of samples, irrespective of their being healthy or diseased, are taken into consideration. When 0 is eliminated, the above grading becomes a 5-point scale. It can be extended to a 10-point scale by suitably modifying the grades.

SAMPLING FOR ASSESSMENT

Certain preliminary steps are essential for better accuracy of assessments and for comparisons. Growth stage of the crop at the time of sampling must be recorded according to prescribed standards. For instance, in cereals (wheat, barley, etc.) there are 10 broad growth stages such as one shoot, tiller initiation, tiller formed, leaf sheaths lengthening, leaf sheaths strongly erect, first node of stem visible, second node visible, last leaf just visible, ligule of last leaf just visible, boot stage, etc. For convenience, the advanced stages of growth can be further subdivided. Similarly, in maize also there are 10 distinct growth stages, such as pre-emergence, emergence, single leaf, two leaf, early whorl, mid whorl, late whorl stages. In legumes, the growth stages are defined as bud, flower, full flower and seed stages. In potato, the early stage of disease intensity is assessed with diagrammatic keys and advanced stages with the help of late blight key tabulated earlier in this chapter. In temperate fruit trees (apple, pear), mostly the growth stages are based on the stages that occur after winter dormancy. Thus, there are pre-bud burst stages, bud burst stage, silver tip stage, green tip stage, pink blossom stage, petal fall stage, etc.

Next step is the sampling procedure which determines to a great extent the accuracy of measurements. Samples should represent the total field, orchard or plantation area under inspection. Samples can be collected from a diagonal in the plot or by any standard statistically accepted method. In cereals (wheat, barley, rice, maize) samples of fertile tillers or stems are selected at random along one diagonal or other appropriate area. In small plots (up to 0.004 ha), 10 tillers constitute a sample with sub-samples of leaf, stem, leaf sheath, ear, etc. In larger plots up to 50 samples should be taken. Complete assessment of the entire sample is done. In potatoes, a length of the row or a small area is chosen for sampling.

When the disease first appears in isolated foci and spreads later, a number of such foci are selected along a diagonal, area of each focus determined and mean calculated. Then the percentage acreage affected by the disease is calculated as below:

Average number of foci per hectare = 5

Average area of foci = 1 sq.m

Percentage ha affected $= \dfrac{5 \times 100}{10000}$

= 0.5%

The same system can be followed for calculating average infection or infection index for a large area. The average index for a field with known acreage can be used as a field rating class.

Calculations

The following example is for explaining the method of calculating infection index derived from records of assessment with the key given earlier.

Host: Rice

Disease: Leaf blast (*Pyricularia grisea*)

Stage of growth: Post-transplanting tillering stage

Number of samples collected from a one hectare field: 50 tillers with total of 200 leaves.

% leaf area covered 1	Grade 2	Number of leaves in the grade (ratings) 3	Disease rating 4	
0	0	75	0	
1	1	25	25	Sum of
10	3	45	135	disease
25	5	30	150	ratings = 495
50	7	20	140	Total ratings =
more than 50	9	5	45	200
				Maximum disease grade = 9

A common formula used to calculate the average infection or infection index is:

$$\% \text{ infection index} = \frac{\text{Sum of all disease ratings}}{\text{Total number of rating} \times \text{Max. disease grade}} \times 100$$

Thus, the index in the above example will be:

[495 divided by (200 × 9)] × 100 = 27.5%

Severity estimates from fairly small areas can be combined to cover large areas (village, district, state) and maps of disease incidence over large areas can be prepared. This overall index can be obtained by using the formula:

$$\frac{\text{Field rating class} \times \text{Number of hectares in the class}}{\text{Total number of hectares}}$$

REMOTE SENSING

Aerial photography can detect objects of land over a large area. It dates back to 1860 but it was not until 1956 that Colwell demonstrated its potential usefulness in plant disease survey. He showed that panchromatic, colour and especially infrared aerial photography could be used to detect rusts and virus diseases of small grains and certain diseases of citrus. Later, infrared photography was used in England for potato late blight disease. The remote sensing by aerial photography has been used for many types of assessments and detection work in agriculture including survey of plant diseases, area covered by them and intensity of the disease in different areas.

The key to distinguishing between healthy and diseased plants in a crop and between diseases is the use of appropriate film-filter combination to catch reflectance from plant surfaces. In some cases, it may be necessary to use two or more such combinations simultaneously (**multiband photography or multispectral remote sensing**). The film types used in this work are panchromatic, infrared normal colour and colour infrared. The infrared films are preferred because of their superior sensitivity to visible light and to near infrared wavelengths (700 to 950 nm). The colour infrared or Ektachrome Aero Infrared (Camouflage Detection Film) has the added superiority in that it can show difference between healthy and diseased patches of plants in colour. The healthy foliage is highly reflective to the wavelengths of 700 to 950 nm and appears red on this film whereas blighted or diseased foliage has low infrared reflectance and does not appear red in the photograph. Identification of disease is possible because reflectance differs in foliage affected by different diseases. Although disease severity of single plant has been assessed accurately by analysis of photographs of the plant, the high cost and inconvenience of taking and analysing colour infrared photographs have limited the application of this method on large scale.

The advantages of aerial photography or remote sensing are many, viz., (i) it reveals pattern of disease incidene, intensity and development which cannot be seen so well, or can not be seen at all, from the ground, (ii) it makes possible study, on an extensive scale, of the disease where it actually occurs in the field, (iii) it frequently poses questions for ground investigations. These questions could not be generated by ground studies. In addition, remote sensing almost compels collaboration between diciplines such as soil physicist and chemist, agronomist and plant pathologist.

The main objectives of remote sensing are detection and identification of diseases, their relation to environment, origin and development of epidemics and quantitative assessment of disease. In addition to the information on epidemiology the technique gives assessment of disease over a vast area without ground parties visiting these areas. Quantification of the assessment is done by recording transmittance of the positive transparancies of the colour infrared photographs with microdensitometers or other types of electronic scanners. The numerical value of transmitance recorded by the densitometer is used as a figure of disease assessment and can be used like the data collected by other methods.

Video Image Analysis

Since early 1980s video image anlysis has been used for assessment of plant disease severity (*cf* Mayee and Datar, 1989). Advances in electronics and computer technology allow video cameras to directly interface with a microcomputer. Rapid, automated, non-subjective estimates of disease severity are made possible by computer-controlled anlysis of video images. Assessment of only those plant diseases with lesions having colour different from healthy tissues is possible by this method. The method is very costly and has limited use in developing countries.

Assessment with inoculum-disease relationship

An alternative to measurements based on symptoms and tissue damage is to count the spores of the fungal pathogens. This has been attempted in diseases where one spore may cause one infection. Although this method is fast, it generally gives wrong estimates, particularly for air-borne pathogens. Infection efficiency of spores differs and environments may disturb efficiency of the spore. Estimation of low severity diseases may be time consuming and may have high standard errot on the mean estimate. Attempts have been made to indirectly estimate low severity diseases by relating incidence to severity. It has been possible to correlate the count of nematode cysts with stunting of the plant and estimate severity.

MEASUREMENT OF LOSS

After the field keys are prepared, plant growth stage defined for disease measurement, and frequencies of assessment determined, next phase of loss appraisal can be taken up by determining the **severity-loss relationship**. The object is to correlate disease severity to reduction in yield and prepare standard tables or arithmetical models which can be used to translate the figures of disease severity into figures for percentage loss.

Determination of potential yield or yield that would have been obtained in total absence of the disease is central problem in estimation of yield loss due to disease. As was mentioned earlier, the yield loss is not only due to disease but other factors also such as bad agronomic management, attack of insect pests, diseases other than the disease under study, etc. Presuming that these deleterious factors are also absent the yield obtained can be taken as potential yield. Once this assessment is available, the actual yield with any measured degree of disease incidence or severity can be expressed as a percentage of potential yield and the yield reduction as the difference between this and 100%.

Rules for estimation of yield reduction from disease measurements to prepare the standard tables are worked out by conducting especially designed multilocational field trials, for years, in different seasons in which the essential controls are plots which are kept free from disease at any cost, by using fungicides, resistant varieties, or other available means. It may also be necessary to create epiphytotics with different characteristics. By comparing the yields from control plots and those from plots having different degrees of the disease, figures for loss for each degree of disease can be worked out. Some models, based on FAO Manual on loss appriasal are given below.

Loose Smut of Wheat
(Ustilago segatum tritici)

Procedure: Sampling is done at the heading stage. In each field 50 plants are taken for disease assessment from 5 sites of 1 m^2 selected at random along a diagonal. Percentage of infected ears (or panicles) of these plants is calculated and yield loss estimated according to the following formula:

$$\text{Loss (kg / ha)} = \frac{Pr}{100 - Pr} \times Pa$$

where Pr is percentage infected ears and
Pa is actual yield (kg/ha)

In large scale surveys, 20 fields for each kind of cereal are assessed in a minimum of 5 locations in each district. Percentage of diseased ears is directly proportional to loss of grain yield.

Same procedure is followed for stinking smut or hill bunt of wheat (*Tilletia tritici* and *T. laevis*) except that the sampling is done at the ripening stage of the crop.

Grain smut of sorghum
(Sphacelotheca sorghi)

Procedure: Crops should be sampled 7–21 days after flowering starts when not less than 10–20% of the spikelets have not yet flowered (i.e., stamens still present inside). For plots, take single sprigs at random from the top, middle and bottom of successive panicles on plants at random. From the sprigs, select at random 500 spikelets, dissect them and determie percentage infected. For surveys of crops, to obtain overall incidence in an area, 50 spikelets from 500 sprigs per field are adequate.

Intensity/loss relationship: The percentage of infected spikelets is directly proportional to the loss of grain yield.

In another method, infection range is determined on cob and area basis and range of yield loss determined for both separately. This study has shown that each unit increase in percentage infection (cob or area basis) causes a decrease of percentage grain yield by 0.9.

Common scab of potato
(Streptomyces scabies)

Procedure: Random samples of all tubers are dug from 3 m length of the ridge. Number of samples vary from 8 in 1.2 to 4 ha fields to 16 in fields of 12 ha. Tubers are graded into:

a) free from scab or less than one eighth surface affected
b) more than one eighth but less than one-quarter surface affected
c) one-quarter or more surface affected.

The disease does not cause any significant loss in yield. It reduces the quality. Substantially affected crop is defined as having more than 1.4 kg/50 kg of tubers in category (c) or with more than 50% of the tubers in categories (b) and (c). Similar procedure can be followed for the black scurf disease (*Rhizoctonia solani*) of potato also.

Stripe disease of barley
(Drechslera graminea)

Procedure: Growth stage for sampling is heading to milky ripe. Samples are taken at random on a diagonal through the field in sample area of 0.4 × 0.4 m. Number of fertile tillers and number of infected tillers are counted and percentage of infected tillers is calculated.

Intensity/loss relationship: Studies in UK have shown that infected plants contribute nothing to yield. Therefore, loss can be calculated as:

% loss = % infection

$$\text{Loss kg / ha} = \frac{\% \text{ infection}}{\% \text{ healthy}} \times \text{yield kg / ha}$$

Cereal cyst nematode (Molya disease)
(Heterodera avenae)

Procedure: Fields are sampled between end of flowering and ripening of grain. Three samples, each consisting of 5–10 plants are collected at random from each field of 0.5 to 2 ha. Where fields show patches of stunted plants, one sample should be collected from such areas. The root system of collected plants is gently washed free of soil and plants are examined for their overall development as well as for the presence of white cysts on roots. The number of cysts per root system is recorded. Based on cyst number and on growth of the plants, the plants are grouped as below:

A = no cysts
B = less than 10 cysts
C = more than 10 cysts but no stunting
D = more than 10 cysts but plants stunted.

The number in each class is recorded and the **nematode index** is calculated according to the following formula:

$$\text{Nematode index} = \frac{\dfrac{\text{No. in class B}}{4} + \dfrac{\text{No. in class C}}{2} + \text{No. in class D}}{\text{total number of samples}}$$

Intensity/ loss relationship: Based on experimental data obtained in nematode control experiments in many north European countries the following approximate relationships have been established:

Crop	Expected yield reduction factor from most severe attack
Wheat	0.50
Barley	0.35
Oats	0.70

To calculate maximum potential yield reduction (MPYR) from a certain area, the following formula can be used:

MPYR (tons/ha) = average yield (t/ha) × nematoed index × expected yield reduction factor × surveyed area (ha)

The average yield data is obtained from existing records.

17

Disease Management—Principles

Measures taken to prevent incidence of a disease, reduce the amount of inoculum that initiates and spreads the disease in the crop and, finally, minimize the loss caused by the disease have been traditionally called the control measures. Since the 1970s there has been a general tendency to consider these measures as management strategies rather than control because the control measures generally manage the population of the pathogens at an innocuous level, not necessarily eliminating them from the ecosystem. Here, the terms disease management and disease control are used interchangeably.

MANAGEMENT PLANNING AGAINST A DISEASE OR FOR A CROP

The principles of plant disease management can have two approaches, viz., management of a single disease of the crop or planning for overall health of the crop. Management planning for a disease is directed against a specific disease which is causing heavy losses, such as late blight in the potato crop, without taking into consideration other fungal, bacterial or virus diseases of the same crop. Management of crop health involves an integrated plan in which all diseases of any significant nature for the crop are taken into consideration, although major stress may be against the most common and severe diseases. Obviously, the second approach, though difficult, is of more practical value for the farmer because he is interested in increasing the productivity of the crop and, therefore, prefers a plan that can provide safeguard against all possible diseases occurring in the area.

A single crop is attacked by several diseases in the same season. Wheat shows presence of smuts and bunts along with the attack of rusts. Sugarcane shows attack of smut in early season, soon followed by red rot and then the wilt. Potato is attacked by several viruses, late and early blight, black scurf and stem canker, root knot etc. There are specific measures reommended for management of each disease. It is not essential that measures against one disease will check the other diseases also. While deciding management practices for a specific disease the nature of the causal agent is a basic consideration. Different diseases of the same crop are caused by pathogens of different nature. Therefore, in planning for health of the crop, the procedures for management are arranged in such a manner that maximum number of diseases of the crop are partly or wholly managed by minimum number of operations and repetition of operations for each separate disease is avoided. Certain principles of plant disease management help in reducing incidence of several diseases. The multiple disease-management methods can reduce cost of plant protection in the crop and the farmer has the ease in applying them.

SOME CONSIDERATIONS IN PLANT DISEASE MANAGEMENT

The aim of disease management is to check the reduction in economic gain from a crop. If the control measures fail to increase economic gain, even if disease incidence is reduced, no farmer is likely to accept the recommendations for plant disease management. Therefore, the cost benefit ratio is a very important consideration in the application of control measures.

The procedures for plant disease management are a part of the general cultural practices for raising a crop. It is necessary that the procedures for disease control should fit in the general schedule of operations for crop production. There may be points of conflict between agronomists and pathologists. A flat recommendation of 120-60-60 NPK may not be necessarily beneficial for crop yield. There are examples where 100 kg N/ha gives as good yield as 150 kg N/ha. Where a disease is favoured by high nitrogen level, the agronomist should agree to lower the recommended dose of nitrogen. Similarly, if 5 irrigations are recommended for a crop but disease is favoured there is justification for reducing the number of irrigations.

For success of any management planning its adoption on large contiguous area is also necessary. Under field conditions dissemination of pathogens takes place through the agency of wind, water, insects and shifting of soil, etc. There is no method for raising an effective and permanent barrier against these agencies of transmission of pathogens. If only one former in the area rigidly implements principles of disease control in his crop but neighbouring farmers growing the same crop neglect them and the disease develops in their crop, its transmission to the neighbouring field where control procedures were adopted may easily take place as soon as the effect of a particular treatment has disappeared. To protect his crop the farmer will have to spend more money, time and energy in repeating the control methods more frequently. When control measures, such as chemical sprays on standing crops, are followed on a large contiguous area under a crop the frequency of application of treatments is less, chances of success of the treatment are better and the cost of control is reduced.

A special feature of plant diseases, especially of the annual field crops, is that the disease is detected when damage to the entire plant or its organs has already taken place. Plants do not forewarn of their sickness although the intelligent farmers is aware of the possibility of occurrence of serious disease in the particular crop. The annual crop plants have no system of tissue rejuvenation. An organ once destroyed can not be usually regenerated. Cure of such diseased organs is not possible. Object of disease management in plants is only to prevent spread of the disease from such organs to healthy organs or from a diseased plant to a healthy plant. Only in perennial plants such as fruit trees diseased organs can be cut and removed thus saving the life of the tree. Structural strength and long life span enables the trees to self heal the wounds and continue their life processes normally. In short term, tender field crops this type of surgery is not possible. Therefore, in plant disease management most of the procedures are preventive rather than curative. These preventive measures are applied before and/or during the appearance of the disease in the field.

DISEASE CYCLE-DISEASE CONTROL RELATIONSHIP

Knowledge of following aspects of diseases development is essential for effective and economic control:

1) Identification of the cause of the disease.
2) Mode of perennation and dissemination of the infectious agent of the disease.
3) Host-parasite relationship and means of secondary spread.
4) Effect of environment on pathogenesis in the plant and spread of the disease in the plant population.

The knowledge of these stages of the disease-cycle helps in selecting control measures which directly or indirectly destroy the pathogen or suppress its growth.

Correct knowledge of the cause of the disease avoids new problems and reduces expenses. Abnormal conditions of the plant may be due to many causes. Yellowing of leaves in rice or stunted growth and yellowing of leaves in wheat may not necessaribly be due to viral, bacterial or fungal infection. It may be manifestation of nutritional deficiency. If fungicides, nematicides, bactericides or insecticides are used without ascertaining the cause it may be only a waste of resources.

The knowledge of medium of survival of the pathogen helps in selection of control procedure to attack it in its most vulnerable stage when numerically it is not much compared to its active stage when it may produce high amount of spores. In the disease cycle of loose smut of wheat the fungus causing the disease is internally seed-borne and primary infection occurs through spores dispersed by wind during flowering stage of the crop. From this information, it can be deduced that soil treatment, crop rotatin, sprays of fungicides, etc., will be ineffective methods for control of this and similar diseases. Only seed treatments that can destroy the internally seed-borne fungus can control this disease and prevent production of spores of the fungus for fresh infections in the crop. In red rot of sugarcane, the disease is carried by seed setts and, for a brief period, the pathogen can survive in soil. Choice of control measures in this case emphasise selection of healthy canes for seed and crop rotation. In polycyclic diseases in which many cycles of spore production occur in the standing crop, the strategy for management involves the prevention of entry of the pathogen in the crop and prevention of spore formation. Once spore formation starts, the control becomes costly and difficult.

The knowledge of environmental relations of the pathogen and disease development strengthen the chemical control measures and provide the farmer with some forewarning of possible attack of a disease. It also helps in assessing the possible severity of the disease if it appears and, thus, directs suitable chemical protection measures. The knowledge of environmental relations of the pathogen also helps in designing strategies that can adversely affect survival of the pathogen in soil or in seed.

A study of the disease-cycle reveals that primary inoculum is seed and/or soil-borne or is brought by wind from external sources of survival into the main crop. Prevention of primary infection or initiation of the disease in the crop depends on the management of this primary inoculum. If primary inoculum is stopped from becoming active, there should be no disease in the crop. In other word, if seed and soil are free from inoculum and entry of the pathogen is prevented in the standing crop there should be no disease. This ideal situation can hardly exist anywhere. Only in such diseases where there is a single source of survival and entry of the pathogen in the field, complete prevention of the disease is possible by creating the above ideal situation. Loose smut of wheat is an example. Nevertheless, integration of different approaches to disease management nearly meet the goal of achieving the three ideal conditions. For most diseases in which the pathogen is seed and/or soil-borne and the disease has secondary spread in the field (polycyclic diseases) or diseases in which the inoculum comes from outside, as in cereal rusts, integrated principle or management are always recommended.

BASES OF DISEASE MANAGEMENT PRINCIPLES

The conventional approach to disease management involves the **immunization-prophylaxis system**. It is based on the fact that cure of a diseased plant is not possible because the disease becomes visible only after injury to the plant has taken place. Therefore, preventive measures are most important. These preventive measures involve induction of resistance in the plant (immunization) and protection of the plant by prophylactic measures. The approach is the backbone of all the methods recommended under integrated pest control, subsistence or sustainable methods, etc.

I. Immunization: 1) Induction of resistance
        a) Genetic manipulation
        b) Systemic acquired resistance
    2) Chemotherapy. Use of systemic fugicides and

antibiotics

II. Prophylaxis: 1) Legislation: quarantines, seed inspection and certification for exclusion of the pathogen.
2) Protection:
   a) Chemical
   b) Cultural
3) Eradication:
   a) Rotation
   b) Sanitation
   c) Eradication of alternate and collateral hosts
   d) Chemical

The groups of management methods listed above can be classified under **manipulation of the disease traingle**. In the chapters on epidemiology and effect of environments it was mentioned that a plant disease is the function of interaction between three forces, the host, the pathogen and the environment. None of these alone can operate to cause a disease. This constitutes the disease triangle Since the three are intimately related in the causation of a disease the principles of management should tackle all the three as shown in

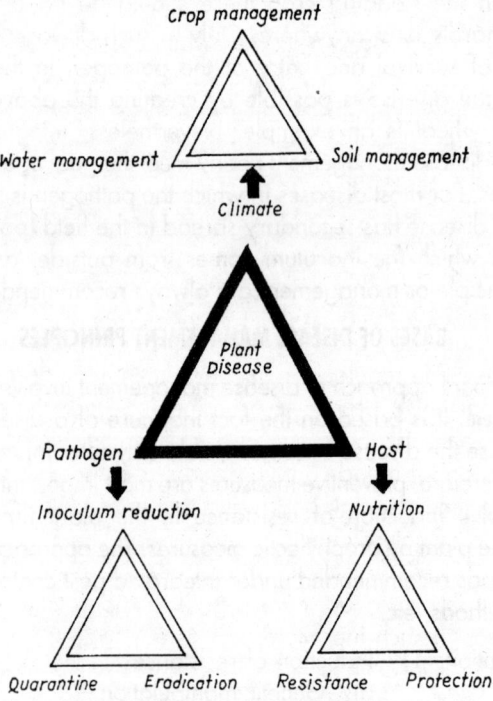

Figure 34. Management of the Disease Triangle.

the following figure.

On the same basis, an appropriate approach for disease management will be integration of methods directed against the pathogen, in favour of the host and for modification of the environment. This forms the basis for **integrated disease management**. Management of the pathogen involves the practices directed to prevent entry of inoculum by legislation (exclusion) and reduce inoculum and eradicate inoculum (eradication). Management of the host involves the practices directed to improve plant vigour and induce resistance through breeding, biotechnology (genetic engineering), nutrition, and enabling the plant to acquire systemic resistance during its life time. It also includes chemical protection against attack of the pathogen and avoiding disease by planting at a time when, or in areas where, inoculum is ineffective due to environmental conditions or is rare or absent (avoidance). The management of the environment involves water management, soil management and crop management. The chemical methods of plant disease management cover both the aspects, the inoculum reduction and eradication (management of the pathogen) and protection of the host (host management). Similarly, cultural practices not only cover the aspects of pathogen management but also management of the host as well as the environment.

EXCLUSION OF THE PATHOGEN

The principle of exclusion applies to management of the pathogen. The aim is to prevent entry of a pathogen in a field or area (state, country) supposedly free from that pathogen. Following types of activities are involved.

1. Quarantine: Plant quarantine aims at preventing entry of pathogens from infested areas into non-infested areas at international or national level. If in a particular area some disease is present in serious form and is likely to be disseminated by propagating materials, the government passes necessary regulations to stop entry of such materials from the infested area. These are known as quarantine regulations. For implementation of these regulations at the international level, proper check is maintained at the points of entry (airports and seaports). Suspected material is kept under quarantine for a specific period and if found contaminated it is either destroyed or effectively treated. Quarantine regulations are justified only when the inoculum cannot be disseminated by natural agencies and imported host material is an effective medium of spread of some known dangerous disease. Such regulations may become necessary when dormant organs of a pathogen are likely to be introduced through such inert materials as soil, wood, packing cases, etc.

As a principle, quarantine is one of the most effective weapons against plant diseases. However, as a common practice its effectiveness is doubtful. These regulations can obstruct trade. In countries where food grains are imported from abroad quarantines usually have no meaning. Such plant materials cannot be treated with toxic chemicals and there are chances that the imported grain is used as seed by some ignorant farmer. If a new pathogen is introduced in this manner it can become a serious threat to the crop when environments favour the development of the disease. One of the best examples demonstrating loopholes in quarantine regulations is that of citrus canker. The disease had been introduced into USA (in 1910), South Africa and many other countries through planting material from south and southeast Asian countries. In the USA and also in Australia, the disease had been eradicated through mass destruction of diseased or suspected to be diseased plants in the nursery and orchards. Strict quarantine was imposed against entry of citrus plants and even fruits from abroad. In spite of these regulations, the disease reappeared in USA in early 1980s. The reappearance was traced to entry of canker affected citrus in spite of quarantine. In Australia also fresh outbreaks of the disease were reported in 1981, 1984 and 1991. These were suspected to have originated from illegal importation of citrus into isolated home gardens in one part of the country. Domestic quarantine in large countries such as India is practically impossible although there is need to implement it to prevent spread of such diseases as banana bunchy top and potato wart. Check of the imported material at the seaports and airports is not so difficult. But there is no check on the road transport. Pathogens may travel with dust on vehicles moving through infested areas. In spite of these limitations quarantine is accepted as a necessary procedure in the fight against global spread of destructive diseases.

2. Inspection and Certification: The crops grown exclusively for seed are periodically inspected for presence of diseases that are disseminated by seed. Necessary precautions are taken to remove the diseased plants. The produce is then certified as seed. There may be further checks at the seed processing plants for presences of inoculum in the seed. The badly affected plots and seed lots are usually rejected. The method is supposed to prevent regional and inter-regional spread of seed-borne pathogens.

3. Seed Treatment: Seed tubers, grafts, bulbs and other propagative materials can be given heat, gas or chemical treatments to exclude the pathogen present in or on them. The method is used for exclusion by eradication. Seed treatment is mandatory for seed agencies supplying certified seed. The treatments reduce loss in germination and development of the disease in the field. Generally, seed treatments are prescribed for imports also and the exporting agency has to provide a certificate regarding this for quarantine purposes.

4. Eradication of Insect Vectors: For effective exclusion of pathogens that can gain entry into a new area through insect vectors or carriers, particularly insects having long flight range, a check on these vectors is necessary. Since the flight of insects cannot be checked, the crop should be given insecticidal cover before arrival of the vectors on the plant surface.

AVOIDANCE OF THE PATHOGEN

The principle of avoidance involves tactics that prevent contact between the host and the pathogen, presuming that the pathogen has crossed the barriers placed by exclusion or it is already present in the area and can attack the host. Avoidance is not applicable to diseases in which host is in a susceptible stage for a long time or showers of inoculum continue for months. Under avoidance, following strategies are included:

1. Choice of Geographic Area: Selection of geographic area for any crop is made on the basis of suitability of climate for the crop. The same climate may be suitable for the activities of the pathogen also. Many fungal and bacterial diseases are more severe in wet areas than in dry areas. Crops susceptible to such diseases in wet areas are not profitable. These can be grown in dry areas with the help of irrigation. Bean anthracnose is common in wet areas where seeds produced are generally infected. For seed production of bean dry areas are always preferred. Smut and ergot of pearl millet are serious in areas where rains occur for long durations during flowering of the crop. Cultivation of this crop in such areas is therefore not profitable.

2. Selection of Field: Successful cultivation of a crop depends, to a great extent, on selection of a proper field. If a disease caused by a soil-borne pathogen has been located in a field the field is not put to the same crop for some time. In such diseases as bacterial wilt of potato, wilt of pigeonpea, ergot and smut of pearl millet, ear cockles of wheat and root knot nematodes the infested field can be avoided.

In selection of fields management of drainage is also important. Poor drainage aggravates many diseases. In fruit orchards the choice of land for starting an orchard is most important because the trees are perennial and occupy the land for decades. If proper selection of land is not done, the trees may show signs of degeneration after few years when roots grow deep in the soil. Hard pan present in the subsoil obstructs root growth and nutrient availability. Apple orchards planted on land cleared from oak forest usually show serious incidence of collar rot.

3. Choice of Time of Planting: In many diseases the incidence or disease severity is most serious when the susceptible stage of the plant growth coincides with favourable conditions for the pathogen. This coincidence can be avoided by alteration in date of planting. It helps in avoiding the critical period. Thus, late sown winter crops escape incidence of root rot and wilt favoured by high temperatue and moisture that usually occur after the summer rainy season.

4. Disease Escaping Varieties: In different crops, certain varieties escape damage by disease because of their growth characters, not due to their genetic resistance to the disease. In India, varieties of pea that mature early (by January) generally escape much damage from powdery mildew which becomes serious in January or later. Groundnut varieties with erect habit suffer less from damage by leaf spots.

5. Selection of Seed and Planting Material: Diseases which are carried by seed or vegetative planting material and spread the infection in the field require proper selection of seed to avoid multiplication of the pathogen in the field and contaminate the healthy crop. Planting of disease-free seed in pathogen-free soil is often the most effective method of control of certain diseases.

ERADICATION OF THE PATHOGEN

The principle of eradication aims at removal of the inoculum already present in the field or the crop. Total eradication being not possible, the aim is to reduce the inoculum density to a level where it cannot cause significant damage. This is attempted through biological means, crop rotation, eradication of diseased plants or plant organs and physical and chemical treatments.

1. Biological Control of Pathogens: The biological control aims at eradication and reduction of inoculum and protection of plant surfaces through the activity of other microorganisms. It includes activities that enhance microbial numbers and quality on plant surfaces in soil or on the leaf. The antagonistic components of the microbial population may have biostatic or biocidal effect against the pathogen. Even non-antagonistic organisms can provide resistance to the host against pathogens.

2. Crop Rotation: When the same crop is raised year after year on the same land the soil-borne pathogens of that crop easily perennate in the soil and increase their population. After some time the soil becomes so heavily infested that it becomes unfit for cultivation of that crop. On the other hand, when immune, resistant or non-host crops are grown for a definite duration after a susceptible crop in the field it is expected that the pathogen will be weakened, starved and killed. It is also possible that different crops modify the chemical and biotic environment of the soil against the pathogens. Crop rotation is one of the oldest methods of fighting soil sickness and root diseases. The method is more effective against pathogens which have limited host range and restricted survival ability in soil. It is not effective against pathogens that have high degree of competitive survival ability, persisting in soil for long in absence of the host and those that have a large host range. In systems for sustainable agriculture recommended for intensive farming, rotation is generally not recommended and is substituted with other biological and chemical methods.

3. Removal and Destruction of Diseased Plants or Plant Organs: The presence of diseased plants in the field or orchard is a source of continuous release of inoculum. Therefore, as far as practicable such plants or their affected organs should be removed and destroyed to reduce the amount of inoculum. On the same basis, the removal of alternate or collateral hosts is also recommended. In removal of diseased plants or plant organs some precautions are necessary. Plant organs bearing dispersible pathogen propagules and their vectors should be removed carefully so that they are not dispersed while the plant or its organ is being physically removed.

(i) Roguing: This practice involves removal of diseased plants or their affected organs from the field. If done properly it checks spread of the disease to healthy plants and also helps in production of disease-free seed. In orchards, where removal of the entire tree is not feasible unless it is badly damaged, the affected organs can be cut or scrapped and destroyed by burning. Roguing is employed in such diseases as loose smut of wheat, loose and covered smuts of barley, red rot of sugarcane and many wilt diseases. The method is practical only when the size of the field is not very large and number of diseased plants is not very high.

(ii) Eradication of Alternate and Collateral Hosts: Many diseases especially those with continuous infection chain persist through alternate or collateral hosts of the pathogen. The primary inoculum is produced on and dispersed from these sources. If these wild or un-economic hosts of the pathogen are destroyed the source of primary inoculum is eliminated and chances of intiation of the disease in the crop are reduced. The method applies to diseases caused by fungi, bacteria, viruses as well as nematodes. It is feasible in and around the field and along the irrigation channels and ponds.

(iii) Sanitation: Field sanitation is essential for control of soil-borne and facultative parasites or saprophytes. Many obligate parasites also perennate through dormant structures in plant organs lying in or on the soil. Destruction of crop debris by burning in the field decreases this type of survival of pathogens in the field. Burying the crop debris deep in the soil by soil turning ploughs also inactivates inoculum of many pathogens. Sanitation is very important when diseased crop residue is left on the field as a general practice by the farmers.

4. Heat and Chemical Treatment of Diseased Plants: The pathogen present in the plant or in its special organs can be inactivated or killed by heat or chemical treatments. This approach has been found useful mostly in virus diseases of fruit trees. Heat therapy inactivates viruses in fruit tree seedlings and grafts and destroys the exposed fungal and bacterial propagules. Bare root dip in nematicides or fungicides is a method of sanitizing the seedlings before transplanting.

5. Soil Treatments: The aim of soil treatment is to inactivate or eradicate the pathogens present in the soil. It involves the use of chemicals, heat and such cultural practices as flooding and fallowing.

In chemical treatment of soil, fungicides and fumigant or granular nematicides are generally used. The fungicidal dusts can be used at the time of planting of the crop. Most of the fungicides used are selective in action and destroy only specific fungi. Therefore, by their use the development of other non-target pathogens may increase due to reduced microbial competition. Majority of the soil fumigants that were recommended in the past for nematode control destroyed microorganisms, including the beneficial bacteria, in the soil. Reinfestation of such soil with fungal pathogens is easy and quick. These soil fumigants were applied few weeks before planting of the crop to avoid injury to seed and seedlings. The garnular, contact and systemic nematicides have now replaced fumigants. These can be applied at the time of planting and in soil around roots of fruit trees.

In chemical treatment of soil the soil around the roots must be treated. Therefore uniform treatment of the soil all over the field is essential. As a result, the cost of chemicals required is very high and the treatment becomes expensive. Such treatments are practical for nurseries, small field plots, and for cash crops where higher income proportional to the expenditure is ensured.

For small quantities of soil or for small plots, heat treatment is an efficient method of eradication of pathogens. Burning of crop debris in the field gives partial heat treatment of the top soil. Soil solarization is a novel method of soil treatment to destroy most fungal, bacterial and nematode propagules as well as weed seeds. It is a system of raising temperature of wet soil kept covered with polyethylene sheets which traps the solar heat. The system is highly useful for sanitizing the nursery soil and small field plots. In many countries steam is used for treating glasshouse soil.

Flooding of the field is a method of eradicating fungal and nematodes pathogens from the field. If about 30 cm deep water is allowed to stand in the field for several weeks, the anaerobic or low oxygen conditions and toxins produced by anaerobic bacteria destroy fungal sclerotia and plant parasitic nematodes. Resting structures of many fungal pathogens float on the surface of the water and if the flood water is rapidly drained such structures are washed out of the field. The feasibility of this method depends on availability of water and topography of the land. As a substitute, wet rice culture in the rotation is recommended. Fallows, though not recommended in intensive, sustainable agriculture, are able to restore micronutrients in the soil and reduce the level of inoculum of many soil-borne pathogens.

PROTECTIVE MEASURES

The inoculum of many fast spreading infectious diseases, especially those caused by fungi, is brought by wind from neighbouring fields or any other distant place of survival. Principles of exclusion, avoidance and eradication are generally inffective or not sufficient to prevent development of such diseases. Plant is, therefore, provided some protective cover to face such pathogens. These measures include use of chemical sprays and dusts to create a toxic barrier between the host surface and the propagules of the pathogen and necessary modification of the environment to make it unfavourable for development of the pathogen.

1. Chemical Treatments: The aim of most chemical sprays, dusts and seed treatment is to form a protective toxic layer on the host surface so that when the pathogen comes in contact with the surface it is killed or prevented from growth. The chemicals used for this type of protection are called protective chemicals. When these chemicals destroy the pathogen that has already landed or has established infection they are called eradicant chemicals. The same chemical can be protectant as well as eradicant. The use of systemic fungicides or nematicides is protectant as well as eradicant. It destroys the pathogen present within the plant.

2. Control of Insect Vectors: Many species of insects are important vectors of viral and other diseases. Some viruses are transmitted only by insect vectors. Timely and effective destruction of these vectors is the most important approach to control of such diseases. Although, theoretically, this approach should prove very effective, in practice it has been often found not so encouraging. If a large population of the vector(s) attacks a crop and feeds on the diseased plants which have been sprayed with an insecticide, many of them escape instant death and may visit healthy plants and transmit the virus before being killed. However, later spread of the diseased is significantly reduced if majority of the vectors have come in contact with the toxic chemical. The success of chemical control of insect vectors depends to a great extent on the stage of plant growth and nature of the pathogen. It also depends on the speed with which the chemical can kill the insect. Those chemicals which kill the insects within few seconds are most effective in control of insect-transmitted diseases.

3. Modification of the Environments: Improvement of aeration under crop canopy reduces humidity on leaves and other aerial parts and thereby checks growth of fungi which flourish in humid atmosphere. Reducing the number of irrigations also helps in modification of environment against certain diseases. Mixed cultivation of crops, one of which provides ground cover, often provides low temperature and high soil moisture. These conditions are not favourable for some root pathogens. Root diseases favoured by high temperature are often controlled by irrigation. Aeration through proper ventilation of the store house provides proper environment for storage of plant products, especially those with succulent tissue. Postharvest decay of fruits and vegetables is reduced or prevented by cold storage which provides a modified environment unfavourable for decay causing organisms.

5. Modification of Host Nutrition: Host nutrition often influences development of a disease in the plant. It generally acts through stengthening of the tissues. Many leaf diseases are favoured by high level of nitrogen in the soil. Lowering nitrogen application in such diseases is a method of checking such diseases. In rice application of 100 kg N/ha instead of 120 kg is recommended to prevent losses from leaf spots, blast and sheath blight. Deficiency of potash in plants renders the tissue susceptible to water soaking and susceptible to many diseases. High calcium increases resistance to wilt and soft rot diseases through strengtening of pectic substance in the cell walls and obstructing the activity of pectic enzymes of the pathogen. Intensity of several diseases is decreased by such micronutrients as zinc, boron, manganese, etc.

DEVELOPMENT OF RESISTANCE IN THE HOST

Resistant varieties are a cheap method of disease management. The conventional approach had been to select resistant plants or develop resistance through hybridization. This type of resistance is genetic. With developments in biotechnology it has been possible to introduce resistance genes in the plant by other methods also. Biochemical resistance of non-genetic nature can be developed in plants by chemotherapy or host nutrition. This type of resistance is temporary.

1. Selection and Hybridization: Selection of resistant individual with poor commercial qualities and hybridizing them with suceptible plants of high commercial qualities is the aim of developing resistance through hybridization. The genetic and molecular basis of this type of resistance has been explained in an earlier chapter.

2. Genetic Manipulation through Biotechnology: Manipulation, genetic modification and multiplication of plants through such techniques as tissue culture and genetic engineering is now used in many crops. This has enabled hybridization between species which normally do not hybridize and hybridization of plants that normally do not produce viable seed. Creation of transgenic plants in which resistance genes from sources other than the particular plant species or in which avirulence genes of the pathogen are introduced to impart resistance is now possible.

3. Induction of Acquired Resistance: There are now numerous examples in which the plant acquires localized or systemic resistance during its life time through the effect of chemicals or microorganisms. Use of phosphates and carbonates as foliar therapy have been found to "turn on" the resistance genes which were otherwise lying "turned off". The epiphytic microflora on leaves may cause stress leading to induction of acquired resistance. The rhizobacteria also are known to induce systemic acquired resistance in the foliar parts against many diseases simultaneously.

4. Resistance through Chemotherapy: Temporary physiological resistance in plants can be developed through chemotherapy. Systemic fungicides and antibiotics when applied to the foliage or through the roots persist in the plant for some time and while their toxic level is maintained the pathogen cannot invade the tissue. Systemic nematicides applied to soil and taken up by the plant keep away not only the nematodes but also aphids and leafhoppers for several weeks. This protects the plants against viruses.

5. Resistance through Host Nutrition: Nutrition cannot change a susceptible variety to a resistant variety. But making available major and micronutrients through foliar sprays, seed treatment or soil application is reported to strengthen the tissues that can ward off invasion by the pathogen. Although the effectiveness of this approach is doubtful, a vigorous growth of the plant is always desirable. Vigorous plants with capacity to form new roots and shoots to replace the damaged once tolerate the attack of many diseases.

THERAPY OF DISEASED PLANT

Although cure of the diseased plant or its organ in most crops is not possible, in many crops and fruit trees chemical and physical therapy has been applied to cure the plant by eradicating the pathogen.

1. Chemotherapy: Chemical treatments applied to eradicate the pathogen from the tissues of the diseased plant and thus curing it are included in chemotherapy. Such chemicals are called chemotherapeutants. The chemicals are mainly systemic fungicides and antibiotics. The principle underlying chemotherapy is that the chemicals used are absorbed by leaves and roots and on reaching the site where the pathogen is present they either kill it or incapacitate it by preventing sporulation, growth or both. So long as they remain at a toxic level they provide temporary resistance also. The chemotherapeutants can act also by detoxyfying the toxins produced by the pathogen. In this way, the tissues not invaded by the pathogen are saved and the plant is cured.

2. Heat or Theromotherapy: Plants which can tolerate the thermal inactivation or death point of the pathogen can be treated by heat to destroy the pathogen. These treatments are especially used for seed, tubers, bulbs and grafts. Grafts of fruit trees are exposed to high temperatures for inactivation of many viruses. For inactivation of nematodes from roots of grafts heat therapy has been suggested. Ratoon stunting disease bacterium and many viruses of sugarcane are eradicated by hot water, air or moist hot air treatment of the seed canes.

3. Tree Surgery: Large size fruit trees are cleaned of infection by cutting or scrapping of the diseased part and covering the wound with a fungicidal paste. This removes the infection.

18

Disease Management—The Practices

The principles under the immunization-prophylaxis system detailed in the preceding chapter are given practical shape through measures under cultural practices, biological control, use of chemicals and induction of resistance. These measures cover the aspects of pathogen management, host management and the management of the environment.

I. CULTURAL PRACTICES FOR DISEASE MANAGEMENT

Adjustment of crop management procedures has been an age old practice with the formers for prevention of losses in crops due to diseases and other causes. It is an integral part of subsistence agriculture in developing countries. In spite of developments in chemical methods and resistant varieties this approach has excellent promise even for the future. Cultural practices are now being considered as essential back up procedures for management of resistant varieties and also for chemically protected crops. Management of plant diseases through cultural practices involves the principles of avoidance, exclusion and eradication. Successful use of cultural practices for disease management requires complete knowledge of the nature of the pathogen and its behavior in different conditions of the environment-climate, cropping systems, etc. Often the cultural practices are the only feasible methods of disease control in crops which give low return per unit area or in which resistant varieties are not available.

For development of a disease, contact between the host and the parasite must occur in an environment that is favourable for the pathogen and pathogenesis. Suitable modification in cultural practices can modify the environment that is not favourable for the pathogen but favorable for the host. On this basis disease control by cultural practices is mainly preventive. Many practices reduce the density and activity of the inoculum. Precautions taken under avoidance are also mainly cultural practices. Many cultural practices act against the pathogen through a system of biological control. However, the objective of complete disease management is met only by integrated approach involving cultural practices, chemical treatments and resistant varieties.

1. PRODUCTION AND USE OF DISEASE-FREE PROPAGATING MATERIAL

A large number of fungal, bacterial and virus pathogens are transmitted through true seed or vegetative propagating material. For effective disease control this source of primary inoculum must be taken care of. Seeds carry the pathogens as (i) internally seed-borne infection, (ii) externally seed-borne inoculum, (iii) contaminants with the seed, and (iv) through nursery raised planting stock. Although the infested or infected propagating material can be made pathogen-free by chemical or physical treatments, production of such seed in the field is the first and important step. Following practices are followed to produce and use pathogen-free seed material.

i) Dry climate for seed production: Control of seed-borne diseases favoured by wet climate can be achieved by raising the crop in dry areas. Some examples are anthracnose of bean (*Colletotrichum lindemuthianum*), anthracnose of cucurbits (*Colletotrichum lagenarium*), Ascochyta blight of pea. (*Ascochya spp.*), bacterial blight of legumes (*Pseudomonas syringae* pv. *pisi* and *Xanthomonas axonopodis* pv. *phaseoli*) and black rot of cabbage (*Xanthomonas campestris* pv. *campestris*). For producing seed of such crops dry areas are preferred where leaf wetness is avoided. Reasonably good dry climate can be created in wet areas through management of crop canopy facilitating air circulation and entry of sun rays.

ii) Isolation distance for seed plots: Separation of seed plots from sources of inoculum (commercial crop in the neighbourhood without proper protective steps) helps in production of healthy seed. In the production of certified seed, a particular distance between plots is mandatory.

iii) Inspection of seed plots: Periodical inspection of the crop being raised for seed or orchards producing grafts and seedlings of fruit trees for distribution is an important step. Eradication of diseased plants or plant organs immediately follows the inspection. If the crop is badly diseased the plot is rejected for seed. The procedure is followed in production of seed of wheat, seed tubers of potato, seed material of sugarcane, and seedlings and grafts of citrus and apple.

iv) Drying and ageing of seed: Some pathogens do not tolerate drying of seeds. The pathogen of downy mildew of maize (*Peronosclerospora sacchari*) is present as mycelium in freshly harvested grains. When the seed is properly dried before storage the mycelium collapses. Viability of some seed is longer than the pathogen present in them. Thus prolonged storage often eliminates the pathogen from the seed. *Fusarium solani* f.sp. *cucurbitae,* infecting different cucurbits, is internally seed-borne. When the infected seed is stored for two years before sowing the fungus is killed. Similar eradication of the pathogen causing anthracnose of cotton is also reported. The cotton blight bacterium (*Xanthomonas axonopodis* pv. *malvacearum*) survives in the seed for a year or so. Storage of the seed for more than a year eliminates the pathogen. In such methods proper conditions for storage must be maintained to avoid harm to the seed.

v) Cleaning of seed: Sclerotia and oospores of many fungal pathogens and cockles or cysts of nematodes may be present in the debris mixed with the seed. Common example are ergot and smut of pearl millet, white blisters of crucifers, cyst nematode of sugarbeet, and ear cockles of wheat. Cleaning of seed is done by hot air blast that removes the dust also and by hand. In hand cleaning the seed is submerged in a 20% common salt solution. The debris and nematode cysts or cockles float on the surface and can be skimmed off by hand.

vi) Thermal and chemical treatment of seed: Heat or chemical treatment of seed before storage is a part of cultural practices for healthy seed production. Chemical seed treatment is a compulsory step in the production of certified seed. Thermal treatment of seed is favored in those cases where the pathogen is deep seated and ordinary protective fungicides cannot reach the pathogen. Many systemic fungicides are capable of reaching the internal tissues. Treatment of wheat seed with Vitavax and thiram, and hot air or hot water treatment of sugarcane seed material, before sowing are some examples of seed treatment for eradication of the pathogen. Bare root dip treatment of seedlings with nematicides is practiced to control root knot and other nematodes.

vii) Site and treatment of nursery beds: Diseases like club root of cabbage, root knot of tomato and citrus gummosis are generally carried by seedlings or grafts from the contaminated nurseries. The nursery site should be chosen with care avoiding locations near infested fields and the soil should be periodically treated with chemicals or heat. Soil solarization is feasible method for nursery soil treatment. The other method is to burn a heap of farm trash over the beds. Soil fungicides are also available for chemical treatment.

viii) Adjustment of harvesting time of the crop: Time of harvesting affects cleanliness of the seed. Delayed harvesting of grain crops in temperate regions gives the pathogens more time for contaminating the seed. Grain crops harvested in wet weather often produce contaminated seed. Harvesting of potato when the leaves are still green allows the late blight pathogen to contaminate tubers which carry it to the next season. Such situations can be avoided by suitable alteration in the timing of harvest of the crop.

2. ADJUSTMENT OF CROP CULTURE TO MINIMIZE DISEASE INCIDENCE

Modifications in crop rotation and choice of crops, method of planting, irrigation and fertilizer application, special treatment of soil such as green manuring, organic amendment, and many other crop culture practices have been found to provide reasonably satisfactory control of plant diseases.

1) Crop rotation: Growing crops in rotation has many benefits, such as:
   a) better use of nutrients,
   b) desirable effect on soil texture with deep rooted crops alternating with shallow rooted crops, *viz.,* a cereal with legume,
   c) water economy, in particular, conservation of water in years of fallow.

d) weed control in row crops in which in-growth tillage to remove weeds could be practiced, alternated with crops sown by broadcasting; crops likely to inhibit weeds by their rapid growth and dense foliage alternated with slow growing crops having sparse foliage.

e) suppression of soil borne pathogens.

Most of the diseases caused by soil-borne pathogens can be significantly reduced by crop rotation. Examples are wilt diseases of pigeonpea, chickpea, pea, cotton and linseed, red rot and wilt of sugarcane, ergot and smut of pearl millet, bunts and flag smut of wheat, leaf smut and bunt of rice, bacterial wilt of potato and tomato, and cereal cyst nematode. Some specific effect of succeeding crop on the pathogen of the preceding crop are listed in the following table.

TABLE 17: Effect of Short Term Rotations on Some Pathogens.

Beneficial crop	Pathogen reduced	Preceding crop (host)
Rice	*Verticillium dahliae*	Cotton
Pea	*Gaeumannomyces graminis*	Wheat
Sudangrass	*Ralstonia solanacearum*	Tomato
Maize, wheat or sorghum	*Ralstonia solanacearum*	Potato
Legume cover crops	*Streptomyces scabies*	Potato
Barley	*Meloidogyne incognita*	Cotton
Legumes, sesame and wheat	*Pratynechus indicus*	Rice
Groundnut	*Meloidogyne incognita*	Tomato
Beet	*Pratylenchus penetrans*	Cereals

The success of crop rotation for disease management depends on proper selection of crops in the sequence. The crop(s) grown between the susceptible host crops should be resistant or immune to the pathogen or should be non-host and their root exudates should not directly or indirectly favour survival of the pathogen. In case of pathogens having very large host range such as the root knot nematodes of vegetable crops, choice of the crop from various vegetables is sometimes difficult. The vegetables have to be rotated with cereals like wheat or rice. However, immune or highly resistant varieties of the vegetable crop such as tomato can be included in the rotation.

Rotation does not help against pathogens which have strong saprophytic survival ability in absence of the host. The structure of survival and its longevity in soil should be known to decide the length of rotation or gap between the susceptible crops. A one year rotation for pigeonpea wilt is likely to fail if the plant roots which harbour the pathogen do not completely decompose. In bacterial wilt of potato, only very long rotations can work since the bacterium has unusual longevity in soil.

Although crop rotation is one of the oldest plant disease management practice, being in existence since ancient times, and has been an effective method in subsistence agriculture being followed by farmers having small holdings, it is not encouraged in sustainable agriculture systems which emphasize intensive cultivation with the help of chemicals. The choice of food grain crops being limited, it has its own limitation even with farmers having limited land. Thus, rice-wheat (both cereals) has become the predominant rotation in most parts or India. The ideal should have been a legume or green manure followed by wheat or rice followed by some winter and summer legume.

2. Fallowing: Keeping the land fallow for some time also has been an ancient agricultural practice. It can be a part of the rotation. Fallowing may have to be adopted out of sheer necessity, such as because of limited water supply or failure of a crop soon after planting and no time left for replanting of the same or another crop. It can also be adopted by choice in order to reap specific benefits. In fallows adopted due to necessity the field should be kept free from weeds and repeated turning of the soil should be done to expose the pathogens to elements of weather. Fallowing by choice is largely an economic consideration, taking into account revenue from the crops that could be grown during the fallow period, alternative methods of reducing pathogen population in soil, and benefits expected from the succeeding crop. Even where fallowing is known to give definite advantages in disease control, it is often not accepted because of the economic considerations. It is practicable on large farms practicing extensive farming. In subsistence agriculture with small holding the needs of the farmer may not permit him to keep the land fallow for one season.

Flood fallowing has been found effective in the management of such diseases as Fusarium wilt of banana and root knot nematodes. The method is not recommended in general crop culture. However wet rice culture where water level is maintained throughout the crop season is considered an alternative to flood fallowing.

3. Monoculture: In strict sense, monoculture means cultivation of single or closely allied crop species in annual or seasonal succession, with interruption only by fallow or by a green manure crop or by application of organic amendments, not necessarily after each crop. This excludes diversity and introduces element of danger in the crop-disease system. Plant pathologists view monoculture as a system of perpetuating the diseases. Apart from helping the pathogen survival, it may increase chances of new biotypes of the pathogen through selection pressure. Most soil-borne plant pathogens are favored by monoculture. *Verticillium albo-atrum* and *V. dahliae,* causing wilt of cotton, reach such a high level (600% increase per year) in monoculture that cotton monoculture becomes impossible. Root knot and cyst nematodes also show uninterrupted high rise in their population in monoculture.

However, monoculture has two advantages. For the farmer, it simplifies cropping by permitting full use of equipments for the crop chosen and it leads to specialization in the culture of the particular crop. In convertible soils, monoculture may induce disease suppressiveness in the soil. However, this is possible only when the soil is not put under any other crop and is not subjected to any heat or chemical treatment.

4. Mixed cropping: Mixed cropping is simultaneous cultivation of more than one crop in the same plot. The examples of mixed cropping are wheat + barley, wheat + chickpea, pigeonpea + sorghum, cotton + mothbean. The economic loss is reduced if the main crop is badly damaged by a disease. In addition, mixed cropping has several other advantages such as (i) increased availability of nitrogen for crops mixed with legumes, (ii) more efficient use of solar radiation due to better interception of light by foliage, (iii) shading effect, (iv) better use of soil moisture at various depths and (v) suppression of weeds. Mixed cropping is different from intercropping in which rows of the main crop are interspersed with rows of some other crop. The rows become plots of the main crop. Intercropping is not so effective in disease reduction as mixed crops sown by broadcast. Mixed crops of pigeonpea and sorghum give significant reduction in the incidence of Fusarium wilt of the former. In wheat or barley raised with legumes such as chickpea or pea, there is reduced spread of rusts. Intercropping of chickpea with barley gives significant control of Ascochyta blight of chickpea with maximum produce from the land.

Reduction in disease incidence in a mixed crop is attributed to one or more of the following:

i)   Due to reduction in the number of the susceptible plants there is sufficient spacing between diseased leave or roots and healthy plants. This prevents spread of infection by contact. The reduced susceptible surface causes reduced production of secondary inoculum for spread of diseases like rusts in the field.

ii) The roots of non-host plants may act as physical barriers obstructing movement, if any, of the pathogen. They may release toxic chemical in roots exudate that suppress growth of the fungus. HCN in root exudates of sorghum is highly toxic to the pigeonpea wilt fungus (*Fusarium udum*). A relationship between HCN content and linseed wilt (*F. oxysporum* f.sp. *lini*) is also reported.

iii) By proper selection of crops for the mixture, soil environment can also be changed that is unfavourable for the pathogen. Control of root rot of cotton by growing cotton with mothbean is an example.

iv) The soil-borne pathogens are only randomly present in the field as dormant structures. Many such structures require contact with host roots to become active. In mixed crops chances of such contact are reduced.

One of the limitations of mixed cropping is the economic loss due to reduced plant population of the main crop in no disease situations. However, for farmers with limited land area it has many benefits. He can harvest two food crops (cereals and pulses) in the same season, in addition to avoiding losses from cereal rusts.

Sometimes, mixed crops enhance disease incidence. This happens when both crops directly or indirectly favour a pathogen of the main crop.

5. Adjustment of date of sowing: There are numerous examples where early or delayed sowing of a crop enables it escape critical period of disease incidence. In northern India, when pea and chickpea are planted early (in October) they suffer heavily from root rot and wilt (a complex of *Fusarium, Rhizoctonia* and *Sclerotium*). This is because the high temperature and high soil moisture favour these pathogens. In addition, the early sown crops show much vegetative growth forming a dense canopy which provides high humidity under the plant that favours blight caused by *Sclerotinia* and also Botrytis gray mold. When these crops are planted late (end of November to December) there is little or no root rot and wilt and incidence of blight and gray mold is also reduced. Dense canopy also favours rapid spread of Ascochyta blight in chickpea. The white rust disease of mustards (*Albugo candida*) causes greater damage in crop planted late (after mid-October) than in crops sown early (late September to early October). Thus, suitable modifications of date of planning in areas where the disease is common can reduce losses.

The incidence of leaf spots of groundnut (*Cercosporidium personatum* and *Cercospora arachidicola*) reaches its maximum intensity towards the end of August. Advancing the planting date to June or even earlier if irrigation is available can permit the crop to escape the damage. In South India, rice sown from January to June develops no more than 5% leaf blast and 1% neck blast. In July planted crop the incidence increases to 20% and 25%. Maximum incidence of the disease occurs in crops sown from August to November. Other diseases favoured by late planting are Karnal bunt of wheat (*Neovossia indica*), stem rust of wheat (*Puccinia graminis tritici*), bacterial blight of cotton (*Xanthomonas axonopodis* pv. *malvacearum*), and downy mildew of maize (*Peronosclerospora sacchari* and other species). Suitable alteration in date of planting in all such diseases provides significant relief.

When warm season crops are sown at relatively low temperatures (maize planted in winter) the germination is slow. This gives seed rot fungi enough time to colonize and destroy the seed. In vector-transmitted plant viruses the intensity of disease can be reduced by avoiding the period when the vector population is high. Early planted potato crop reaches the stage of tuberization before the population of vectors reaches the peak in January. The okra yellow vein mosaic virus is not common in crops planted during February-March because the vector population is very low or absent. In crops sown in June or later, the disease is a serious threat due to high vector population.

6. Adjustment of depth of seeding: Varying the depth of sowing enables the host to avoid the pathogen. The shallow sowing is a method of reducing damping off of seedlings. Quicker emergence of seedlings in shallow sowing reduces infection of certain smut fungi which cause seedling infection. Deep placement of seed gives the pathogen more time to invade the seed and seedlings. Stripe disease of barley is favoured by deep placement of seed. Maximum disease occurs when seeds are placed 4–5 cm deep and minimum when planting is done at 2–3 cm. Infection of head smut of maize (*Sporisorium reilianum*) increases with plating depth. In grain smut of sorghum (*Sphacelotheca sorghi*) also conditions that induce slow emergence of seedlings cause more infection. But deep sowing reduces the infection of chickpea by *Ascochyta rabiei* although epidemiologically it may not be very important.

7. Plant spacing-rate of sowing and density of stand: Host density in the field is an important factor associated with most plant diseases. Plant spacing created by rate of seeding or by transplanting of seedlings or grafts affects disease through underground closeness of roots (for soil-borne pathogen) and through creation of a definite density of the foliage (crop canopy structure) for pathogens of aerial parts.

Crop canopy is a major factor determining the crop climate. Dense canopy provides shade, increases humidity under the crop, delays drying of soil under the plants, prevents aeration and radiation and lowers temperature. These conditions favour most foliar diseases including downy mildews, late blight, damping off, etc. The abundance and closeness of susceptible tissue causes rapid spread of the disease. Dense crop canopy results from close planting, high nitrogen doses, time of planting, and varietal characters. Where diseases favoured by high humidity are expected proper spacing between plants should be ensured. In damping off of seedlings in nurseries close seeding is always dangerous. Crowded seedlings remain in a juvenile stage for long and high humidity favours the damping off fungi to cause infection for longer periods. In downy mildew (*Plasmopara viticola*) and bunch rot (*Botrytis cinerea*) of grapes, the dense foliar canopy favours the diseases. Proper training of the vines and removal of excess leaves from near the bunches help in maintaining proper air circulation and penetration of light which reduce humidity and the diseases are significantly reduced.

In recommending plant spacing one should not lose sight of the fact that the overriding consideration in choosing rate of sowing is the economic one. When high seedling mortality is expected due to poor seed vigour dense seeding is desirable. Often densely sown and disease-prone crops have given better yield than crops sown less densely. There are also examples where dense planting is a method of disease reduction. The virus of tomato leaf curl, transmitted by *Bemisia tabaci,* is less severe in a crowded planting than in spaced planting. Same is true for cucumber mosaic, transmitted by *Aphis gossypii* and groundnut rosette transmitted by *Aphis craccivora*. Verticillium wilt of cotton also is less severe in close planting. The incidence of brown rot (*Cephalosporium gragatum*) of soybean is reduced in dense plant population.

The reduction is ascribed to the reduction of effective inoculum per plant in proportion to the increase in the number of plants per unit area in the densely sown field.

8. Management of irrigation: The aim of irrigation is to wet the soil only to the extent that roots easily get water and mineral nutrients. If excess water is added to the soil, it may directly affect activity of the pathogen and/or it may affect disease incidence through the effect on the host. Beneficial effect of irrigation is seen in the control of common scab of potato (*Streptomyces scabies*). Scab attack on potato tubers is prevented by maintaining soil moisture near field capacity during tuberization. The effect is mediated through increasing the activity of antagonistic bacteria on the lenticels. The charcoal rot fungus, *Macophomina phaseolina*, attacks potato tubers when the soil temperature rises and there is water stress. Irrigation lowers the temperature, removes the stress and charcoal rot is avoided.

The disadvantages of excess irrigation and prolonged stay of water around the root zone are many. The juvenile stage of seedlings is lengthened making them susceptible to damping off (*Pythium* spp.). Frequent but low quantity of watering is therefore recommended for avoiding damping off. In citrus foot and root rot caused by *Phytophthora parasitica, P. palmivora* and *P. citrophthora* the number of propagules of the fungi present in the soil increase rapidly immediately after irrigation in irrigated orchards and nurseries. The release of zoospores and their movement is optimal in saturated soils. In continuously wet conditions there is poor development of fibrous roots. If the irrigation is controlled so that soil dries rapidly between irrigations chances of spread of propagules are less and the susceptible feeder roots have time to harden and escape infection. Sprinkler irrigation generally favours diseases of the foliage by increasing leaf wetness and by dispersing propagules of the pathogen by water splashes just like rain water. Excess irrigation increases guttation on leaves. Guttation drops contain nutrients that favour such pathogens as *Helminthosporium* and many bacteria. Timing of irrigation should take into consideration the timing of most active phase of a pathogen. Irrigation should be avoided when the pathogen is likely to attack the host. In Karnal bunt of wheat high humidity during flowering promotes the disease. Where the disease is of regular occurrence, irrigation of wheat at the time of flowering should be avoided.

9. Management of host nutrition: Role of plant nutrition in disease incidence was explained in the chapter on effect of environments. Often suitable modifications in fertilizer doses or exogenous application of micronutrients reduce disease severity. Application of nitrogen more than the normal dose or requirement of the crop causes new succulent vegetative growth of the plant and delays maturity. Pathogens that attack new vegetative growth are favoured by high level of nitrogen. The late blight of potato is severe when due to high dose of nitrogen crop canopy is dense. In such cases, either the dosage should be reduced or spacing should be adjusted to offset the effect of nitrogen. In rice diseases such as false smut (*Ustilaginoidea virens*), blast (*Pyricuaria grisea*), sheath blight (*Rhizoctonia solani*), and bacterial leaf blight (*Xanthomonas oryzae* pv. *oryzae*) heavy nitrogen fertilization promotes the diseases. Since there is no difference in yield of the crops receiving 100, 150 or 200 kg N/ha, the minimum dose should be given. Heavy nitrogen application promotes the incidence of ergot of pearl millet also.

When nitrogen is deficient plant are weak, development is incomplete and maturity is hastened. Pathogens favoured by slow growth of the host are more severe in nitrogen deficiency. In this category are tomato wilt (*Fusarium oxysporum* f.sp. *lycopersici*), bacterial wilt of potato and tomato (*Ralstonia solanacearum*), root rot caused by *Sclerotium rolfsii,* and early blight caused by *Alternaria solani*. The role of chemical form of nitrogen was explained in the chapter on effect of environment.

Micronutrient deficiencies are common in all types of cultivated plants. Calcium is very important for strengthening the middle lamella and prevents soft rot in apples and potato. Zinc deficiency in rice is often a serious problem in many area of India. Zinc deficiency in guava is linked to wilt and decline of the trees. Association of magnesium deficiency with groundnut leaf spots and potassium and zinc with linseed wilt are reported. In maize, the downy mildews are more aggressive in zinc deficiency. Deficiency of boron in soil or its non-availability may result in disintegration of core tissues in apple and other fruits. Application of boron alone or with nitrogen reduces black rot of cauliflower (*Xanthomonas campestris* pv. *campestris*). The deficiency of these micronutrients can be checked by foliar sprays. However, excess of these micronutrients may harm the plant. Therefore, their application should be based on soil and leaf tests.

10. Management of soil acidity and alkalinity: Most fungi, bacteria and nematodes can tolerate the pH range in which their host plants grow. There are only some diseases in which the tolerance limits of the pathogen are narrow while the host can grow outside these limits. This provides a method for control of the specific diseases. However, directly or indirectly soil pH affects the speed of disease cycle and, thereby, the disease intensity in a larger number of cases.

The bacterium of common scab of potato (*Streptomyces scabies*) is sensitive to acidity. The disease is significantly decreased if the soil pH is brought down to below 5.2. Within the pH range of 5.2–8.0 severity of the disease increases with pH. Although lowering of soil pH by application of sulphur had been a successful method, the cost involved is prohibitive. In clubroot of crucifers (*Plasmodiophora brassicae*) raising the pH of field soil to 7 or above by adding lime (calcium carbonate) provides good control of the disease. In addition to lime, calcium sulphate (gypsum) can also be used.

11. Organic amendments of soil: Methods of providing organic sources of nutrition to plants include use of compost, farmyard manure, green manure and special organic amendments such as oil cakes alfalfa meal, wood sawdust, tree bark, etc. While compost and farmyard manure are added in decomposed form, the materials in green manure and special organic amendments are added in undecomposed form and decomposition takes place in the field soil where pathogens as supposed to be surviving. Apart from nutritive value, which is not directly is high as in inorganic fertilizers, the addition of organic matter has the major advantage of promoting diversified microflora in the soil including the rhizosphere. This provides advantages in disease management as explained in the chapter on effect of environment. Organic amendments are one way of promoting naturally occurring biological control of pathogens in the soil.

The reduction in common scab of potato (*Streptomyces scabies*) by green manuring with soybean through prevention of build up of inoculum was the first report, in 1926, providing experimental evidence of benefits of organic amendments. Since then, bulk of research reports in favour of organic amendments has increased at tremendous speed (*cf.* Singh, 2001). The following is a partial list of diseases in which organic amendments have caused suppression of the pathogen. Many other fungal and nematode diseases are

TABLE 18: Pathogens Suppressed by Organic Amendment of Soil.

Pathogen	Host/disease	Amendment
Aphanomyces euteiches	Pea root rot	Curcifer green manure
Phytophthora cinnamomi	Avocado root rot	Lucern, alfalfa meal.
Fusarium oxysporum f.sp. cubense	Banana wilt	Sugarcane bagasse
F. oxysporum f.sp. corianderi	Coriander wilt	Oil-cakes
Gaeumannomyces graminis	Wheat take-all	Rape, pea green manure
Thielaviopsis basicola	Bean root rot	Oat straw, corn stover, lucern hay
Phymatotrichum omnivorum	Cotton root rot	Green manure with pea, *Melilotus*
Macrophomina phaseolina	Cowpea charcoal rot	Margosa oil cake, FYM, barley
	Cotton root rot	straw, lucern hay
Rhizoctonia solani	Patato black scurf	Sawdust, oil cakes wheat straw
Streptomyces scabies	Potato scab	Soybean green manure
Verticillium albo atrum	Potato wilt	Barley straw
Meloidogyne spp.	Root knot	Oil cakes, sawdust green leaves
Heterodera avenae	Cereal cyst nematode	Crucifer green manure, Black mustard oil
Globodera rostochiensis	Potato cyst nematode	Farmyard manure

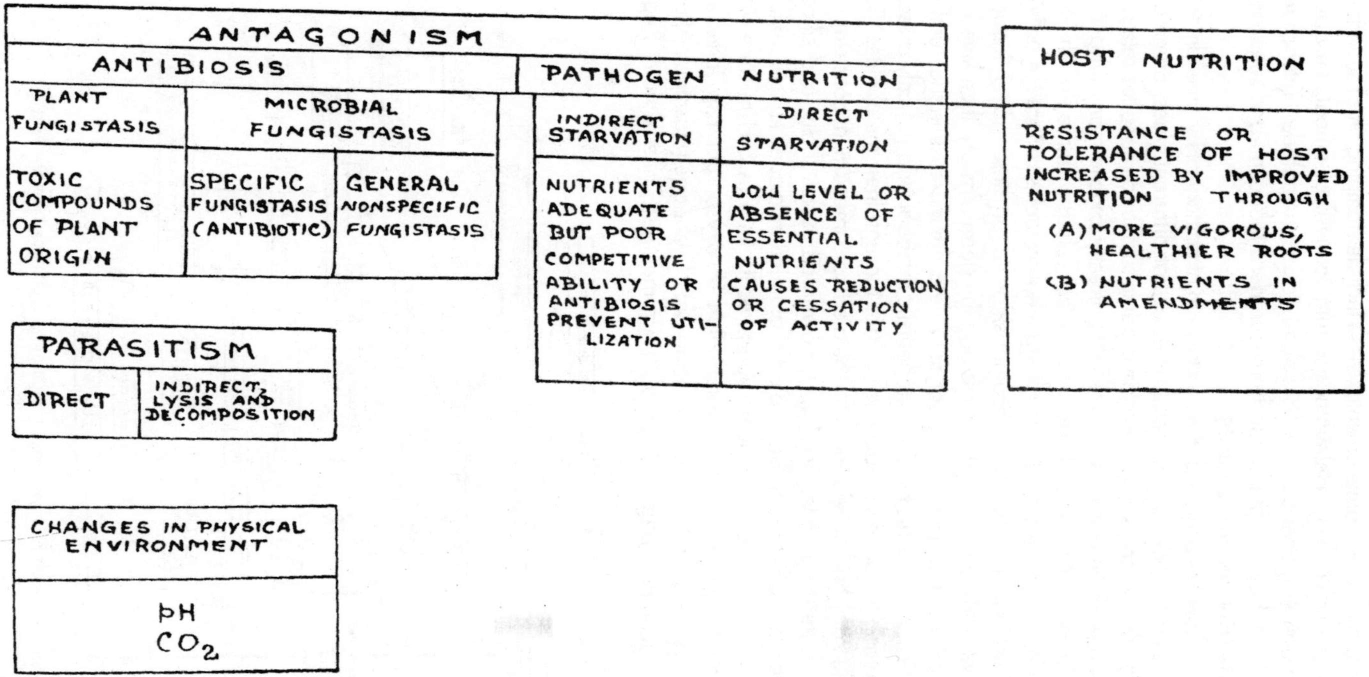

Figure 35. Mechanisms of Disease Control Through Organic Amendments (Slover 1962).

also suppressed by organic amendments. These include damping off of sugarbeet (*Aphanomyces cochlioides*), Helminthosporium foot rot of wheat, Fusarium root rot of bean, pigeonpea wilt (*F. udum*), citrus nematode (*Tylenchulus semipenetrans*), tobacco cyst nematode (*Heterodera tabacum*) and sugarbeet cyst nematode (*Heterodera schachti*).

Different mechanisms operate in the suppression of pathogens by amendments. The decomposition products and metabolites are of different types. Some are directly toxic to the pathogens. The volatile organic sulphur compounds released from the decomposition of crucifer leaves are directly toxic to many fungi and nematodes. Some materials such as lucern meal release ammonia which is toxic to fungi as well as nematodes. In alfalfa meal saponins reduce the activity of *Phytophthora cinnamomi*. The decomposition of organic matter induces germination of dormant fungal structures and the germlings are then killed by lysis (germination-lysis). The most important role is the increase in the number of antagonistic microorganisms. The antagonistic rhizobacteria not only directly inhibit the pathogens but also induce resistance through their metabolites. Sun and Huang (1985) had developed a formulated soil amendment consisting of agricultural waste (baggasse, rice husk), powdered oyster shells, and silicon dioxide, calcium oxide, magnesium oxide, aluminium oxide, ferrous oxide, urea, potassium nitrate and calcium superphosphate. This S-H mixture was effective against Fusarium wilt of watermelon (*F. oxysporum* f.sp. *niveum*), radish wilt (*F. oxysporum* f.sp. *raphani*), clubroot of cabbage (*Plasmodiophora brassicae*), cucumber blight (*Phytophthora melonis*), rice sheath blight

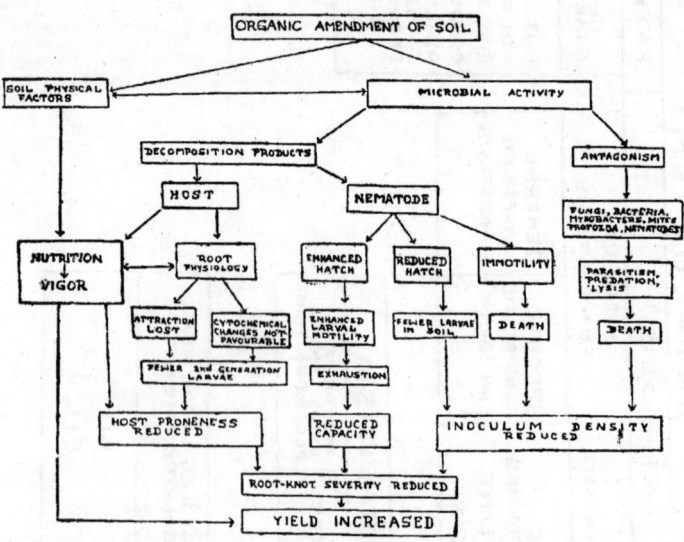

Figure 36. Possible Pathways of Action of Organic Amendments Against Root Knot Nematode (Singh and Ditaramaiah 1973).

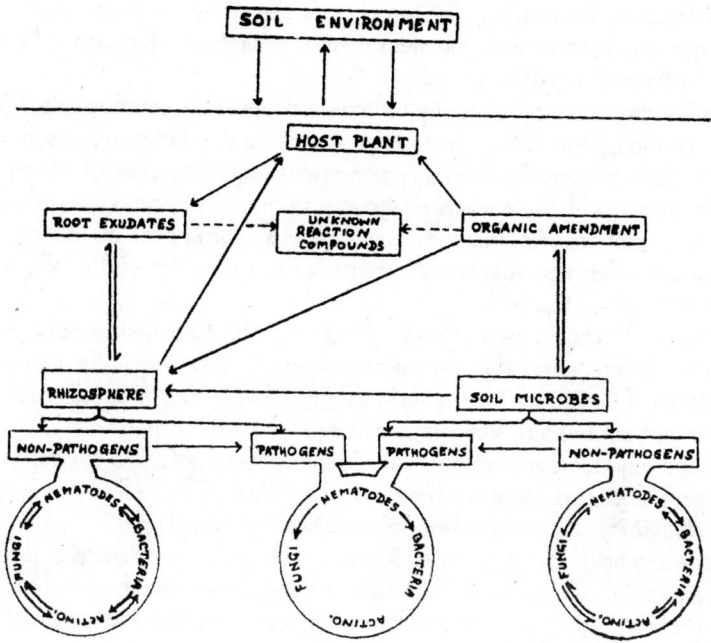

Figure 37. Interaction of Amendments, Rhizosphere and Soil Microflora.

(*Rhizoctonia solani*), and tomato bacterial wilt (*Ralstonia solanacearum*). They were of the opinion that a multitude of factors are involved in disease suppression. It could be due to direct effect of antagonists or specific effects of constituents of the mixture.

12. Management of the top soil: Mulching or covering of top soil with organic residues often helps in reducing plant diseases. Only those materials should be considered for mulching which are not related to the crop. If sugarcane leaves are used as mulch in sugarcane field to suppress weeds, it will perpetuate the pathogens of sugarcane. However, unrelated materials like pine needles, waste paper, wheat straw, etc., have been used as mulch, especially in vegetable crops, with success in reducing diseases, particularly the nematodes. These mulches are known to release inhibitory substances in the underlying soil and also promote development of parasites and predators of nematodes.

Shaping the top soil into ridges or beds is a part of tillage and is a well known and useful means of keeping dry the soil that is in direct contact with collar and root crown. Ridges have proved effective in reducing incidence of *Sclerotium rolfsii* on many vegetables and in groundnut, Pythium fruit rot of cucurbits and *Sclerotinia sclerotiorum* in some vegetables. High riding (15 cm or more) prevents infection of potato tubers by zoospores of the late blight fungus from leaf lesions. Incidence of *Rhizoctonia solani* in many vegetables is also reduced by ridging. The practice is generally recommended where water accumulation around the stem base is a problem. Ridging is harmfull where water deficit exists.

Heat treatment of soil up to the cultivated depth is also management of top soil. Deep ploughing to turn the top soil, exposing it to hot summer sun is one way of killing propagules of many pathogens including cysts of *Heterodera avenae*. Burning of crop debris destroys propagules of pathogens present in the surface soil. For treatment of nursery beds, burning of 30–45 cm thick layer of crop debris, grasses and weeds spread over the bed is effective in sterilization of the top soil

Soil solarization is claimed as a novel technique for treatment of soil by solar heat. It is a hydrothermal treatment. If moistened soil is covered under transparent polyethylene film (about 25 μm thick) during summer to prevent dissipation of the trapped solar heat, the temperature of the soil rises, sometimes 13°C higher than the temperature outside, reaching 50°–55°C at 8 cm depth. This kills most fungal, bacterial and nematode propagules as well as weed seeds.

13. Minimizing influx of inoculum from neighbouring crops: Unprotected crops in neighbouring fields constitute a major source of inoculum for spread of diseases. Some examples of diseases which become serious because the neighborhood crops provide source of inoculum are: leaf curl of tomato, downy mildew of cucurbits and crucifers, leaf spots caused by *Alternaria* and *Cercospora*, late blight of potato and tomato and ergot of pearl millet. Fields having loose smut affected wheat crop may provide inoculum for floral infection of neighbouring fields and introduce the disease in the seed lot. When such dangerous neighbouring fields are owned by other farmers the precautions include:

i)   Isolation distance: The farmers may agree to a definite distance between plots of the same crop. Even a distance of few dozen meters reduces the risk.

ii) If the direction of prevalent wind during the crop season is not likely to carry inoculum from one field to another, the risk is reduced.

iii) Shape of the plot: A crop grown in an elongated field will be able to transmit much more inoculum into a neighbouring field running parallel to its long axis than a field of similar acreage but more or less quadratic in shape.

iv) Early crops should be kept as far away as possible from the main crop and should be given maximum plant protection coverage.

v)   Stubbles likely to carry inoculum must be ploughed deep into the soil well before a new, susceptible crop is sown in the vicinity.

14. Choice of crop variety: Choice of one or more varieties of any crop resistant to a disease is always desired provided the varieties are not susceptible to some disease. Large scale cultivation of a single resistant variety and for many years in the same area is unjustified because it enables the pathogens to develop new races which breakdown the resistance.

3. FIELD AND PLANT SANITATION

Field and plant sanitation is a main part of disease management through cultural practices. This step is essential even if disease or pathogen-free seed or planting material has been used and other recommended cultural practices like crop rotation and alteration in date and method of sowing have been followed. The inoculum present on few plants in the field may multiply in soil or on the plant and in due course may be sufficient to nullify the effect of other cultural practices.

In wilt disease of banana (*F. oxysporum* f.sp. *cubense*), so long as the dead roots and rhizomes of the affected plants are present the fungus continues its growth and production of conidia and chlamydospores. When these substrates are removed there is rapid decline of the pathogen population in the soil. Similarly, the populations of *F. oxysporum* f.sp. *vasinfectum* (wilt of cotton), *F. udum* (wilt of pigeonpea), pathogens of root rot of bean and Verticillium wilt of cotton are also reduced by removal of diseased plant debris. This sanitary precaution makes even a short duration rotation effective in the control of the disease. In addition to the reduction of inoculum in the field, sanitation of irrigation water, agricultural implements, etc., is also important.

1. Management of crop debris: Most crop debris, especially the roots, is left in the field in which the crop had grown. Infected crop debris not only serves as medium for survival of the pathogens, it serves as a substrate for their multiplication and increase in the quantity of inoculum. Destruction of this source of survival and multiplication helps in the control of many diseases of roots and foliage. The fungi causing downy mildew of maize, sorghum, pea, pearlmillet and grapevines, white blisters of crucifers and ergot diseases of cereals and pearlmillet are examples in which the oospores and sclerotia survive in the crop debris left in the field. The fungus causing early blight of tomato survives as dormant conidia in the crop debris. Rotting potato tubers left in the field after harvest of the crop are a source of survival of the wilt and soft rot pathogens. Apple leaves bearing scab lesions and falling on the ground in November are the most important source of survival of the apple scab pathogen which lives as saprophyte in these leaves, produces perithecia and liberates inoculum for initiation of the disease in the next spring.

The straw of rice harbours resting structures of many pathogens such as true sclerotia of false smut fungus and sclerotia of *Sclerotium oryzae* (stem rot) and *R. solani* (sheath blight). Survival of pathogens through crop debris are more common in cold climate areas where decomposition of the debris is slow than in warm and wet areas where the debris decomposes early. The pathogen of Ascochyta blight of chickpea survives in crop debris during winter in cold climate regions but is killed during summer in tropical and subtropical areas.

There are different methods of destruction of the debris. Burning of debris of a diseased field crop is one method. Deep ploughing after crop harvest buries the debris deep in the soil. The practice of burning rice crop residue followed by stubble disc ploughing is the most effective method of minimizing inoculum of *Sclerotium oryzae*. Benefits of deep ploughing have been reported for sclerotial fungi such as *Sclerotinia sclerotiorum* and *Sclerotium rolfsii*. Deep ploughing during summer followed by roguing of diseased plants during the crop season is reported to reduce the oospores of *Peronosclerospora sorghi* (downy mildew of sorghum) in soil. The cereal cyst nematode *Heterodera avenae* is highly susceptible to desiccation. Summer ploughing helps in destruction of its cysts. Fallows and flooding also help in destruction of the debris in absence of a host of the pathogens. In apple orchards, spray of urea on the fallen leaves promotes their early decomposition. In some countries, thermal treatment of apple orchard floor by specially designed propane flamers (temperature raised to 150°–200°C) has been demonstrated to destroy the apple scab pathogen in its saprophytic stage.

2. *Management of the diseased plant:* Regular removal of diseased plants from a population (**roguing**) is an important sanitary precaution and is effective in reducing the amount of initial inoculum and spread of diseases. It is one of the effective recommendations in the management of virus diseases of field crops. However, the practice is effective only when population of the vectors is low. By spraying an insecticide before roguing this problem can be solved. The removal of potato haulms before vector population rises is also a form of roguing for seed production. For the control of loose smut of wheat roguing of smutted plants is recommended but once the smutted ears have appeared the practice is not very effective. In many wheat varieties appearance of smutted heads is preceded by yellowing of flag leaves. This indicator facilitates effective roguing in such varieties. Red rot of sugarcane, wilt of pigeonpea and cotton, smut of sugarcane and downy mildew of sorghum and maize are other diseases in which roguing is recommended. Sometime, roguing or destruction of healthy looking plants around the diseased plants may have to be done such as in citrus canker.

A number of important pathogens of fruit trees and vines are capable of survival on dormant parts of the plant, mostly on or between bud scales. The examples are downy mildew (*Plasmopara viticola*) and powdery mildew (*Uncinula necator*) of grapes and powdery mildew of apple (*Podosphaera leucotricha*). Mummified fruits hanging on the trees or fallen on the ground are an important source of inoculum of the brown rot fungi (*Monilinia* spp.). Destruction of such infected or dead organs (**defoliation** and **pruning**) is important for sanitation of the trees or vines. The graft transmissible diseases such as those caused by viruses and fastidious prokaryotes are perpetuated and spread through **budding** and **grafting**. The precautions that are recommended include (i) stock and scion must be free from the disease and (ii) they should be given chemical or heat treatment before planting them in the orchard.

3. Management of irrigation water: Irrigation water is often a means of transport of inoculum through the field and to different fields in the area. Transfer of sporangia and zoospores of *Phytophthora* in citrus orchards, sporangia of *Phytophthora* and *Pythium* in papaya plantations, sporangia of downy mildew fungi, conidia of the red rot fungus (*Colletotrichum falcatum*), cells of bacteria causing leaf blight and leaf streak of rice, bacterial blight of bean (*Xanthomonas axonopodis* pv. *phaseoli*), angular leaf spot of cotton (*X. axonopodis* pv. *malvacearum*) and many other pathogens occurs through irrigation and drainage water in and outside the field. Nematodes and cysts of nematodes are also moved by water in and out of the field. The passage of irrigation water through infested field should, therefore, be avoided. Pond water is often used for irrigation. The ponds receive drainage water from the surrounding fields which brings inoculum of pathogens present in the fields. When the ponds dry, the weeds growing in them become alternate hosts of many pathogens thus increasing the amount of inoculum. The rice sheath blight fungus (*R. solani*) has very large host range among such weeds. Pond water has been implicated in the epidemiology of this disease.

4. Crop-free period and crop-free zone: The pathogens attacking crops of secondary importance and having a narrow host range can be controlled by maintaining a crop-free period of definite duration. This requires cooperative effort by farmers in the urea. The pathogen is starved and eliminated in absence of the host. Similarly, when the host crop is not grown in a wide zone surrounding the infested area the disease is checked. In bunchy top of banana this is a recommended practice.

5. Creating barriers by non-host and dead hosts: Many diseases depend for their spread on nearness of healthy roots to infected roots. Presence of non-host roots creates a barrier between the two. One of the control measures recommended for bacterial wilt of banana (*Ralstonia solanacearum*) and spreading decline of citrus (*Radopholus similis*) is to destroy the healthy plants around the diseased plants. This checks the movement of the pathogen.

6. Decoy crops, trap crops and antagonistic crops: Decoy crops or cover crops are non-host crops sown with the purpose of making soil-borne pathogens waste their infection potential before the susceptible main crop is grown. This is effected by activating the dormant propagules of fungi, larvae of nematodes and seeds of parasitic plants in the absence of the host. Most of the decoy crops are of low economic value. Small scale farmers may not like to use them.

Trap crops are highly susceptible host crops of the pathogen, planted to attract it but destined to be harvested or destroyed before the pathogen completes its life cycle. Such crops are effective in checking the populations of root knot and cyst nematodes and some parasitic weeds such as *Orobanche*. In Germany, planting of somewhat resistant potatoes and harvesting them before the cyst nematode matures is recommended. Oats have been used in England as a trap

TABLE 19: Decoy Crops for Reduction of Pathogen Population in Soil.

Pathogen	Host	Decoy crops
Plasmodiophora brassicae	Crucifers	Rye-grass, Reseda odorata
Spongospora subterranea	Potato	Datura stromanium
Meloidogyne incognita	Tomato	Castor
Meloidogyne spp.	Eggplant	Sesamum
Heterodera avenae	Oats	Maize
Orobanche spp.	Tomato,	Sunflower, safflower
	Tobacco	Chickpea, lucerne
Striga asiatica	Various	Sudangrass

crop for cereal cyst nematode. Oats are harvested early for fodder. *Solanum nigrum*, a weed stimulates hatching of eggs of *Globodera rostochiensis* (potato cyst nematode). The second stage larvae enter the roots of the weed but cyst development is poor. *Crotalaria* spp. and Bermuda grass prevent complete morphogenesis of the root knot nematodes. Flax (linseed) is used as a trap crop for *Orobanche ramosa* and *O. aegyptiaca*. It pinpoints areas of infestation in the field. The parasite damages the linseed plants but does not produce flowering stalks. This eliminates seed of the parasite for the next crop.

Antagonistic or enemy crops produce some toxic compounds that directly destroy the nematodes in soil. They are not hosts of the pathogen. Some grasses, certain varieties of mustard, marigold, species of *Crotalaria* and asparagus have been listed as enemy plants.

7. Management of weed, collateral and volunteer host plants: The non-specialized pathogens having a very wide host range among weeds, collateral and volunteer host plants carry over the pathogens from one season to next and also provide a base for multiplication of the inoculum which may reach epidemic proportion. Powdery mildew of cucurbits (*Sphaerotheca fuliginea*) persists during winter on wild or out of season cucurbit plants. Same is true for viruses of cucurbits. Kans grass (*Saccharum spontaneum*) is a collateral host of sugarcane smut (*Ustilago scitaminea*) and sugarcane downy mildew (*Peronosclerospora sacchari*) while *Saccharum munja* is also host of many sugarcane pathogens. These are perennial grasses present around cultivated fields in India. Removal of such hosts is an important sanitary precaution. In some diseases such as in ratoon stunting disease of sugarcane (*Clavibacter xyli* subsp. *xyli*) weeds add to the destruction of the crop by overgrowing the latter.

8. Management of insect vectors: Sanitation of the field involves the control of insect vectors also as they are a source of bringing inoculum from outside and spreading inoculum within the field.

9. Harvesting time and practices: The choice of suitable time and method of harvesting often requires many sanitary precautions not only to protect the produce from deterioration (fruits and vegetables) but also for reducing chances of carry over of the pathogen from the current season to the next within the field and through seed or propagating material to other fields in the next season. Some examples of pathogens that are dispersed in the field during harvest are bunts of wheat, flag smut of wheat (*Urocystis agropyri*), potato cyst nematodes (*Globodera rostochiensis* and *G. pallida*) and phanerogamic plant parasites such as dodder and *Orobanche*. The danger of potato tubers getting contaminated with the late blight fungus at the harvest time was mentioned earlier.

Harvesting of fruits without proper sanitary precautions is one of the major causes of fruit spoilage during transport and storage. Nail or thorn injury to fruit skin during harvest predisposes the fruit to infection of decay causing fungi, the spores of which are present in the atmosphere in the orchard. This is further enhanced by harvesting in cloudy and humid weather. There is evidence for mangoes that infection of stem-end rot fungi (*Diplodia natalensis, Lasiodiplodia theobromae*) usually occurs when the fruits are harvested in cloudy weather and the pedicels are removed on the orchard floor.

The risk of the produce becoming infected at the harvest time can be lessened in the field crops by (i) first removing the badly affected plants or affected earheads, (ii) chosing a suitable method of harvesting so that disease carrying parts are not spread in the field or go with the seed, and (iii) harvesting under conditions unfavourable for the pathogen. Thus, lifting of potato tubers in dry weather almost removes the danger of infection of zoospores of the late blight fungus. In orchards, picking fruits only when they are not over-ripe and the weather is dry with good sunshine reduces the losses from post-harvest decay. Some virus diseases of vegetable crops are spread by contact during periodical picking of fruits such as in tomato and okra. Obviously, the worker has to avoid contact with diseased plants while picking from healthy plants. Similarly, picking of fruits, such as apples, requires cleanliness of hands of workers who should ensure that fruits are not injured in any way.

II. BIOLOGICAL CONTROL

Disease management through cultural practices that enhance microbial diversity and intensity in the soil often causes suppression or destruction of the inoculum of pathogens by the microbiota. This role of microorganisms is a part of natural or biological control of pathogens. Biological control is defined as the reduction of inoculum density or disease producing activities of a pathogen or parasite in its active or dormant state by one or more organisms, except man, accomplished naturally or through manipulation of the environment, host or antagonists or by mass introduction of one or more antagonists. A simplified definition is "biological control is the reduction of the amount of inoculum or disease producing activity of a pathogen accomplished by one or more organisms other than man (Cook and Baker, 1983). Reviews of progress in the study of biological control were published by Cook (1993) and Whipps

(1997). Although there is strong evidence for natural occurrence of biological control on the phylloplane and in the rhizosphere, there are not many examples of field application of the method. It has been more amenable to its natural occurrence than to mass introduction or manipulation of antagonists except where such plant organs as seed, fruits and seedling roots have been treated with known antagonists.

The entities involved in biological control are (i) avirulent or hypovirulent individuals or populations within the pathogenic species itself, (ii) the host plant manipulated genetically, by cultural practices or with microorganisms toward greater or more efficient resistance to the pathogen and (iii) antagonists (antibiosis, competition, parasitism and predation), defined as microorganisms that interfere with survival and disease producing activities of the pathogen. The approaches to biological control include (i) biological control of inoculum in the soil, (ii) biological protection of the plant surfaces (seed, fruit and roots) and (iii) cross protection or induction of resistance.

1. Destruction of surviving propagules: The destruction of survival structures of fungi in soil is by many mechanisms employed by antagonists. Parasitic activity of *Trichoderma viride* against *Rhizoctonia solani* was reported by Weindling in 1932. Since then, the species of *Trichoderma* have received the most attention as fungal antagonists of not only soil-borne root pathogens but also some pathogens of the foliage and fruits (viz., *Botrytis cinerea*). Formulated products of *Trichoderma* have been developed for their mass application. The mode of action of *Trichoderma* spp. includes all the attributes of a successful antagonist such as competition for space and nutrients on the same energy source as the pathogen, production of enzymes such as glucanase and chitinase to lyse fungal cell walls, antibiotic and inhibitory metabolite production, predatory effect or direct parasitic action against the pathogen and indirect effect resulting in modification of the environment adverse to the pathogen. The natural populations of these antagonists are enhanced by organic amendments of the soil.

Sclerotia of many fungal pathogens are destroyed by parasitism of *Sporodesmium sclerotivorum*, *Coniothyrium minitans* and other fungi (*cf.* Singh 2001, p. 78) Introduction of *Trichoderma viride* into soil infested with *Verticillium dahliae* before sowing cotton suppresses the Verticillium wilt. Reduction of charcoal rot of cowpea (*M. phaseolina*) by addition of *T. viride, T. harzianum* and some species of *Penicillium* and *Aspergillus* to the soil is also reported. Oospores of *Phytophthora cinnamomi, P. cactorum, P. parasitica, P. megasperma* var. *sojae, Pythium* spp., and *Aphanomyces euteiches* are parasitised by many fungi including chytridiales which reduce the number of oospores.

Conidia of *Helminthosporium sativum* and chlamydospores of *Thielaviopsis basicola* are killed by vampyrellid amoebae (viz., *Arachnula impatiens*) through perforations caused in the protective walls. These amoebae also cause depressions and perforation in pigmented hyphae of *Gaeumannomyces graminis* and *R. solani.* Only dormant hyphae are destroyed in this manner.

Examples of microorganisms destroying phytonematodes in the soil are numerous (Davis *et al,* 1991; Singh, 2001). The amoeba *Arachnula impatiens* captures an active larva of *M. incognita* with fine filopodia, makes several holes in the cuticular sheath and completely empties contents of the larva within 2–3 hours. The amoeba *Vampyrella vorax* completely surrounds the whole body of a larva and takes 12–24 hours for completion of feeding. Certain species of collembola rapidly eat away eggs of *M. javanica* in soil amended with oil-cakes. Predaceous fungi were implicated in the reduction of cereal cyst nematode (*Heterodera avenae*) and beet cyst nematode (*Heterodera schachtii*) in soil green manured with chopped cabbage leaves. The nemato-phagous fungi *Verticillium chlamydosporum, Cylindrocarpon destructans,* and *Phialophora heteroderae* destroy eggs and cysts of *Heterodera schachtii, H. avenae* and *Globodera rostochiensis. Nematophthora gynophila* parasites eggs of *H. avenae* and is the major factor limiting the natural multiplication of this nematode in soil.

Mycophagous nematodes in the soil are responsible for bringing down inoculum levels of fungal plant pathogens and reduce some root diseases. The nematode *Aphelenchus avenae* is a special feeder of hyphae of *R. solani.* Multiplication of this nematode differs with the anastomosis group (AG) of *R. solani.* Certain combinations of AG and the nematode have potential for biological control of the fungus.

These naturally occurring enemies of plant pathogens are active in a soil where they have established. Generally, antagonists added to the soil as pure culture fail to perform. The antagonists have their own ecological characteristics and establish themselves only in environments where they were a resident microflora. If they are resident microflora in a particular soil environment their strength in the field can be increased by suitable amendments such as addition of chopped cabbage leaves against *Heterodera.* The basic objective should be to maintain the natural antagonistic potential of the soil. However there are reports where exogenous application of antagonists to the field soil have controlled specific diseases. The examples are *T. harzianum* isolated from rotting sclerotia used against *S. rolfsii* and broadcast application of sclerotia destroying fungi to soil against *Sclerotinia sclerotiorum. Streptomyces griseus* inhibits conidial and chlamydospore germination and growth of germ tubes of *Fusarium udum* (pigeonpea wilt). It is naturally present in rhizosphere of resistant cultivars. Other antagonists of this pathogen include *Aspergillus niger, A. terreus, A. flavus, Trichoderma* spp. and species of fluorescent *Pseudomonas.* These are favoured by organic matter in the soil and wet and warm conditions.

Siderophores produced by fluorescent pseudomonads cause suppression of *Fusarium* spp. in soil through competition for iron. The extent of suppression is reduced when iron is added to the soil. Mutants of *Pseudomonas* not producing siderophores cause low level of suppression by competition for carbon sources. Field control of Pythium damping off and Aphanomyces root rot of pea by introducing *P. cepacia* or *P. fluorescens* in soil is reported (Parke *et al*, 1997). *P. fluorescens* and *Streptomyces graminofasciens* introduced into soil check root hair infection process and club-formation in the clubroot disease of crucifers caused by *Plasmodiophora brassicae* (Bhattacharya and Pramanik, 1998).

2. Prevention of Inoculum Formation: The above mentioned biological destruction of such propagules as sclerotia, oospores, cysts and eggs helps in prevention of next generation of inoculum formation. Other logic behind prevention of inoculum formation are (i) preventing a pathogenic fungus from colonizing plant residue in soil where it could multiply the inoculum, (ii) encouraging the development of antagonists on aerial parts of the plant where they could destroy the inoculum or prevent sporulation and (iii) using decoy or trap crops that could be destroyed before the inoculum is formed and released in soil.

Many fungal pathogens such as *Pythium aphanidermatum*, *Phytophthora parasitica* and *Armillaria mellea* fail to colonize host plant residue in soil if the latter are precolonized by saprophytic antagonists such as *Fusarium roseum* "Culmorum". Colonization of sclerotia of *Sclerotinia sclerotiorum* by saprophytes, especially when the sclerotia are injured or weakened by some treatment, prevents apothecia formation and release of ascospores. *Fusarium roseum* "Sambucinum" and *F. heterosporum* are highly potential biocontrol agents against *Claviceps purpurea* (ergot of cereals). The conidia of the hyperparasites are applied through aerosols on the crop during flowering. They colonize the honey dew and prevent sclerotia formation. Similar parasitization of honey dew and immature sclerotia of *Claviceps fusiformis* (ergot of pearl millet) is also reported. A strain of *Pseudomonas fluorescens* is reported to inhibit teliospore germination of *Tilletia laevis* (common bunt of wheat) resulting in no sporidia formation.

Biocontrol of active pathogen body on the host surface to prevent inoculum formation and dispersal has great potential for powdery mildews and rusts in which inoculum production is of compound interest rate type. Several fungi such as *Ampelomyces*, *Tilletiopsis*, *Cladosporium* and *Verticillium* and some insects are hyperparasites of these pathogens. *Ampelomyces quisqualis* (*Cicinnobolus cesatii*) is a mycoparasite of *Uncinula necator* (grape powdery mildew), *Podosphaera leucotricha* (apple powdery mildew), *Erysiphe cichorasearum*, *Sphaerotheca fuliginea* (cucurbits powdery mildew) and *Sphaerotheca pannosa* (rose powdery mildew). It is a naturally occurring pycnidial parasite and perpetuates in plant buds and host cleistothecia. *Fusarium proliferatum*, as a biocontrol agent, can reduce sporulation of *Plasmopara viticola* (grape downy mildew) by 97%. In vineyards weekly sprays of conidial suspension of the hyperparasite have given significant control of the disease. A tydeid mite, *Orthotydeus lambi*, is an effective biocontrol agent against *Uncinula necator*. It reduces mildew colonies on the leaves and prevents cleistothecia formation (English *et al*, 1999). Suppressive soils, explained earlier, prevent inoculum formation through the antagonists associated with them.

3. Reduction of Vigour or Virulence of the Pathogen: This approach involves suppression of pathogenicity factors in the pathogen. It can be accomplished through modifications of soil environment and factors inherent (or carried) in the pathogen itself. Thus, altered sex ratio in phytonematodes (*Meloidogyne, Heterodera, Globodera*), determined by the availability of food and other environments, hypovirulence in some fungi and similar phenomena are candidates for means of this type of biological control. In phytonematodes, the intensity of infection depends on preponderance of female larvae. Under conditions of insufficient food, most larvae tend to become males. In extracts of soils amended with oil-cakes and sawdust, it was found that larvae of *Meloidogyne* become very active with jerking motion as if affected by some irritant. Since fitness of larvae to cause penetration of the host root depends on their strength derived from the lipid reserves in their body the hyperactivity exhausts the energy reserve and larvae become weak and fail to cause penetration.

Hypovirulent strains of fungal and bacterial pathogens occur in nature. In fungi hypovirulence is supposed to result from infection or the presence of one or more dsRNA determinants in the hyphae. In those fungi where vegetative compatibility is known and anastomosis can occur, these determinants of hypovirulence can be transmitted to virulent strains. A dsRNA mycovirus-like agent is found in some strains of *Rhizoctonia solani*. Vegetative compatibility within the anastomosis groups (AG) of this fungus is common. A mixture of normal and hypovirulent strains significantly reduces damping off caused by the fungus. Similar mycovirus infection is common in and has been found associated with decline of the wheat take-all fungus *Gaeumannomyces graminis*.

4. Biological Protection of Planting Material: Biological control has been more successful in protection of the planting material than through mass incorporation in the soil. Crown gall has provided an example of commercial biological control of a plant disease. Crown gall is caused by *Agrobacterium radiobacter* pv. *tumefaciens*. The tumor inducing factor is contained in its plasmid. The plasmids in non-gall forming strains lack the tumor inducing factor. Plasmids of *A. radiobacter* pv. *radiobacter* (K84) carry genes for production of a bacteriocin (Agrocin 84). This bacteriocin is highly effective against many pathogenic strains of the pathovar *tumefaciens*. The bacteriocin producing strains have been used for control of crown gall on a field scale. The antagonist is used as cell suspension for dip treatment of nursery stock, seedlings and seeds.

There are many examples of biological control achieved by protective covering of seed, rhizomes, tubers and fruits with propagules of an antagonist. Seed coating with cell or spore suspension of antagonists enhances their number in the rhizosphere. Rhizobacteria that are only antagonistic or both antagonistic and plant growth promoting (PGPR) can be introduced in the soil by this method. *Bacillus subtilis,* some species of *Pseudomonas, Penicillium, Trichoderma* and *Chaetomium globosum* are often as effective as seed protectant chemicals such as thiram and captan in protecting seed of maize, pea and soybean against seed rot fungi. Bacteria are better colonizer of seed surface than fungi. When pea seed is coated with spores of *Penicillium oxalicum* or any other aggressive colonizer of pea seed, the antagonists exhaust the seed exudates before the pathogens (*Pythium* and *Fusarium*) can utilize them. Aphanomyces root rot of pea is also checked by seed treatment with *P. oxalicum* or *Trichoderma* spp. Many commercial products based on *Trichoderma* have been developed. These can be stored for some time and can be used in the same manner as protectant fungicides.

Bacillus subtilis is a strong antagonist of many fusaria including *Fusarium udum* (pigeonpea wilt). Seed treatment with cell suspension of *B. subtilis* controls various diseases caused by *R. solani, Helminthosporium* in rice and damping off of tomato. Production and harvest of *B. subtilis* in quantities necessary for field scale use is easy. It forms endospores, hence it can be formulated as dust or wettable powder without losing efficacy. Commercial products based on this antagonistic bacterium came up for sale in 1980s. The bacterium stimulates germination, causes better emergence of seedlings, enhances plant nutrition, reduces cankers caused by *R. solani* AG-4 in groundnut, increases root growth and finally yield.

Association of fluorescent pseudomonads with suppression of many root diseases and take-all of wheat was reported in 1983. A similar control of wilt disease of chickpea (*F. oxysporum* f.sp. *ciceris*) by a strain of *Pseudomonas fluorescens* was reported in 1993. In a talc based formulation the bacterium survived for 8 months in storage. On seeds treated with the formulation the bacterium survived for 6 months (*cf.* Singh 2001, p. 93). Along with seed treatment, the application of the powder to the rhizosphere further improved the control of wilt. Biological control of blast (*Pyricularia grisea*) and sheath blight (*R. solani*) of rice by seed treatment with and soil application of *P. fluorescens* and *P. putida* is reported from India and Philippines. The bacteria have long survival on the roots and also in the phylloplane when applied through roots. Pea seed coating with *P. cepacia, P. fluorescens* or *B. subtilis* suppresses Aphanomyces root rot and Pythium seed rot.

The use of bacteria in root and rhizosphere treatment not only provides protection against major soil-borne pathogens through competition for nutrition and space and antibiotic production but also by induced resistance (Weller, 1988; Van Loon *et al*, 1998; Sticher *et al*, 1997). Many of these bacteria stimulate plant growth even if there is no disease. The increased plant productivity in no-disease situations results largely from the suppression of deleterious microorganisms and often through metabolic changes in the plant. Bacterization of potato seed tubers is an example. Not only bacteria but antagonistic fungi such as *T. harzianum* also are involved in growth promotion of plants. The fungus produces a growth regulatory factor that increases the rate of seed germination and dry weight of shoots irrespective of pathogen control (*cf.* Singh, 2001).

Often the rhizobacteria without being antagonistic have given better disease control than such antagonists as *Trichoderma* and *Pseudomonas fluorescens* (Jubina and Girija, 1998). Rhizobacteria obtained from host roots from widely located fields have given control of clubroot of crucifers. Indigenous rhizobacteria (strains of *Pseudomonas fluorescens, P. putida* and *P. alcaligenes*) used for cotton seed treatment not only improve germination and seedling growth but also suppress the cotton blight bacterium on leaves (Mondal *et al,* 1999) suggesting induction of systemic resistance.

Bacterial antagonists which have been used for commercial production of marketable formulations include different strains of *Bacillus subtilis, P. fluorescens, P. syringae, Agrobacterium radiobacter, P. cepacia, P. putida, P. aeruginosa, Sterptomyces griseoviridis* and non-pathogenic *Ralstonia solanacearum.*

5. Biological Protection of Foliage and Blossoms: There are many examples where presence of natural microflora on the foliage has been cited to explain the reduction of disease incidence. Brown leaf spot of rice (*Drechslera oryzae*), leaf spot of rye (*Helminthosporium sativum*), fire blight of apple and pear (*Erwinia amylovora*), Alternaria spots of tobacco and many other foliage diseases are less severe when the normal epiphytic microflora is allowed undisturbed. The epiphytic flora suppresses pathogens through competition for nutrition and space, antibiosis and hyperparasitism or through induction of resistance in the host.

Strawberry gray mold rot (*Botrytis cinerea*) is significantly reduced by the presence of *Gliocladium roseum*, a versatile antagonist which suppresses the pathogen by more than 90% on leaves, petals and stamens. This is better than control obtained with the fungicide captan. Weekly sprays of suspension of propagules of the mycoparasite are recommended. Botrytis gray mold of chickpea is also suppressed by spray of spore suspension of *T. viride* especially when combined with a spray of the fungicide Ronilan (Agarwal and Tripathi, 1999). *Athelia bombacina* and *Chaetomium globosum* are promising biocontrol agents against apple scab (*Venturia inaequalis*). When sprayed with calcium nitrate as foliar fertilizer they suppress the leaf scab. Strains of *B. subtilis* and *Aspergillus terreus* from phylloplane of citrus suppress the citrus canker bacterium.

Successful introduction of epiphytic microflora can be achieved by spraying their propagules provided they have been selected or developed from the same site or similar plant surfaces. There are methods that promote growth of such beneficial organisms on the foliage. In Germany, leaves sprayed with water extracts of composted horse manure-straw-soil mixture were found to develop resistance to downy mildew (*Plasmopara viticola*). Later, in attempts to provide biological control of grapevine powdery mildew (*Uncinula necator*) it was revealed that the control of mildew by water extracts of compost or fermented cowdung was through enhanced microbial activity on the leaf surface. Populations of *Bacillus, Pseudomonas, Serratia, Penicillium* and *Trichoderma* were enhanced on the host surface. In India spray of fermented margosa cake on citrus had been reported to give significant control of canker during 1950s. Pruning wounds and other cut surfaces on the trees can also be protected by application of antagonists, hypovirulent strains or materials that can promote development of antagonistic microflora.

6. Biological Control of Post-harvest Fruit Decay: In the recent past there have been successful demonstrations of biological control of post-harvest decay of fruits. Bacterial and fungal antagonists used as an alternative to fungicides for control of fruit rots of citrus, apple, peaches, pears and cherries include *Bacillus subtilis* (against brown rot caused by *Monilinia* spp.), *Enterobacter cloecae* (against Rhizopus rot), *Pseudomonas cepacia, P. syringae, Trichoderma* spp., *Acremonium brevae,* and several species of yeasts. As a biocontrol agent (fruit dip in cell suspension) *B. subtilis* can suppress ten citrus fruit pathogens by antibiotic production. These include *Lasiodiplodia theobromae,* a very common cause of fruit rots in the tropics and subtropics. *Trichoderma viride, P. cepacia* and *P. fluorescens* are effective against various types of Penicillium rots of citrus. The yeast, *Candida guilliermondii,* when applied before infection can protect citrus fruits against *Penicillium italicum* and *Geotrichum candidum* (sour rot). Guava fruit rots caused by *Lasiodiplodia theobromae, Colletotrichum gleosporioides, Pestalotiopsis versicolor, Phomopsis psidii* and *Rhizopus arrhizus* are effectively checked by *Trichoderma* spp. Treatment of apple and pear fruits with *P. cepacia* checks the rots caused by *Penicillium* spp. and storage decay of apples is reduced by fruit treatment with the yeast *Sporobolomyces roseus.*

7. Cross Protection and Induced Resistance: Cross protection, induced resistance or induced systemic resistance (ISR), systemic acquired resistance (SAR) and localized acquired resistance (LAR) are terms used to denote the condition in the plant in which biological or chemical treatments or stress trigger latent mechanisms of resistance (Kuc, 1995). The resistance becomes systemic in the plant (SAR) or remains localized at the site of treatment (LAR). In the latter case, the signal that propagates the enhanced capacity throughout the plant is lacking (Van Loon *et al,* 1998). The induced resistance can be triggered by non-pathogens, avirulent form of the pathogen, incompatible races of the pathogen, by a virulent pathogen under circumstances when infection is stalled owing to environmental conditions or by substrates that promote growth and activity of rhizobacteria. It can also be triggered by exogenous application of water extracts of compost and organic and inorganic chemicals. Systemic acquired resistance is reported in large number of plant species but systemic resistance induced by rhizobacteria is not reported among monocots. SAR confers quantitative protection simultaneously against a broad spectrum of fungi, bacteria, viruses and occasionally insects and nematodes. In cucumber, strain 89B-27 of *Pseudomonas putida* induces resistance against anthracnose, bacterial leaf spots, Fusarium wilt, bacterial wilt, cucumber mosaic virus and the insect *Diabrotica undecimpunctata*. In the same plant species, the strain 99–166 of *Serratia marcescens* induces resistance against Fusarium wilt, anthracnose, cucumber mosaic virus, angular leaf spot and the insects *Diabrotica undecimpunctata* and *Acalymna vittatum* (cf. Van Loon *et al*, 1998).

Cross immunity or cross protection is well known in virus diseases where it is one of the few established control measures. When a mild or avirulent strain of the virus is introduced into the host it may induce resistance to a virulent strain of the same or serologically related viruses. The explanation is that the surplus of coat protein made by the first virus prevents uncoating of the virulent virus thereby preventing its multiplication. In citrus tristeza virus the cross protection has been proved and employed. Cross protection or induced resistance against bacterial wilt caused by *Ralstonia solanacearum* in many plant species by prior inoculation with non-pathogenic, avirulent, bacteriocin producing or incompatible strains of the bacterium or heat-killed cells of the bacterium is reported. The efficacy of heat killed cells suggests disease suppression by means other than antagonism.

In wilt disease of mint caused by *Verticillium dahliae* inoculation of roots at least 2 days earlier with *V. nigrescens* reduces the incidence of wilt. Inoculation with mild or avirulent strains of *F. oxysporum* f.sp. *vasinfectum* provides protection to cotton against Fusarium wilt. In *F. oxysporum* f.sp. *ciceris* (wilt of chickpea) seven races are known. Races 0 to 1 are less virulent while race 5 is most virulent. Seed inoculation with races 0 and 1 provides protection against race 5. Similar protection by use of avirulent races is reported in linseed rust (*Melampsora lini*), watermelon anthracnose (*Colletotrichum lagenarium*) and bean anthracnose (*Colletotrichum lindemuthianum*). In these examples the inducer of resistance is mainly the strain of the pathogenic species. Non-parasitic organisms, not related to the pathogen, even if not antagonistic, also induce resistance to one or more diseases of the same host.

The rhizobacteria, particularly the pseudomonads, are common inducers of resistance (Schippers, 1988; Van Loon *et al*, 1998) even against foliar pathogens. The antagonistic sterptomycetes in rhizosphere of pigeonpea and banana are known to suppress wilt. The plant growth promoting rhizobacteria applied to seed or soil remain localized at the root surface and can induce resistance in the leaves or the stem. The role of vesicular arbuscular endomycorrhizae and ectomycorrhizae in acting as barrier to invasion by fungi and nematodes was mentioned earlier. This is also a biological control.

III. HOST RESISTANCE FOR DISEASE MANAGEMENT

Varietal resistance is a type of biological control in which the host itself plays the role of an antagonist. Resistant varieties can be the most simple, practical, effective and economical method of plant disease management. They not only ensure protection against disease but also save the time, energy and money spent on other measures of control. In many diseases such as vascular wilts, rusts and viruses, resistant varieties appear to be the only practical method of control. In crops of low cash value, chemical methods of control are often too expensive. In such crops varieties resistant to common and important diseases can be an acceptable recommendation for the farmers.

Disease Escape, Tolerance or Endurance and True Resistance

True resistance denotes incompatibility between the host and the pathogen at the genetic level. It implies that so long as this incompatibility exists the host should ward off infection by the pathogen. Disease escape is not true genetic resistance to the pathogen with which the plant may be compatible. The escape is due to other characters of the plant which may be genetically determined. A host or a population of host plants is defined as having tolerance if the signs and symptoms it manifests are visually similar to those on a susceptible variety but the damage is less than in susceptible variety. Technically, tolerance is not resistance since, by definition, the tolerant plants must look susceptible. But, practically, tolerance is resistance that is usually polygenically inherited and general in nature. One of the factors that determines less damage is the high yielding capacity even if infected. There is a general trend to favour tolerant varieties over truly resistant varieties because, if properly managed, such varieties last for longer periods than resistant varieties.

Types of Genetic Resistance

The terms monogenic, polygenic, vertical, horizontal, specific and general resistance were introduced in the chapter on genetic and molecular basis of resistance. According to the genetic concept a host is defined as having *specific resistance* or susceptibility if there is a differential interaction among genotypes of the host and genotypes of the pathogen. A host is defined as having *general resistance* or susceptibility if there is no known differential interaction among the genotypes of the host and the pathogen. Epidemiologically, a population of host plants is defined as having *discriminatory resistance* or susceptibility if it affects the epidemic by discriminating among races of the pathogen, i.e. by favouring or rejecting certain races of the pathogen population. A population of the host is defined as having *dilatory resistance* if it affects the epidemic by reducing the rate of development of pathogen population.

The various terms used to express resistance of a cultivar to a disease are divided into two basic types: specific and general resistance:

Specific resistance	*General resistance*
Race specific	Race nonspecific
Vertical	Horizontal
Major gene	Minor gene
Monogenic	Polygenic
Oligogenic	Multigenic
Multiple allele	Multiple gene
Qualitative	Quantitative
High	Low, Moderate
Seedling	Adult plant
	Generalized, uniform, field, partial, permanent and durable resistance.
	Late and slow rusting

When the defence mechanism is controlled by a single gene pair it is known as **monogenic resistance**. Usually the gene pair is dominant as in leaf blight of maize (*Helminthosporium carbonum*), blast of rice (*Pyricularia grisea*) and wilt diseases of cabbage, pea, and tomato caused by different form species of *F. oxysporum*. A recessive gene pair can also provide resistance such as in powdery mildew of pea (*Erysiphe polygoni*) and bacterial pustules of soybean (*Xanthomonas axonopodis* pv. *glycines*).

When the resistance is contributed by combined action of more than one minor defence mechanism, each one of which individually cannot resist infection, and these mechanisms are determined by one or more groups of supplementary genes, the resistance is known as **generalized** or **polygenic**. Usually the determinant gene pairs are minor or recessive.

Varieties with monogenic resistance remain completely free from attack of one or two races of the particular pathogen in different environments. However, the resistance is stable only until new races of the pathogen have not developed. Since the resistance is due to a single gene pair in the host the pathogen can easily develop a new race capable of breaking resistance by a single step mutation and possession of a single gene for virulence and aggressiveness. However, in some soil-borne diseases monogenic resistance is known to last for very long periods. Cabbage yellows (wilt caused by *F. oxysporum* f.sp. *congulitinans*) is an example.

The polygenic resistance is liable to disappear with change in environments but is more tolerant to evolution of new races of the pathogen. Since the resistance is due to multiple gene action the pathogen has to evolve many new races possessing suitable combination of genes for virulence to break the resistance. Therefore, polygenic resistance, though unstable in different environments, remains useful for a longer period in a fixed environment in which it was developed. As a principle, resistance developed by combination of many dominant and recessive genes is more desirable. A combination of disease escape, disease endurance and generalized resistance is considered a better method of disease control than resistance contributed by one or more specific genes.

Development of Resistant Varieties

In the conventional approach to crop improvement for disease resistance three common methods were employed: selection, mutation and hybridization. The last is the most important since only in this method manipulation of genes is possible. Selection and mutation are often utilized for creating variability in the process of hybridization.

i) **Selection:** Crop improvement by selection of desired individuals is an old practice. From good cultivated species and varieties those individuals in the populations are selected which have exhibited good growth and other desired qualities and have given good yield. These are individuals that have survived the natural elimination of weak and disease susceptible individuals from the population. The seeds of such individuals are multiplied. The undesirable individuals in the progeny are discarded. If the desired characters are found stable the progeny is used as a new variety or is used as one of the parents for hybridization. In fields where natural incidence of a pathogen occurs regularly or in artificially created sick plots having populations of a pathogen with a large number of its races, mass selection of seeds can be made from the most highly resistant individuals that survive. The method is simple but improves plants only slowly and does not work well in cross-pollinated crops where there is no control over the source of pollen. Pure line or pedigree selection involves separate propagation of highly resistant individuals and their progeny and repeated inoculations to test for resistance. This method is easy and most efficient with self-pollinated crops but most difficult with cross-pollinated crops.

ii) **Mutation:** Development of resistant varieties by mutation depends on chance. Mutant individuals are those whose genetic configuration has changed due to some chemical or physical shock or effect. The change is inherited. Mutations happen in nature and can also be induced by chemical (colchicine) or physical (radiation) mutagens. Mutation has generally been used to create variability for use in hybridization.

iii) **Hybridization:** Hybridization involves sexual act between two individuals (parents) one of which is the cultivated variety having good commercial qualities but lacking resistance to some specific disease while the other individual is the source of resistance lacking desired commercial qualities. Thus, availability of source of resistance is the first requirement in a hybridization programme. The source of resistance may be obtained by selections from a variety or species cultivated in the area, related wild plants, other species, rejected breeding materials, etc. A common source of resistance is wild relatives of the crop in the area of origin of the species where the host and its wild relatives and the pathogen have coexisted for long. The late blight fungus (*Phytophthora infestans*) and species of *Solanum* (*S. tuberosum, S. demissum,* etc.) have coexisted in the highlands of Central and South America. Most of the resistant material for this disease has been obtained from these areas (Mexico, Guatemala). Tomato (*Lycopersicon esculentum*) and its wild relatives such as *L. peruvianum* and *L. pimpinellifolium* also have their origin in Peru and Chile in South America. These two wild relatives of cultivated tomato have provided sources of resistance to many diseases of tomato. Chickpea (*Cicer arietinum*) had its original habitat in Asia Minor (Turkey and Syria). This area almost exclusively grows Kabuli type of chickpea and most of the varieties showing resistance to Ascochyta blight belong to the Kabuli type. Resistance to wildfire, black root rot and mosaic was introduced in tobacco from wild species of *Nicotiana*. Wild *Avena strigasa* has provided resistance against crown rust in oats (*Avena sativa*). Resistance to sugarcane mosaic is found in such wild species of *Saccharum* as *S. spontaneum* and *S. barberi*.

The same methods used to breed for any heritable character (crossing and back-crossing) are used for breeding for disease resistance also and depend upon the mating system of the plant (self or cross-pollinated) and same laws of inheritance apply in this case also. However, breeding for disease resistance is more complicated because (i) it can be assayed only by exposing the plant to the pathogen (another biological entity) which interacts with the plant and (ii) the existence of physiologic races in the pathogen which may not allow stability of resistance in the progeny.

Breeding for Resistance using Biotechnology

Biotechnology is defined as the manipulation, genetic modifications and multiplication of living organisms through novel technologies, such as tissue culture and genetic engineering, resulting in the production of improved or new organisms and products that can be used in a variety of ways.

In tissue culture, sections (explants) of shoot or root tip, shoot nodes, germinated seedlings and in some plants sections of leaves are properly trimmed, sterilized, repeatedly washed in sterile distilled water and then placed in test tubes containing suitable culture medium in liquid or solid form. These explants produce numerous shoots, roots or boths. Complete plants are then finally set out in greenhouse pots.

During regeneration the tissues multiply and form a mass of unorganized, undifferentiated mass of cells called callus. If the culture is kept in constant slow motion, single cells can be obtained. Suitable treatment with enzymes can remove the cell wall leaving only membrane bound protoplast. Protoplasts from different sources can be induced to fuse to generate a new type of plant.

The potential of the tissue culture methods have further increased with the use of recombinant DNA and other technologies (broadly called ***genetic*** *engineering*) which include detection isolation, modification, transfer and expression of single genes or group of related genes from one organism to another. This forms *transgenic plants* that may possess enhanced resistance to pathogens and insect pests (Broglie *et al*, 1991).

The tissue culture methods have helped in production of plantlets of potato, sugarcane and banana in which propagation by seed is restricted or seeds are not formed. It has also opened the way to use interspecific hybridization. The recombinant DNA technology using suitable gene vectors (*Agrobacterium* or certain DNA viruses) has helped introduction of bacterial genes into the plant to make them resistant to insect attack and introduction of virus protein genes and pathogen avirulence genes into the host plant to provide resistance to specific diseases.

Testing for Resistance

Testing of newly developed resistant varieties for sufficient time is necessary before the variety is released for the farmers.

Following precautions are essential:

i) The variety has been exposed to epiphytotic conditions. For this the variety must be tried in hot spots of the disease.

ii) Multilocational trials: The variety is grown in selected areas of the region or the country and effect of local environments, especially temperature, on expression of resistance is studied. This decides for which area the variety will be suitable.

iii) Physiologic races of the pathogen prevalent in different regions have been used for inoculation on the variety and its resistance to different races under each set of environmental conditions is determined.

iv) The stability of resistance and other desirable characters of the variety have been confirmed at different locations in successive generations of the crop for several years.

Effect of Environments on Expression of Resistance

The expression of resistance may be enhanced or suppressed by one or other external conditions, especially temperature and moisture. The examples of red rot resistance at different seasonal temperatures, rice blast under low temperature conditions, cabbage yellows, Fusarium wilt of tomato and pea, bacterial wilt of tomato and root knot were cited earlier in the chapter on effect of environments (temperature and moisture).

Causes of Failure of Resistance

There are very few varieties which have maintained their complete resistance to a disease for very long. This can be due to many avoidable and unavoidable causes. Among avoidable causes are laxity in screening, improper back-crossing and such other failures. Among unavoidable causes the most important is the continuous variability in the pathogen, especially when there is selection pressure due to cultivation of resistant varieties. Prolonged cultivation of a single genotype in the area helps in development of new races of the pathogen.

Management of Resistant Varieties

The resistance of varieties is maintained for a longer period when more than one resistant genotypes are cultivated in the area. The same resistant variety should not be cultivated for many years. Varietal diversification in time and space should use race-specific and race non-specific resistance. The resistant varieties backed by suitable cultural practices and low level of chemical protection last longer than when resistance is used as the only method of disease control. The present trend is to combine generalized resistance with a fungicide spray schedule to reduce the number of sprays to be given to the crop.

Induction of Resistance by Means Other than Gene Manipulation

The role of nutrition, including calcium and other micronutrients in strengthening the tissues to fight pathogens and role of systemic fungicides and antibiotics in providing temporary resistance was mentioned in different chapters and sections of this book. The induction of systemic resistance by biological means was also mentioned. Systemic acquired resistance is characterized by accumulation of salicylic acid (SA) and pathogenesis-related proteins or PR (Ryals et al, 1998; Sticher et al, 1997, Van Loon et al, 1997). Exogenous application of salicylic acid induces resistance in grapes against downy mildew and powdery mildew and many other diseases. SA acts as a signal for SAR. It activates the resistance genes or primes them to become active when challenged by the pathogen. These genes are present in the host in a "turned off" position.

Foliar application of carbonates, phosphates and salicylic acid or the application of inorganic chemicals through seed are known to reduce some diseases. Foliar fertilizers therapy is now considered a concept in integrated pest management (Chen 1998; Fuhr et al, 1998, Kassemeyer et al, 1998; Reuveni and Reuveni, 1998). A single spray of 0.1 M solution of phosphate induces a systemic protection in cucumber, mangoes, grapes and nectarines against powdery mildew. In China, spray of copper ammonium on citrus during autumn and summer has given significant control of citrus canker (Chen, 1998).

IV. DISEASE MANAGEMENT THROUGH TOXIC CHEMICALS

Use of toxic chemicals in plant disease control is both host management (protection of the host) as well as pathogen management (eradication of the

344

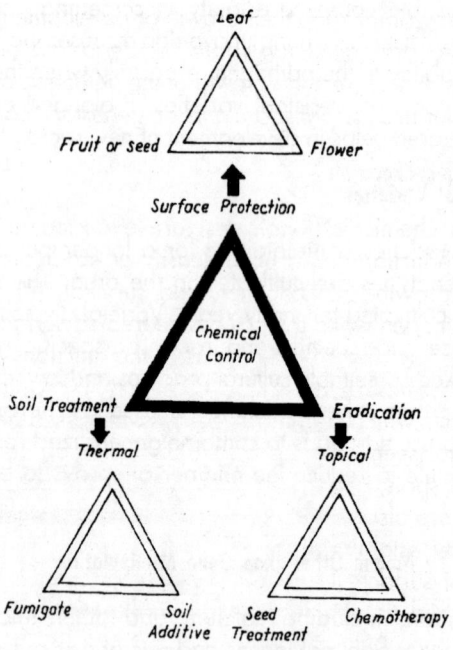

Figure 38. Chemical Control of Plant Diseases.

pathogen). It supplements the cultural practices and host resistance in the management of diseases. Application of chemicals is a highly visible and effective technique for plant disease management but at the same time it has become controversial because of the rising cost involved and its polluting effect on the environment. However, worldwide, the disease controlling chemicals (fungicides) are not as extensively used as insecticides and weedicides. The consumption of pesticides in India is much lower (435 g/ha) compared to USA and Europe (1.5–1.9 kg/ha) and Japan (10 kg/ha).

Aims of Chemical Control

The aims of use of chemicals in plant disease control are (i) to create a toxic barrier between the host surface or tissues and the pathogen and (ii) to eradicate the pathogen present at a particular site on or in the host including seed, foliage and roots. The chemicals oppose the germination, growth and multiplication of the pathogen or directly destroy it through toxicity.

According to type of the pathogen against which they are used, the antipathogen chemicals are called *fungicides, bactericides, nematicides* or *viricides*. The chemical used for eradication of flowering plant parasites are called *herbicides* or *weedicides*. However, all antipathogen chemicals actually do not kill the pathogen. They may simply stop their growth or development without killing them. In that case they are called *fungistatic, bacteriostatic* or *nematostatic*. When the toxicants do not prevent the growth of the pathogen but inhibit spore production they are termed as *antisporulants* or *genestatic* compounds.

Functions of Antipathogen Chemicals

The antipathogen chemicals (fungicides) are expected to perform three major functions: (i) reduction in inoculum density or eradication of inoculum from the source of growth, multiplication and survival, (ii) inactivation or destruction of the pathogen when it lands on the treated surface and (iii) cure of the diseased plant (therapy). On the basis of the functions performed, the fungicides can be classified as:

1) Protectant chemicals which are effective only when used before infection. The fungicides used for seed treatment against damping off and for sprays on the crop against leaf spots, blights, etc., come under this category. However, chemotherapeutants are also used as protectants. **Contact protective fungicides** kill the pathogen already established on the host surface or when it comes in contact with the host surface. Thus, they act as eradicant also. The **residual chemicals** are established as a fixed layer on the host surface (sprays, dusts, pastes) and destroy the active pathogen.

2) Eradicant chemicals eradicate the dormant or active pathogen from the host. Such chemicals can remain effective on or in the host for some time and function as protectants also.

3) Chemotherapeutants can specifically eradicate the pathogen from the host after it has caused infection. In this manner they are capable of curing the plant. Mostly the chemotherapeutants are systemic in action. A chemotherapeutant is thus protectant as well as eradicant.

Pesticides are available in the market as wettable powder, flowables, suspensions and slurries, emulsifiable concentrates, solutions, dusts and granules. The wettable fungicides or materials available in solution or emulsifiable form are used as spray with suitable equipments. Slurries and dusts are mostly used for seed treatment.

Factors Affecting Efficacy of Fungicides

The efficacy of fungicides used in the field depends on a number of factors operating independently or jointly. The factors may be the characters of the fungicide, prevailing environment at the time of application and timing and method of application of the fungicides.

Physical and chemical factors may be innate characters of the toxicant and/ or may be related to the environment in which it is used. The physical properties of the fungicide include its particle size, polarity, solubility, differential water lipid solubility, adherence to leaf surface (tenacity), molecular size and shape

346

and vapour pressure. Chemical properties include differential reactivity between the compound and essential metabolic systems of the pathogen and the host as well as stability under different environmental conditions.

Prevailing environment influences efficacy of fungicides. Toxicity of chloranil (Spergon) is decreased in alkaline medium or under strong light due to its degradation. While mancozeb retains its efficacy in all wetting regimes, triforine (Saprol) loses its efficacy under conditions of increased precipitation. It is effective only in light mist. Fungicides that can resist weathering are helped by dew or light rains in redistribution providing protection to new growth of the host.

Materials that alter the physical and chemical properties of a fungicide may alter its efficacy. When copper sulphate is used alone it is a poor foliage fungicide and many cause phytotoxicity. When it is used with lime it becomes a good fungicide and the phytotoxicity disappears. The mixture increases its tenacity and prolongs its effectiveness. Often, adjuvants (surfactants, spreaders, stickers) are added to a fungicide. These increase the coverage capacity and tenacity. Wetting agents improve coverage but reduce tenacity. Use of surfactant when the fungicide is to be applied to easily wetted leaves (such as potato) should be avoided since it decreases retention of the fungicide on the foliage. The adjuvants, insecticides or any other material added to the fungicide must be compatible with it. Often two fungicides may become mutually antagonistic. On the other hand, synergistic effects are also noted. Mixture of mancozeb with systemic fungicides is an example.

One hundred per cent coverage of the susceptible surface is desired. Fungal spores or bacterial cells are very minute bodies. A very small gap left uncovered by the fungicide may permit the pathogen to survive and cause infection. Many fungal pathogens sporulate on the lower surface of the leaf (viz., downy mildew fungi and late blight fungus) and many flourish on the lower part of the stems. The fungicidal spray must cover these areas. Correct timing and frequency of fungicide application against foliage diseases is the most important factor determining success of fungicides. The timing can be determined by experience (calendar based sprays) but the best way to decide timing is disease forecasting which was discussed earlier. Concentration of the fungicide used is not as important as the frequency of application. Sprays applied with a lower concentration but at short intervals are more effective than sprays applied with high concentration but at long intervals.

FUNGICIDES, NEMATICIDES AND ANTIBIOTICS IN USE

On the basis of brand names the number of fungicides in the market all over the world runs into several hundred but if they are considered on the basis of chemical groups the number is not so large. There are now 113 active ingredients registered as fungicides worldwide (cf. Knight et al, 1997). All the brand names are not equally safe, effective and popular. Salts of toxic metals and organic acids, organic compounds of mercury and sulphur, quinones and heterocyclic nitrogen compounds have been the major fungicides in use. Mercury compound are now mostly withdrawn because of persistence in soil and pollution of ground water.

1. SULPHUR FUNGICIDES

The use of sulphur in plant disease control is probably the oldest. Fungicides based on sulphur continue to be in maximum use even today. Inorganic sulphur is used as elemental sulphur or as lime-sulphur mixture in dust form or as wettable powder which is in maximum use. **Elemental sulphur** dust is prepared by find grinding of the mineral. Tenacity and fungicidal efficacy depends on the particle size which should be within the range of 47–75 μm (200–300 mesh). **Wettable sulphur** can be suspended in water and used as spray material for control of powdery mildews and rusts. In India the commercial products of wettable sulphur are sold as Elosal, Cosan and Sulfex.

Lime-sulphur mixture can be prepared by boiling the two together (10 kg lime and 7 kg sulphur in 250 litres of water). Chemically the mixture is calcium polysulphide. Standard commercial formulations of lime sulphur are also available. It is mainly used in orchards to control such diseases as peach leaf curl, apple powdery mildew, anthracnose and brown rot. The dosage is 10–15 litres of the mixture in 500 litres of water. High concentration is used on dormant trees but on trees with opening buds and leaves the concentration is reduced. Phytotoxicity is a problem with sulphur. The toxicity is common at high temperature (above 30°C) than at lower temperatures for sulphur dust and above 26°C for lime sulphur. The higher toxicity of lime sulphur is due to its solubility.

The organic compounds of sulphur **(dithiocarbamates)** are highly effective protectant fungicides and the most popular spray materials in use today. The first useful dithiocarbamate fungicide was **thiram** (tetramethyl thiuram disulphide) which continues to be extensively used for seed treatment, usually at the rate of 250 g/quintal seed. Thiram has some antibacterial activity also. It has an interesting side effect. When used as foliar spray it acts as repellent for animals like deer and thiram treated seeds repel rodents.

The metallic dithicarbamates are prepared by reacting methylated dithiocarbamic acid with metals. Examples are **Ferbam** with iron and **Ziram** with zinc. In 1943, fungitoxic properties of disodium ethylene bisdithiocarbamates were discovered which initiated the development of the most widely used organic fungicides known as ethylenebisdithiocarbamaetes (EBDC). **Nabam** (disodium ehtylene bisdithiocarbamate) sold as Dithane A–40, Dithane D–14 and Parzate liquid was the first product in this group. It is effective against potato and tomato blights. When sodium in nabam is substituted with zinc **Zineb** (zinc ethyelene bisdithiocarbamate) is formed. It is sold as Zineb, Dithane Z–78, Lonacol, etc. It was extensively used against late and early blights of potato and tomato, blast of rice, helminthosporiose of rice, anthracnose and ripe rot of chillies, and leaf blight and downy mildew of maize. **Maneb** (manganese ethylene bisdithiocarbamate) contains manganese in place of zinc and is sold as Dithane M–22 or Manzate. In **mancozeb** (Dithane M–45 or Indofil M–45) manganese and zinc both are present. This fungicide is used against late and early blights of potato and tomato, anthracnose diseases caused by *Colletotrichum*, and downy mildews of many crops. Dithane S–31 contains manganese and nickel. The dosage of application of most dithiocarbamates varies from 2 to 3 kg/ha.

Another member of the dithiocarbamate group is **vapam** (sodium methyl dithiocarbamate) which is available in liquid form and is used for soil treatment. Vapam is effective against many fungi including species of *Fusarium* and *Pythium* and also against some important nematode parasites.

2. COPPER FUNGICIDES

The fungicidal activity of copper was mentioned as early as 1807 but its large scale use as a fungicide started in 1885 after the discovery of Bordeaux mixture. Together with sulphur the copper fungicides had been most extensively used in plant disease control before the advent of organic fungicides. Copper fungicides are all inorganic compounds. They are prepared from copper sulphate, copper carbonate, cuproud oxide and copper oxychloride.

Bordeaux mixture is prepared in a non-metallic container by 5 lb (2.5 kg) copper sulphate, 5 lb (2.5 kg) slaked lime dissolved in a small quantity of water and then making up the volume to 50 gallons (25 litres). This was 5-5-50 formulation. Variations can be made but quantity of lime must be equal to or more than copper sulphate (5-10-50). When the fungicides is used on copper sensitive plants amount of lime is further increased. The preparation can not be stored. Bordeaux mixture has been used against a very large number of diseases including blights, leaf spots and downy mildews. **Burgundy mixture** contains sodium carbonate in place of copper sulphate and is less effective. **Cheshnut compound** contains 2 parts copper sulphate and 11 parts ammonium carbonate. It is suspended in water and can be used as spray material.

The fixed or insoluble copper compounds have the copper ions fixed to the molecules chemically more firmly so that it is only slightly soluble. Therefore these compounds are less phytotoxic and addition of lime is not required. Due to ease in application and availability these fungicides became more prevalent than Bordeaux mixture. **Copper oxychloride** is the main compound among copper fungicides. The formulations available in market contain 4–50% metallic copper. The formulations with 4–12% copper (viz., Blimix 4%) are used as dust. The preparations having 50% copper are Blitox-50, Blue copper and Fytolan. These coper fungicides are used against the disease in which Bordeaux mixture was used.

3. QUINONE FUNGICIDES

Quinones naturally occur in plants and animals and are the source of colouration. They are also produced by oxidation of phenolics. They have antimicrobial properties and may be responsible for resistance to disease in plants. Quinone fungicides had limited, specialized uses. **Chloranil** (Spergon), a benzoquinone and **dichlone** (Phygon), a naphthoquionone, were two common fungicides in this group. Only dichlome is still used for seed treatment against externally seed-borne pathogens.

4. BENZENE FUNGICIDES

Chlorothalonil (tetrachloro-isophthalonitrile) is sold as Bravo, Daconil and Termil. It is a broad specturm fungicide used against a wide variety of fungal diseases of field and vegetable crops as well as orchard fruit trees. The diseases include late and early blights of potato and tomato, leaf spots of groundnut, downy mildews of different crops, Botrytis gray mold of grapes, peach leaf curl, rusts, anthracnose and apple scab. Chlorothalonil is relatively persistent in different environments. During rains much of the deposit is removed but the remaining residue persists for long.

Dichloran (dichloro nitroaniline, DCNA) is sold as Botran and is used against fruit and vegetable diseases as spray or soil treatment material. It is effective against sclerotia producing fungi such as *Botrytis, Monilinia, Sclerotium* and *Sclerotinia* and also against *Rhizopus*. Botran is especially effective as post-harvest dip or spray in suppressing fruit rots caused by above fungi.

Dinocap or Karathane (methyl heptyl dinitrophenyl) is an acaricide but most commonly used for control of powdery mildew in fruit orchards.

Pentachloronitrobenzene (PCNB) was highly effective against clubroot disease of crucifers and rots caused by sclerotial fungi. Due to its long persistence in soil and polluting effect it is now a withdrawn fungicide.

Biphenyl (diphenyl) is a polynuclear aromatic compound widely used to prevent post-harvest rot of citrus fruits. It is particularly effective in suppressing fruit rots caused by *Penicillium digitatum, P. italicum, Diplodia natalensis, Botrytis cinerea* and *Phomopsis citri*. It does not affect bacteria, yeasts, Phycomycetes and resistant strains of *Penicillium* and *Diplodia*. Biphenyl is volatile and when used for impregnating paper wraps or packaging material it diffuses to the stored fruits. Treatment with biphenyl leaves undesirable odor on fruits but it is temporary.

5. HETEROCYCLIC NITROGEN COMPOUNDS

This group has some of the best fungicides also known as phthalamides and dicarboximides such as captan, captafol, dyrene, glyodin, iprodione and vinclozolin.

Captan is sold under the trade names of Captan, Orthocide, Vancide etc. Primarily meant for seed treatment against *Pythium* it has also been used as soil drench and foliar spray. Captan had been extensively used for control of apple scab. It is relatively ineffective against downy mildews, powdery mildews and rusts. Captan is unstable at alkaline pH. **Folpet** is closely related to captan and is sold as Phaltan, Orthophaltan, etc. In addition to diseases controlled by captan, this fungicide is effective against powdery mildews and some rusts.

Captafol is sold as Difolatan, Difosan and Sanspor. It is a foliar spray as well as seed treatment fungicide and controls downy mildews also. Difolatan has high resistance to weathering and has very low phytostoxicity permitting it to be used in massive dosages. It has been used against apple scab, late blight of potato and tomato and several other foliar diseases.

The new fungicides in the dicarboximides group are iprodione, procymidone, vinclozolin, etc. **Iprodione** is sold as Rovral or Chipko-26019. It is a broad spectrum fungicide, mainly preventive. Rovral has been effectively used against diseases caused by *Botrytis, Monilinia, Sclerotinia, Alternaria, Helminthosporium* and *Rhizoctonia*. It can be used as spray material as well as for seed treatment and for fruit dip. **Vinclozolin** sold as Ornalin, Eonilan or Vorlan is used as a foliar fungicide against sclerotia forming fungi such as *Botrytis, Monilinia* and *Sclerotinia*. These two fungicides have shown good effficacy against Botrytis diseases including bunch rot of grapes. Ronilan has been found effective against Botrytis gray mold of chickpea. Resistance to these fugicides in strains of *B. cinerea* and *Monilinia fructicola* is reported. To avoid resistance development, these fugicides can be used in mixture with mancozeb.

6. ORGANO TIN COMPOUNDS

The organic tin fungicides include Du-ter, Brestanol and Brestan. **Du-Ter** is chemically triphenyltin hydroxide. It is very effective against *Cercospora, Helminthosporium, Alternaria, Pythium, Phytophthora* and *Rhizoctonia*. **Brestan** is chemically triphenyltin acetate (TPTA) and is effective against *Cercospora, Alternaria, Septoria* and many other fungi. **Brestanol** is triphenyltin chloride with activity similar to Du-Ter and Bestan. Due to phytotoxicity to the host these fungicides have restricted use.

SYSTEMIC FUNGICIDES

The discovery of the systemic fungicides in 1966 is a major landmark in the history of fungicidal management of plant diseases. A systemic fungicide is one which is taken up and translocated within the plant as a result of which the later becomes fungitoxic. In this respect systemic fungicides resemble antibiotics which are also systemic. However, antibiotics are basically of microbial origin and differ from systemic fungicides in translocability.

The uptake and translocation of almost all the systemic fungicides is passive.

This is a disadvantage because the active ingredient can move only with the sap stream. Thus, with few exceptions, when a systemic fungicide is applied to roots, the active ingredient passes intercellulary into the xylem vessels in which it is swept along with the sap stream towards the foliage (apoplast transport). When applied to the foliage, it passes into xylem and spreads to distal parts of the leaves but not towards roots. Coming from the roots, these compounds move into all organs that transpire, that is, they have functional stomata. They will not pass into flowers or fruits. They accumulate in mature rather than in newly formed leaves and cannot be removed from senescent leaves to newly formed leaves. Thus, actively growing plants need a continuous supply of the fungicide via soil or repeated foliar application to protect new tissues.

1) Oxathiins, Carboxins and Carboxamides: The first systemic fungicides that were developed in 1966 were the carboxin (**Vitavax**) and oxycarboxin (**Plantvax**). Both chemicals were translocated from roots in sufficient quantity to control rust on primary leaves of bean inoculated with *Uromyces phaseoli*. Vitavax has become the most popular fungicide for seed treatment to control loose smut of wheat and barley, replacing the solar or hot water treatment. It also gives satisfactory control of bunt (*Tilletia tritici* and *T. leavis*) and flag smut (*Urocystis agropyri*) of wheat. Other Basidiomycetes are also affected by Vitavax and Plantvax. Both have some antibacterial activity. They have proved better than antibiotics in the control of bacterial blight of cotton (Verma, 1995). Sprays of 0.15–0.3% Vitavax are reported to suppress bacterial leaf streak of rice also. In combination with thiram or copper oxyquinolinate the efficiency of both compounds is increased.

2) Benzimidazoles: This group of systemic fungicides has made more impact than the oxathiins because of their broad spectrum activity. The suppression of fungi in the xylem had not been possible earlier but these fungicides can do it. Most benzimidazoles are converted on the plant surface to methyl benzimidazole carbamate (MBC, Bavistin) which is the active ingredient. The group includes such fungicides as benomyl, carbendazim, thiabendazole and thiophanate. The list of fungi controlled by this group is very large and covers most classes of fungi except the oomycetes. In addition to fungicidal property benzimidazole are also miticidal and anti-nemic. Addition of oil or surfactants to benzimidazole spray mixture increases its penetration into the plant. However, development of resistance to these fungicides is very common.

Benomyl is marketed as Benlate, Tersan and many other names. It has been used for foliar spray, seed treatment and fruit dip against fruit rot pathogens. Benomyl is effective against powdery mildew fungi (*Erysiphe, Sphaerotheca, Uncinula* and *Podosphaera*), *Venturia inaequalis* (apple scab), species of *Penicillium* (fruit decay), *Cercospora, Fusarium, Verticillium, Cephalosporium, Colletotrichum, Gloeosporium, Botrytis, Monilinia, Rhizoctonia, Thielaviopsis* and *Ceratocystis*. Oomycetes, dark coloured fungi such as *Alternaria* and

Helminthosporium, some Basidiomycetes and bacteria are not affected. Remission of symptoms of sandal spike (MLO) is also reported.

Carbendazim (MBC) has been in the market under the name of Bavistin which has same fungicidal properties as benomyl but is more stable. It is mainly used as seed dressing material for control of seed-born smuts and wilt causing fungi. As spray material it is used against apple scab and rice sheath blight. Sprays of a mixture of Bavistin and Ledermycin have provided control of the citrus greening disease caused by the bacterium *Liberobacter asiaticum*.

Thiabendazole (TBZ) is sold under the name Mertect. This is also a broad spectrum fungicide, same as benomyl, but quantitatively less effective. It is taken up and translocated without hydrolysis to leaves and stems.

Thiophanates are actually not benzimidazoles because they are based on thiourea. They are sold as Topsin and Cercobin. They are specially effective against powdery mildews. **Thiophanate methyl** (Topsin-M or Cercobin-M) is more effective than the ethoxy compound.

3) Acylalanines or Acylanilides were the first group of systemic fungicides effective against Oomycetes (*Pythium, Phytophthora, Sclerospora, Peronospora*). The group includes metalaxyl (Ridomil), furalaxyl (Fungorid) and banalaxyl.

Metalaxyl is sold under the name Ridomil for foliar spray and Apron for seed treatment. It is highly water soluble and is readily translocated from roots to aerial parts but its lateral movement is slight. It is practically non-toxic to fish and bees and is compatible with most other fungicides, insecticides and acaricides in common use.

Metalaxyl is highly effective against *Pythium, Phytophthora* and the downy mildew fungi. It has been used as spray material against downy mildew of grapevines and Koleroga or Mahali disease of areca palm and as seed treament and foliar spray against downy mildews of maize, sorghum and pearl millet as well as white blisters of crucifers. Foliar sprays have little effect against seed and root infection since almost no amount of the fungicide is translocated to the roots. It is recommended against rhizome rot of ginger through corn treatment. As soil drench it has been used against Phytophthora root rot of citrus and collar rot of apple. The fungicide has curative effect. Metalaxyl does not kill the target fungi but is fungistatic affecting different stages of the life cycle of the pathogens. In soil, the fungicide persists for long.

The highly selective metalaxyl can promote diseases caused by non-target pathogens. As side effect metalaxyl induces mild shift in host resistance expression. It increases potato tuber resistance to *Alternaria solani*, enhances lytic capabilities of antagonistic microorganisms in rhizosphere and promotes growth of endo- and ectomycorrhizae. Resistance to this fungicide is most common in *Phytophthora infestans*, some in downy mildew fungi and only one in *Pythium*. This has necessitated use of Ridomil with protectant dithiocarbamates. Against late blight of potato Ridomil MZ (metalaxyl + mancozeb) or Ridomil ZM (metalaxyl + Ziram) are recommended. Ridomil MZ is used against downy mildew of grapevines. Sprays of metalaxyl + copper oxychloride are effective against downy mildew of hop (*Pseudoperonospora humuli*). Mixture of metalaxyl with daconil is effective against *Sclerophthora macrospora* in maize. In some diseases combination of metalaxyl with vitavax has been used. Metalaxyl can be combined with furadon or aldicarb for protection against fungal and nematode or insect attack.

Cyprofuram (Vinicur) is also a acylanilide and effective against oomycetes. It is applied as foliar spray and for soil surface and seed treatment.

4) Sterol Biosynthesis Inhibiting Fungicides: A groups of sterol biosynthesis inhibiting fungicides has come in common use during the last 30 years. It includes pyrimidines, imidazoles, triazoles, piparazines and morpholines.

The hydroxypyramidines include **diamethirimol** (Milcurb), **ethirimol** (Milstem) and **bupirimate** (Nimrod). The substituted pyrimidines include **fenarimol** (Rubigan), **nuarimol** (Trimidal) and **triarimol**. Diamethirimol is effective against powdery mildew of cucurbits (*Sphaerotheca fuliginea*). Ethirimol is effective against powdery mildew of barley. These compounds check powdery mildew of wheat also. Diamethirimol is applied to soil to be taken up by the plant while ethirimol can be applied through soil as well as seed. Bupirimate also is effective against powdery mildew of many hosts including apple and roses. Applied on consecutive days at 2-week intervals it is reported to give effective control of apple powdery mildew. Protected plants are more vigorous with greater root weight than the unprotected plants. Fenarimol and nuarimol are effective against several leaf spot fungi, rusts and smuts in addition to powdery mildews. Fenarimol could control apple scab by only 4 sprays given at tight cluster, pink bud, petal fall and 10 days after petal fall stages (Wilcox *et al*, 1992). Triarimol also has systemic and curative effect on apple scab disease. It has been found effective against *Ustilago striiformis* and *Urocystis agropyri*.

The triazoles include some of the best SI fungicides such as **triadimefon** (Bayleton), **triadimenol** (Bayton), **bitertanol** (Bacor), **boutrizol** (Indar or RH-124), **propiconazole** (Tilt), **etaconazole** (Vangard), **myclobutanil** (Eagle, Nova, Rally, Prothane, Enzone and Systhane), **difenoconazole, cyproconazole, tebuconazole** and **fluquiconazole.** These fungicides are generally used as spray materials but seed and soil treatment is also common. They show long term protective and curative activity thus facilitating fewer applications and longer intervals between sprays. A single application of a triazole between bloom and two weeks after scatter in grapevines for control of downy mildew was recommended by Pearson *et al*, (1994). Bayleton has been used against cereal rusts, Karnal bunt and powdery mildews in India. It is effective against ergot (*Claviceps*) also. Propiconazole (Tilt) is very effective against cereal rusts even with one spray. Boutrizol has been highly successful chemical against leaf rust (*Puccinia recondita*) and stripe rust (*Puccinia striiformis*) in many countries. Myclobutanil and flusilazole can control apple scab with only four sprays. Cyproconazole and tebuconazole are very effective against groundnut leaf spot (*Cercospora* *arachidicola* and *Cercosporidium personatum*) and against soil-borne diseases caused by *Rhizoctonia* and *Sclerotium*. Difenoconazole is highly effective against *Alternaria solani* on tomato, *C. arachidicola* on groundnut and apple scab.

Morpholines include the fungicides **dodemorph** (Meltatox) and **tridemorph** (Calixin). These are preventive and eradicant foliar fungicides effective against powdery mildews and leaf spots on cereals. Calixin has been commercially used against powdery mildew of wheat, pea, cucurbits and apple. It is also effective against cereal stripe rust, Shigatoka of banana and pink disease of rubber and other plantation crops. **Triforine** (piparazine) is a foliar fungicide sold as Saprol, Sella or Funginex and used against powdery mildews, leaf and fruit spots, fruit rots and anthracnose.

5) Organic Phosphates: The systemic fungicides based on organic phosphate include fosetyl-Al (sold as Aliette), Kitazin, edifenphos (Hinosan) and pyrazophos (Afugan).

Aluminium-*O*-ethyl phosphonate or phosethyl-Al commonly called **Fosetyl-Al (Aliette)** was develop as a fungicide against oomycetes after metalaxyl. It can be applied as foliar spray, soil drench, root dip, post-harvest fruit dip and soil mix. The fungicide is highly phloem mobile (unusual for systemic fungicides) and is translocated from leaves to roots. In soil it persists for several months. A single application through soil protects citrus roots against *Phytophthora* spp. for 2–3 months. Single pre-plant dip treatments protects pineapples against *Phytophthora* for 18 months. Although the fungicide gives excellent control of Phytophthora diseases often it does not show *ir vitro* activity against the pathogen. In such cases it probably acts by triggering induced resistance in the host. Like metalaxyl this fungicide is also compatible with insecticides.

Kitazin and **Hinosan** are highly effective against the rice blast fungus *Pyricularia grisea*. Hinosan has been extensively used in India against this disease. Granules of Kitazin added to water in rice fields provide a reservoir of the fungicide to be absorbed and translocated by the plants.

6) Miscellaneous Systemic Fungicides: Pyracarbolid is closely related to oxathiins and is more effective than the latter. It controls rusts, smuts and *Rhizoctonia solani*. **Chloroneb** (Demosan) is used as seed and soil fungicide. It does not leach from soil and is effective against seedling blight of cotton, bean, beet, etc. **Ethazol** (Turban, Terrazole, Koban) is a seed and soil fungicide effective against damping off, seed and stem rot caused by *Pythium* and *Phytophthora*. When combined with PCNB or Topsin M it has wide spectrum effect against *Fusarium* and *Rhizoctonia* also. **Imazalil** (Fungaflor) has excellent curative and preventive properties and is effective against powdery mildew, leaf spot, vascular wilts and fruit rots caused by Ascomycetes and Deuteromycotina. **Cyprodinil** belongs to the anilopyrimidine group and was developed in 1994 against a variety of diseases caused by Ascomycotina and Deuteromycotina. It is recommended against scab and powdery mildew of apple and Botrytis diseases. **Carpropamid**, a new generation carboximide fungicide, is a melanin biosynthesis inhibitor and checks rice blast by preventing melanization of appressoria of *Pyricularia grisea*. It has no *in vitro* activity against the pathogen.

Strobilurins are a new class of fungicides derived from natural antibiotic found in the symbiotic fungus *Strobilurus tenacellus*. The fungicides in this class have a very broad and balanced spectrum of activity as foliar fungicides having favourable toxicological profile, rapidly dissipating from the soil and surface water and are unlikely to harm beneficial microorganisms. They have protective and curative action against fungi in all the taxonomic groups. So far, they have not shown resistance development in the sensitive fungi.

ANTIBIOTICS

The antibiotics are defined as metabolites of microorganisms which, in very dilute concentration, have the capacity to inhibit growth of, or even destroy other microorganisms. Antibiotics have systemic action in plants moving in both direction, from leaves to roots and from roots to the foliage. They are not only eradicants but also protectants providing temporary resistance in the host. Phytotoxicity is one major disadvantage with antibiotics and has limited their large scale use. If the dosage increases beyond the permissible limit plants are damaged. The other disadvantage is their narrow spectrum of antipathogen action.

The antibiotics used in plant disease control mainly belong to groups known as streptomycin, tetracyclines, polyenes cycloheximide and griseofulvin. **Streptomycin** was the first antibiotic used in plant disease control. It was used against fire blight of pear (*Erwinia amylovora*) in 1953. Streptomycin formulations are available under the name Agrimycin, Phytomycin, Orthostreptomycin, Streptocycline, etc. Agrimycin and Streptocycline are mixed preparation of streptomycin and tetracyline. Maximum use of streptomycin as streptomycin sulphate has been against bacterial diseases of fruit trees such as fire blight of apple and pear and citrus canker. Streptomycin and Agrimycin have been used against bacterial blight of cotton. Streptocycline is used in combination with copper oxychloride for the control of bacterial leaf blight of rice.

Antibiotics in the tetracycline group are **tetracycline** (Acromycin), **oxytetracycline** (Terramycin) and **chloretetracycline** (Aureomycin). These antibiotics are bacteriostatic and bactericidal. The sensitive groups of bacteria include the fastidious bacteria, spiroplasma and MLOs.

Cycloheximide (Actidione, Actispray, Actidione RZ) was a very strong antifungal antibiotic but because of its phytotoxicity it could not become popular. **Aureofungin**, a heptaene antifungal antibiotic, is recommended for the control of rice leaf spot, rice blast, barley stripe disease and covered smut of barley through seed treatment and against powdery mildew, anthracnose and downy mildew of grapes, powdery mildew of apple, leaf rust of wheat and citrus gummosis through foliar sprays. Post-harvest decay of mango and guava fruits is also checked by fruit dip in aureofungin.

Griseofulvin had been found very effective against *Botrytis cinerea, Sphaerotheca pannosa, Pseudoperonospora cubensis* and many other common plant pathogens. However, under field conditions it had failed to perform well and is no more a promising fungicide.

NEMATICIDES

Nematicides belong to two groups: volatile soil fumigants and non-fumigants (contact and systemic nematicides). The fumigants consist of compounds belonging to halogenated hydrocarbons and isothiocyanate groups while the non-fumigants are mostly organo-phosphorus and carbamates. The former directly kill the nematode larvae while the latter do not cause direct kill. Eggs are generally not affected by nematicides being protected in cysts or in crop debris.

The **soil fumigants** such as methyl bromide, ethylene dibromide (EDB), D-D mixture, methyl isothiocyanate, etc. posses high vapor pressure, get dissolved in soil moisture in concentrations high enough to kill nematode larvae and disperse through the soil pores. They are non-selective and can kill even beneficial bacteria in the soil. The method of application involves injection of the fumigants into the soil with the help of suitable equipments, make the soil compact or provide water seal and let the fumigant act for 2–3 weeks before planting the crop. The halogenated hydrocarbon group includes methyl bromide, ethylene dibromide, Toluene, D-D mixture, DBCP (Nemagon) and chloropicrin. The isothiocyanate group includes fumigants that release methyl cyanate such as Vapam, Dazomet, Mylone or Basamid.

The non-fumigant nematicides have gradually replaced the fumigants. They are available in granular form and can be applied to row crops at the time of planting or to soil around the standing trees. The contact non-fumigants include fensulphothion (Dasanit or Terracur P), thionazin (Nemafos, Zinophos), diazinon (Basudin) and ethoprop or ethoprofos (Mocap). The systemic nematicides in this group are phenamiphos (Nemacur), phorate (Thimet), aldicarb (Temik) and carbofuran (Furadan).

MODE OF ACTION

The basis of fungicidal action is selective toxicity of chemicals to various forms of life. A majority of fungicides kill or inhibit the fungi responsive to them through direct effect on fungal cells or spores after entering them. Solubilization of fungicides on the host surface is facilitated by free water, carbon dioxide and ammonia in rain water or dew, guttation fluid and other exudates from the plants, spores exudates, and ability of the spores to accumulate fungicides from very dilute solutions. Killing of fungal cells by fungicides may be brought about by (i) injurious effects on cell-walls and on cell division, (ii) effect on the permeability of the cell membrane, (iii) effect on enzyme system of the fungal cells (iv) chelation and precipitation of chemicals and (v) by antimetabolism.

Unspecifically Acting Fungicides: The protectant fungicides produce the above effects at the point of entry of the pathogen. Inorganic copper, sulphur, mercury and organic compound of mercury and sulphur have fairly nonspecific modes of action. They are inherently toxic to fungi, plants and even animals. Within the body of the fungus, the protectant fungicides exhibit fairly nonspecific chemical reaction with different proteins (enzyme systems). The attack is almost simultaneous on several essential cell components resulting in multisite action.

Copper and mercury ions are toxic to all cells because they form complexes with different cell components resulting in inactivation of essential enzymes. Cytoplasmic membrane is attacked first. There is a rapid loss of potassium ions from the fungal cell. Since these ions forms chelates with cell components they accumulate in the cell. Almost all the organic protectant fungicides cause similar multisite action through inactivation of enzymes (mainly sulfhydryl group of proteins).

The Specifically Acting or Systemic Fungicides: Once in contact with the pathogen the chemotherapeutants appear to affect pathogens in the same manner as the protectant fungicides. However, the systemic fungicides are much more specific and affect only one or two functions in the pathogen. The action of systemic fungicides can be any of the following:

(i) *Damage to cell membrane:* The polyene antibiotics (nystatin, pimaricin, aureofungin, filipin, etc.) and some fungicides such as dodine have the cell membranes as their specific site of action. The fungi that have none or very little sterols in their cell membranes and obtain sterols from external sources (as in oomycetes) are affected by polyene antibiotic which forms complexes with sterols. This results in membrane damage and severe leakage of potassium ions.

(ii) *Effects on enzyme systems:* The mitochondrial respiration is affected by oxathiins (Vitavax, Plantvax) through inhibition of the necessary enzyme system. Calixin interacts with the respiratory chain, attacking the mitochondrial electron transport system. Dexon also interferes with the respiratory system of the pathogen.

(iii) *Inhibition of protein, RNA, DNA synthesis and nuclear division:* Antibiotics are mainly responsible for inhibition of protein synthesis. Metalaxyl and related compounds affect ribosomal RNA of the fungi and interfere with protein synthesis. The benzimidazole fungicides act as spindle poison binding to the protein subunits of spindle microtubules. It inhibits mitosis.

(iv) *Action on other systems:* The organic phosphorus fungicides Kitazin and Hinosan primarily act by inhibiting chitin synthesis in cell walls of Ascomycotina and Basidiomycotina. Fungi that have no chitin in the cell walls, as in oomycetes, are not affected by these fungicides. Triazole fungicides inhibit ergosterol biosynthesis. Ergosterol is a cellular component that plays a crucial role in the structure and function of cell walls.

The fungicides that have multisite action (broad spectrum protectants) do not allow the fungi to develop resistance against them because the fungus will have to develop many changes in its genetic configuration. The site specific fungicides (systemic fungicides) affect only few sites in the pathogen which can easily develop resistance by one or two mutations.

NON-CONVENTIONAL CHEMICALS IN DISEASE CONTROL

Mineral and vegetable oils have been used in plant disease control. Large scale use of mineral oil for control of Shigatoka disease of banana was prevalent in the past. Oils of sunflower, olive, maize and rapeseed have antifungal properties and have shown efficacy against powdery mildew of apple. Mechanically emulsified rape oil was comparable to dinocap (Karathane) giving 99% control when applied 1–7 days after inoculation (Northover and Schneider, 1993). Significant control of grapevine powdery mildew with rape oil derivative is reported from Australia (Azam, et al, 1998). Applying a coat of mustard oil to ripe mango fruits gives 90% control of Aspergillus rot. Groundnut oil coating prevents papaya fruit decay. Similarly, aqueous extracts of many plants are also reported to provide control of fungi, bacteria and viruses. Some herbicides are known to directly affect a disease. The herbicide propanil has antifungal activity against Drechslera oryzae. Foliar application of benthiocarb suppresses sheath blight of rice.

Bibliography

1. Adams, M.J. 1991. Transmission of plant viruses by fungi *Ann. Appl. Biol.* 118: 479
2. Adams, P.B. 1990. The potential of mycoparasites for biological control of plant diseases. *Annu. Rev. Phytopathol.* 28: 59
3. Agarwal, A. and H.S. Tripathi. 1999. Biological and chemical control of *Botrytis* gray mold of chickpea. *J. Mycol. Pl. Pathol.* 29: 52
4. Agnihothrudu, V. 1955. Incidence of fungistatic organisms in the rhizosphere of pigeonpea (*Cajanus cajan*) in relation to resistance and susceptibility to wilt cause by *Fusarium udum*. *Naturwissenschaften* 43: 373
5. Agnihotri, V.P., N. Singh, H.S. Chaube, U.S. Singh and T.S. Dwivedi (eds.). 1989. *Perspectives in Plant Pathology*. Today and Tomorrow Printers and Publishers, New Delhi, India
6. Agnihotri, V.P., A.K. Sarbhoy and D.V. Singh. 1997. *Management of Threatening Plant Diseases of National Importance*. Malhotra Publishing House, New Delhi, India
7. Agrios, G.N. 1980. Escape from disease, pp. 18-38. In: *Ref. No. 260*
8. Agrios, G.N. 1988. *Plant Pathology*, 3rd Ed. 1997. 4th Ed. Academic Press, New York
9. Aist, J.R. 1983. Structural responses as resistance mechanisms, pp. 33-70. In: *Ref. No. 37*
10. Akai, S. and M. Fucutomi. 1980. Preformed internal physical defences, pp. 139-160. In: *Ref. No. 260*
11. Akhtar, M. and M.M. Alam. 1993. Utilization of waste materials in nematode control. A review. *Bioresour. Technol.* 45: 1
12. Akino, S. and A. Ogoshi, 1995. Pathogenesis and host specificity in *Rhizoctonia solani*, pp. 37-49. In: *Ref. No. 481*, Vol. 2
13. Alam, M.M., M.A. Siddiqui and A. Ahmad. 1992. Antagonistic plants, pp. 41-50. In *Nematode Biocontrol* (*Aspects and Prospects*). M.S. Jairajpuri *et al* (eds.). CBS Publishers, Delhi
14. Alba, A.P.C., S.D. Guzzo, M.F.P. Mahlow and W.B.C. Moraes. 1983. Common antigens in extracts of *Hemileia vastatrix* uredinospores and of *Coffea arabica* leaves and roots. *Fitopathologia Brasileira* 8: 47
15. Alghisi, P. and F. Favarun. 1995. Pectin degrading enzymes and plant parasite interactions. *Eur. J. Plant Pathol.* 101: 365
16. Alstrom, S. 1991. Induction of disease resistance in common bean susceptible to halo blight bacterial pathogen after seed bacterization with rhizosphere

360

pseudomonads. *J. Gen. App. Microbiol.* 37: 495

17. Alstrom, S. 1995. Evidence of disease resistance induced by rhizosphere pseudomonads against *Pseudomonas syringae* pv. *phaseolicola. J. Gen. Appl. Microbiol.* 41: 315

18. Altman, J. and C.L. Campbell. 1997. Effect of herbicides on plant diseases. *Annu. Rev. Phytopathol.* 15: 361

19. Ameisen, J.C. 1996. The origin of programmed cell death. *Science* 272: 1278

20. Amin, K.S. and C. Venkatarao. 1979. Rice blast control by nitrogen management. *Phytopath. Z.* 96: 140

21. Anderson, A.J. 1982. Performed resistance mechanisms. pp. 119-137. In: *Ref. No. 392,* Vol. 2

22. Anderson, A.J. 1984. The biology of glycoproteins as elicitors. In: *Ref. No. 335*

23. Anderson, T.R. and Z.A. Patrick. 1980. Soil vampyrellid amoebae that cause perforations in conidia of *Cochliobolus sativum. Soil Biol. Biochem.* 12: 159

24. Apple, J.L. 1977. The theory of disease management, pp. 79-102. In: *Ref. No. 259*

25. Arntzen, C.J., M.F. Haugh and S. Bobick. 1973. Induction of stomatal closure by *Helminthosporium maydis* pathotoxin. *Plant Physiol.* 52: 569

26. Arora, D.K. and A.K. Pandey. 1989 Soil solarization for control of soil-borne diseases: Theory and applications, pp. 429-438. In: *Ref. No. 5*

27. Asada, Y., W.R. Bushnell, S. Ouchi and C.P. Vance (eds.). 1982. *Plant infection-The Physiological and Biochemical Basis.* Springer-Verlag, Berlin.

28. Autrey, L.J.C., A. Dookun, S. Saumatly, *et al.,* 1991. Soil transmission of the ratoon stunting disease bacterium, *Clavibacter xyli* subsp. *xyli. Sugarcane* 6: 5

29. Azam, M.G.N., G.M. Gurr and P.A. Magarey. 1998. Efficacy of a compound based on canola oil as a fungicide for control of powdery mildew caused by *Uncinula necator. Aust. Plant Pathol.* 27: 116

30. Babcock, M.J., E.C. Eckwall and J.L. Schottel. 1993. Production and regulation of potato scab inducing phytotoxin by *Streptomyces scabies. J. Gen. Microbiol.* 139: 1579

31. Backman, P.A. 1978. Fungicide formulations: Relationship to biological activity. *Annu. Rev. Phytopathol.* 16: 211

32. Bagyaraj, D.J., A. Manjunath and D.D.R. Reddy. 1979. Interaction of vesicular arbuscular mycorrhiza with root knot nematodes in tomato. *Plant Soil* 51: 397

33. Bailey, J.A. 1982. Mechanism of phytoalexin accumulation, pp. 289-318. In: *Ref. No. 36*

34. Bailey, J.A. 1983. Biological perspectives of host-pathogen interactions, pp. 1-32. In: *Ref. No. 37*

35. Bailey, J.A. 1995. Plant-pathogen interactions: a target for fungicide development, pp. 233-244. In: *Ref. No. 171*

36. Bailey, J.A. and J.W. Mansfield. 1982. *Phytoalexins.* Wiley, New York, 333 pp.

37. Bailey, J.A. and B.J. Deverall (ed.). 1983. *The Dynamics of Host Defence.* Academic Press

38. Baker, B., P. Zambryski, B. Staskawicz and P. Dinesh Kumar. 1997. Signalling in plant-microbe interactions. *Science* 276: 726

39. Baker, C.J. and B.W. Orlandi. 1995. Active oxygen in plant pathogenesis. *Annu. Rev. Phytopathol.* 33: 299

40. Baker, D.R., J.G. Fenyes and W.K. Moberg. (eds.). 1991. *Synthesis and Chemistry of Agrochemicals*, Vol. II. ACS, Washington

41. Baker, K.F. 1987. Evolving concepts of biological control of plant pathogens. *Annu. Rev. Phytopathol.* 25: 67

42. Baker, K.F. and S.H. Smith. 1966. Dynamics of seed transmission of plant pathogens. *Annu. Rev. Phytopathol.* 3: 153

43. Baker, K.F. and W.C. Snyder (eds.). 1965. *Ecology of Soil-borne Plant Pathogens.* Univ. California Press. Berkeley

44. Baker, K.F. and R.J. Cook. 1974. *Biological Control of Plant Pathogens.* Freeman, San Francisco

45. Baker, R. 1981. Biological control: Eradication of plant pathogens by adding organic amendments to soil, pp. 137-157. In: *Handbook of Pest Management in Agriculture*, Vol. 2. D. Pimental (ed.). CRC Press, Boca Raton, Fl. USA

46. Bakker, A.W. and B. Schippers. 1987. Microbiol cyanide production in the rhizosphere in relation to potato yield reduction and *Pseudomonas* spp. mediated plant growth promotion. *Soil Biol. Biochem.* 19: 451

47. Bakker, P.A.H.M., R. Van Peer and V. Schippers. 1991. Suppression of soil-borne plant pathogens by fluorescent pseudomonads: mechanisms and prospects, pp. 217-230. In: *Biotic Interactions and Soil-borne Diseases.* A.B.R. Beemster *et al.*, (eds.). Elsevier, Amsterdam

48. Bakker, P.A.H.M., J.M. Raaijmakers and B. Schippers. 1993. Role of iron in the suppression of bacterial plant pathogens by fluorescent pseudomonads, pp. 131-142. In: *Ref. No. 256*

49. Baldwin, B.C. and W.G. Rathwell. 1988. Evolution of concepts for chemical control of plant disease. *Annu. Rev. Phytopathol.* 26: 265

50. Banwart, W.L. and J.M. Bremner. 1976. Evolution of volatile sulphur compounds from soils treated with sulphur containing organic materials. *Soil Biol. Biochem.* 8: 439

51. Barras, F., F. von Gijsegem and A.K. Chatterjee. 1994. Extracellular enzymes and pathogenesis of soft rot *Erwinia*. *Annu. Rev. Phytopathol.* 32: 201

52. Beachy, R.N., S. Loesch-Fries and N.F. Turner. 1990. Coat protein mediated resistance against plant viruses. *Annu. Rev. Phytopathol.* 28: 451

53. Beardmore, J., J.P. Ride and J.W. Granger. 1983. Cellular lignification as a factor in the hypersensitive resistance of wheat to stem rust. *Physiol. Plant Pathol.* 22: 209

54. Beckman, C.H. 1987. *The nature of Wilt in Plants.* APS Press, St. Paul, Minn, USA

55. Bell, A.A. 1981. Biochemical mechanisms of disease resistance. *Annu. Rev. Plant Physiol.* 32: 31

56. Beresford, R.M. and D.W.I., Manktelow. 1994. Economics of reducing fungicide use by weather-based disease forecasts for control of *Venturia inaequalis* in apples. *N.Z.J. Crop Hortic Sci.* 22: 133

57. Bhattacharya, A. and A.K. Roy. 1998. Induction of resistance in rice plants against sheath blight with non-conventional chemicals. *Indian Phytopath.* 51: 81

58. Bhattacharya, I. and M. Pramanik. 1998. Effect of different antagonist rhizobacteria and neem products on clubroot of crucifers *Indian Phytopath.* 51: 87

59. Biggs, A.R. M.M. El-Kohli, S.El-Neshawy and R. Nickerson. 1997. Effect of

calcium salts on growth, polygalacturonase activity and infection of peach fruit by *Monilinia fructicola Plant Dis.* 81: 399

60. Billing, E. 1980. Fire blight (*Erwinia amylovora*) and weather: a comparison of warning systems. *Ann. Appl. Biol.* 95: 365

61. Billing, E. 1982. Entry and establishment of pathogenic bacteria in plant tissues, pp. 51-70. In: *Ref. No.* 430

62. Bird, P.M. 1988. The role of lignification in plant disease, pp. 523-535, In: *Ref. No.* 478

63. Bishop, C.D. 1988. Pathogenesis in the fungal vascular wilts. pp. 256-271. In: *Ref. No.* 478

64. Blad, B.L., J.R. Steadman and A. Weiss. 1978. Canopy structure and irrigation influence on white mold disease and microclimate of dry edible beans. *Phytopathology* 68: 1431

65. Blakeman, J.P. 1991. Foliar plant pathogens: epiphytic growth and interactions on leaves. *J. App. Bacteriol.* 70(S): 49

66. Blakeman, J.P. and P. Atkinson. 1981. Antimicrobial substances associated with the aerial surfaces of plants. In: *Microbial Ecology of the Phylloplane.* Academic Press, New York

67. Blakeman, J.P. and N.J. Fokkema. 1982. Potential for biological control of plant disease on the phylloplane. *Annu. Rev. Phytopathol.* 20: 167

68. Boller, T. 1991. Ethylene in pathogenesis and disease resistance, pp. 293-314. In: *The Plant Hormone Ethylene.* A.K. Mattoo (ed.). CRC Press, Boca Raton, Florida

69. Boller, T. 1995. Chemoreception of microbial signals in plant cells. *Annu. Rev. Plant Physiol. Plant Mol. Biol.* 46: 189

70. Boller, T. and F. Meins Jr. (eds.). 1992. *Genes Involved in Plant Defence.* Springer-Verlag, New York

71. Bordoloi, D.K. and S.K. Addy. 1985. Common antigens between *Xanthomonas campestris* pv. *oryzae* and rice leaves to resistance and susceptibility. *Indian Phytopath.* 38: 112

72. Bos. L. 1977. Seed-borne viruses, pp. 39-69. In: *Plant health and Quarantine in International Transfer of Genetic Resources.* L. Ciarappa and W.B. Hewitt (eds.). CRC Press

73. Bourke, P.M.A. 1970. Use of weather information in the prediction of plant disease epiphytotics. *Annu. Rev. Phytopathol.* 8: 345

74. Bowler, C. and N-H, Chau. 1994. Emerging themes of plant signal transduction. *Plant Cell* 6: 1529

75. Bozarth, R.P. and R.E. Ford. 1988. Viral interactions: Induced resistance (Cross Protection) and viral interference among plant viruses, pp. 551-567. In: *Ref. No.* 478

76. Bracker, C.E. and L.J. Littlefield. 1973. Structural concepts of host-pathogen interface, pp. 159-318. In: *Ref. No.* 94

77. Bridge, J. 1996. Nematode management in sustainable and subsistence agriculture. *Annu. Rev. Phytopathol.* 34: 201

78. Briggs, S.P., G.S. Johal. 1994. Genetic patterns of plant host-parasite interactions. *Trends Genet.* 10: 12

79. Britten, R.J. and Davidson. 1969. Gene regulation for higher cells-a theory. *Science* 165: 349

80. Broadbent, P. and K.F. Baker. 1975. Soil suppressive to *Phytophthora* root rot in Eastern Australia, pp. 152-157. In: *Ref. No. 88*

81. Broglie K., I. Chet, M. Holliday *et al.* 1991. Transgenic plants with enhanced resistance to fungal pathogen, *Rhizoctonia solani. Science* 254: 1194

82. Brown, D.J.F., W.M. Robertso and D.L. Trudgill. 1995. Transmission of viruses by plant nematodes. *Annu. Rev. Phytopathol.* 33: 223

83. Brown, M.E. 1974. Seed and root bacterization. *Annu. Rev. Phytopathol.* 12: 181

84. Brown, R.H. and B.R. Kerry (eds.). 1987. *Principles and Practices of Nematode Control in Crops.* Academic Press, Sydney

85. Brown, W. 1965. Toxins and cell wall dissolving enzymes in relation to plant disease. *Annu. Rev. Phytopathol.* 3: 1

86. Browning, J.A. and K.J. Frey. 1969. Multiline cultivars as a means of disease control. *Annu. Rev. Phytopathol.* 7: 355

87. Browning, J.A., M.D. Simons and E. Torres. 1977. Managing host genes: Epidemiological and Genetic Concepts, pp. 191-212. In: *Ref. No. 258*

88. Bruehl, G.W. (ed.) 1975. *Biology and Control of Soil-borne Pathogens.* Am. Phytopath. Soc., St. Paul, Minn. USA

89. Buddenhagen, I.W. and A. Kelman. 1964. Biological and physiological aspects of bacterial wilt caused by *Pseudomonas solanacearum. Annu. Rev. Phytopathol.* 2: 203

90. Burr, T.J., M.N. Schroth and T. Suslow. 1978. Increased potato yields by treatment of seed pieces with specific strains of *Pseudomonas fluorescens* and *Pseudomonas putida Phytopathology* 68: 1377

91. Bushnell, W.R. and J.B. Rowell. 1981. Suppressors of defence reactions: a model for roles in specificity. *Phytopathology.* 71: 1012

92. Butt, D.J. and X.M. Xu. 1994. Vintem TM - a computerized apple scab warning system for use on farms. *Norwegian J. Agric. Sci. Suppl.* 17: 247

93. Buyer, J.S. and J. Leong. 1986. Iron transport-mediated antagonism between plant growth promoting and plant deleterious *Pseudomonas* strains. *J. Biol. Chem.* 261: 791

94. Byrde, R.J.W. and C.V.V. Cuttings. 1973. *Fungal Pathogenicity and Plant's Response.* Academic Press, New York

95. Cahill, D. and G. Weste. 1983. Formation of callose depostis as a response to infection with *Phytophthora cinnamomi. Trans. Brit. Mycol. Soc.* 80: 23

96. Campbell, C.L. and L.V. Madden. 1990. *Introduction to Plant Disease Epidemiology.* John Wiley and Sons, New York

97. Campbell, R.N. 1996. Fungal transmission of plant viruses. *Annu. Rev. Phytopathol* 34: 97

98. Carlson, G.A. and C.E. Main. 1976. Economics of disease-loss management. *Annu. Rev. Phytopathol* 14: 381

99. Caron, M. 1989. Potential use of mycorrhizae in control of soil-brone diseases. *Can. J. Plant Pathol.* 11: 177

100. Carver, T.L.W. 1988. Pathogenesis and host-parasite interaction in cereal powdery mildew, pp. 351-381. In: *Ref. No. 478*

101. Carver, T.L.W., S.M. Ingersonn-Morris, B.J. Thomas and R.J. Zeyen. 1995. Early interactions during powdery mildew infection. *Can. J. Bot.* 73: 5637

102. Castor, L.L., J.E. Ayers, A.A. MacNab and R.A. Krause. 1975. Computerized

364

forecasting system for Stewart's bacterial disease on corn. *Plant Dis. Rep.* 59: 533

103. Chakrabarti, S.K., G.S. Shekhawat and A.V. Gadewar. 1995. Phenotypic reversion from afluidal to fluidal colony types in the strains of *Pseudomonas solanacearum. Indian Phytopath.* 48: 353

104. Chakraborty, B.N. 1988. Antigenic disparity, pp. 477-484. In: *Ref. No. 478*

105. Chakraborty, B.N. and R.P. Purkayastha. 1983. Serological relationship between *Macrophomina phaseolina* and soybean cultivars. *Physiol Plant Pathol.* 23: 197

106. Chakravarty, A.K. and M. Shaw. 1977. A possible molecular basis for obligate host-pathogen interactions. *Biol. Rev.* 52: 147

107. Chan, M.K.Y. and R.C. Close. 1987. Aphanomyces root rot of pea. 3. Control by use of cruciferous amendments. *N.Z.J. Agric. Res.* 30: 225

108. Chang, C.J. and I.M. Lee. 1995. Pathogenesis of diseases associated with mycoplasma-like organisms, pp. 237-246. In: *Ref. No. 481*, Vol. 1

109. Chao, W.L., E.B. Nelson, G.E. Harman and H.C. Hoch. 1986. Colonization of the rhizosphere by biological control agents applied to seed. *Phytopathology* 76: 60

110. Charudattan, R. and J.E. De Vay. 1972. Common antigens among varieties of *Gossypium hirsutum* and isolates of *Fusarium* and *Verticillium. Phytopatology.* 62: 230

111. Charudattan, R. and J.E. Devay. 1981. Purification and partial characterization of an antigen from *Fusarium oxysporum* f.sp. *vasinfectum* that cross reacts with antiserum to cotton (*Gossypium hirsutum*) root antigen, *Physiol Plant Pathol.* 18: 289

112. Chaube, H.S. 1989. Suppressive soils and plant pathogens, pp. 409-428. In: *Ref. No. 5*

113. Chavan, S.B., S.V. Khandge, M.C. Varshneya and J.D. Patil. 1995. Influence of weather parameters on conidia formation in powdery mildew of grape. *Indian Phytopath.* 48: 40

114. Chen, W.Y. and E. Echandi. 1984. Effect of avirulent bacteriocin producing strains of *Pseudomonas solanacearum* on control of bacterial wilt of tomato. *Plant Pathology* 33: 245

115. Chen, Z.S. 1998. Control of canker of citrus with copper ammonia WC. *J. Zhejieng For. Coll.* 15: 108

116. Chet, I. (ed.). 1987. *Innovative Approaches to Plant Disease Control.* Wiley-Interscience, New York

117. Chet, I. (ed.). 1993. *Biotechnological Prospects for Plant Pathogen Control.* John Wiley, New York

118. Christ, B.J. 1990. Influence of potato cultivars on effectiveness of fungicidal control of early blight. *Am. Potato J.* 67: 419

119. Chun, D. and J.L. Lockwood. 1985. Reduction of *Pythium ultimum, Thielaviopsis basicola* and *Macrophomina phaseolina* in soil associated with ammonia generated from urea. *Plant Dis.* 69: 154

120. Clarke, A.F. and R.N. Knox. 1984. Cell recognition in flowering plants. *Q. Rev. Biol.* 53: 3-28

121. Cohen, Y. and M.D. Coffey. 1983. Systemic fungicides and the control of oomycetes. *Annu. Rev. Phytopathol* 24: 311

122. Cole, G.T. and H.C. Hoch (eds.). 1991. *Fungal Spore and Disease Initiation*

365

in Plants and Animals. Plenum, New York

123. Coley-Smith, J.R., A. Ghaffar and Z.U.R. Javed. 1974. The effect of dry conditions on subsequent leakage and rotting of fungal sclerotia. *Soil Biol. Biochem.* 6: 307

124. Collins, G.B. and J.G. Petolino. 1984. *Application of Genetic Engineering to Crop Improvement.* Nijhoff/Junk, Dordrecht

125. Collmer, A. and N.T. Keen. 1987. The role of pectic enzymes in plant pathogenesis. *Annu. Rev. Phytopathol* 24: 383

126. Conway, W.C., C.E. Sams, R.G. McGuird and A. Kelma. 1992. Calcium treatment of apples and potatoes to reduce postharvet decay. *Plant Dis.* 76: 329

127. Conway, W.C., C.E. Sams, G.A. Brown, W.D. Beavers, R.B. Tonias and L.S. Kennedy. 1994. Pilot test for the commercial use of postharvest pressure infiltration of calcium into apples to maintain fruit quality in storage, *Hort Technol.* 4: 239

128. Cook, R.J. 1993. Twenty five years of progress towards biological control. In: *Ref. No.* 256

129. Cook, R.J. and K.F. Baker. 1983. *The Nature and Practice of Biological Control of Plant Pathogens.* Am Phytopath. Soc., St Paul, Minn., USA

130. Cook. R.J., M.G. Boosalis and B. Doupnik. 1978. Influence of crop residue on plant diseases, pp. 147-163. In: *Crop Residue Management Systems,* Am. Soc. Agron. Spec. Publi. 31

131. Costa, A.S. and G.W. Muller. 1980. Tristeza control by cross protection. A U.S.-Brazil cooperative success. *Plant Dis.* 64: 538

132. Costanho, B. and E.E. Butler. 1978. *Rhizoctonia* decline. A degenerative disease of *Rhizoctonia solani.* II. Studies on hypovirulence and potential use in biological control. III. The association of double standed RNA with Rhizoctonia decline. *Phytopathology* 68: 1605

133. Coyler, P.D. and M.S. Mount. 1984. Bacterization of potatoes with *Pseudomonas putida* and its influence on postharvest soft rot disease. *Plant Dis.* 68: 703

134. Cruickshank, I.A.M. 1980. Defence triggered by the invader: Chemical defence, pp. 247-267. In: *Ref. No. 260*

135. Curl, E.A. 1963. Control of plant diseases by crop rotation. *Bot Rev.* 29: 413

136. Curl, E.A. 1982. The rhizosphere: Relation to pathogen behaviour and root disease. *Plant Dis.* 66: 624

137. D'Abbabbo, T. 1995. The nematicidal effect of organic amendments: A review of the literature, 1982-1994. *Nematol. Medit.* 23: 299

138. D'Arcy, C.J. and L.R. Nault. 1982. Insect transmission of plant viruses, mycoplasma-like and rickettsia-like organisms. *Plant Dis.* 66: 99

139. Daly, J.M. 1984. The role of recognition in plant diseases. *Annu. Rev. Phytopathol.* 22: 73

140. Daly, J.M. and I. Uritani. 1979. *Recognition and Specificity in Plant-parasite Interactions.* Univ. Tokyo Press, Tokyo

141. Darvill, A.G. and P. Albersheim. 1984. Phytoalexins and their elicitors- a defence against microbial infection in plants. *Annu. Rev. Plant Physiol.* 35: 243

142. Dastur, R.H. and J.G. Bhatt. 1964. Relation of potassium to Fusarium wilt of flax. *Nature* 201: 1243

143. Datar, V.V. 1999. Bioefficacy of plant extracts against *Macrophomina*

366

phaseolina, the incitant of charcoal rot of sorghm. *Indian Phytopath.* 52: 251

144. Datnoff, L.E. G.H. Snyder and C.W. Deren. 1992. Influence of silicon fertilizer grades on blast and brown spot development and on rice yields. *Plant Dis.* 76: 1011

145. Daub, M.E. 1986. Tissue culture and the selection of resistance to pathogens. *Annu. Rev. Phytopathol.* 24: 159

146. Dautry-Varsat, A. and H.F. Lodish. 1984. How receptors bring proteins and particles into cells. *Sci. Am.* 250: 52

147. Davidse, L.C. and M.A. de Ward. 1984. Systemic fungicides. *Plant Pathology* 2: 191

148. Davis, J.R., R.E. Dole and R.H. Callihan. 1976. Fertilizer effect on common scab of potato and the relation of calcium and phosphate phosphorus. *Phytopathology* 66: 1236

149. Davies, K.G., F.A.A.M. De Ley and B.R. Kerry. 1991. Microbial agents for the biological control of plant parasitic nematodes. *Trop. Pest Manage.* 37: 303

150. Day, P.R. 1974. *Genetics of Host-parasite Interaction.* Freeman, San Frasisco

151. Day, P.R. (ed.). 1977. *The Genetic Basis of Epidemics in Agriculture.* New York Academy of Science

152. Deacon, J.W. 1983. *Microbial Control of Plant Pests and Diseases.* Van Nostrand Reinhold, UK

153. De Boer, S.H. 1982. Survival of phytopathogenic bacteria in soil, pp. 285-306. In: *Ref. No. 392,* Vol. I

154. DeFeyter, R., Y. Yang and D.W. Gabriel. 1993. Gene-for-gene interaction between cotton R genes and *Xanthomonas campestris* pv. *malvacearum avr* genes *Mol. Plant-Microbe Interact.* 6: 225

155. Dekker, J. 1980. Chemotherapy, pp. 307-325. In: *Ref. No. 260*

156. Delaney, T.P. 1997. Genetic dissection of acquired resista.ce to disease. *Plant Physiol* 113: 5

157. Delanye, T.P., S. Uknes, B. Vernooij, *et al.* 1994. A central role of salicylic acid in plant disease resistance. *Science* 266: 1247

158. Delp, G.J. (ed.). 1988. *Fungicide Resistance in North America.* APS, St. Paul, Minn

159. De Meyer, C. and M. Hofte. 1997. Salicylic acid produced by the rhizobacterium *Pseudomonas aeruginosa* 7NSK2 induces resistance to leaf infection by *Botrytis cinerea* on bean. Phytopathology 87: 588

160. Denarie, J., F. Debelle and C. Rosenberg. 1992. Signalling and host range in nodulation. *Annu. Rev. Microbiol.* 46: 497

161. Denny, T.P. and S.R. Black. 1991. Genetic evidence that extracellular polysaccharide is a virulence factor in *Pseudomonas solanacearum. Mol. Plant-Microbe Interaction* 4: 198

162. De Vay, J.E. and H.E. Adler. 1976. Antigens common to hosts and parasites. *Annu. Rev. Microbiol.* 30: 147

163. Deverall, B.J. 1977. *Defence Mechanisms of Plants.* University Press, Cambridge

164. De Wit, P.J.G.M. 1992. Molecular characterization of gene-for-gene systems in plant-fungus interactions and the application of avirulence genes in control of plant pathogen. *Annu. Rev. Phytopathol.* 30: 391

165. De Wit, P.J.G.M. 1992. Functional models to explain gene-for-gene relationships in plant-pathogen interactions, pp. 25-47. In: *Ref. No. 70*

166. de Wit, P.J.G.M. 1995. Fungal avirulence genes and plant resistance genes: unravelling the molecular basis for gene-for-gene interactions. *Adv. Bot. Res.* 21: 147

167. de Wit, P.J.G.M. 1997. Pathogen a virulence and plant resistance: a role for recognition. *Trends Plant Sci.* 2: 452

168. de Wit, P.J.G.M. and P.H.M. Roseboom. 1980. Isolation, partial characterization and specificity of glycoprotein elicitors from culture filtrate, mycelium and cell walls of *Cladosporium fulvum*. *Physiol. Plant Pathol.* 16: 391

169. Dickman, M.B. and S.S. Patil. 1988. The role of cutinase from *Colletotrichum gloeosporioides* in the penetration of papaya, pp. 175-182. In: *Ref. No. 478*

170. Dimond, A.E. 1970. Biophysics and biochemistry of the vascular wilt syndrome. *Annu. Rev. Phytopathol.* 8: 301

171. Dixon, G.K., L.G. Copping and D.W. Holloman. 1995. *Antifungal Agents: Discovery and Mode of Action. Bios,* Oxford

172. Dixon, R.A. 1986. The phytoalexin response: elicitation, signaling and control of host gene expression. *Biol. Rev.* 61: 239

173. Dixon, R.A. and C.J. Lamb. 1990. Molecular communication in interactions between plants and microbial pathognes. *Annu. Rev. Plant Physiol Plant Mol. Biol.* 41: 339

174. Dixon, R.A., M.J. Harrison and C.J. Lamb. 1994. Early events in the activation of plant defense responses. *Annu. Rev. Phytopathol.* 32: 479

175. Doi, Y., M. Teranaka, K. Yora and H. Asuyama. 1967. Mycoplasma or PLT group-like organisms found in the phloem elements of plants infected with mulberry dwarf, potato witches' broom, aster yellows, or Paulownia witches' broom. *Ann. Phytcpath. Soc. Japan* 33: 259

176. Doke, N. and K. Tomiyama. 1980. Suppression of the hypersensitive response of potato tuber protoplasts to hyphal wall components by water soluble glucans isolated from *Phytophthora infestans. Physiol. Plant Pathol.* 16: 177

177. Doke, N.Y. Miura, L.M. Sanches *et al,* 1996. The oxidative burst protects plants against pathogen attack: mechanisms and role as an emergency signal for plant biodefence-a review. *Gene* 179: 45

178. Doubly, J.A., H.H. Flor and C.D. Clagett. 1960. Relation of antigen of *Melampsora lini* and *Linum usitatissimum* to resistance and susceptibility. *Science.* 131: 229

179. Dubos, B. 1987. Fungal antagonism in aerial agrobiocenases. In: *Ref. No. 116*

180. Duijff, B.J., J.W. Meijer, P.A.H.M. Bakker and B. Schippers. 1993. Siderophore-mediated competition for iron and induced resistance in the suppression of Fusarium wilt of carnation by fluorescent *Pseudomonas* spp. *Neth. J. Plant. Pathol.* 99: 277

181. Duniway, J.M. 1976. Water status and imbalance, pp. 430-449. In: *Ref. No. 245*

182. Duncan, L.W. 1991. Current options for nematode management. *Annu. Rev. Phytopathol.* 29: 469

183. Dunez, J. 1987. Perspectives in the control of plant viruses, pp. 297-324. In: *Ref. No. 116.*

184. Durner, J.J. Shah and D.F. Klessig. 1997. Salicylic acid and disease resistance in plants. *Trends Plant Sci.* 2: 226

185. Dusenbery, D.B. 1992. *Sensory Ecology: How Organisms Acquire and Respond to Information.* Freeman, New York

186. Dutta Mazumdar, S.K., N. Singh and V.P. Agnihotri. 1990. Behaviour of *Colletotrichum falcatum* under waterlogged conditions. *Indian Phytopath.* 43: 227

187. Ebel, J.J. and E.G. Cosio. 1994. Elicitors of plant defence responses. *Int. Rev. Cytol.* 148: 1

188. Elad, Y. and R. Baker. 1985. The role of competition for iron and carbon in suppression of chlamydospore germination of *Fusarium* spp. by *Pseudomonas* spp. *Phytopathology* 75: 1053

189. Elad, Y. and I. Chet. 1987. Possible role of competition for nutrition in biocontrol of *Pythium* damping off by bacteria. *Phytopathology* 77: 190

190. English, J.T., C.S. Thomas, J.J. Marios and W.D. Gubler. 1989. Microclimate of grapevine canopies associated with leaf removal and control of Botrytis bunch rot. *Phytopathology* 79: 395

191. English, J.T., M.L. Caps, J.F. Moore, J.Hill and M. Nakova. 1999. Control of powdery mildew of wild and cultivated grapes by the tydeid mite. *Biological Control* 14: 97

192. Enyedi, A.J., N. Yalpani, P. Silverman and I. Raskin. 1992. Signal molecules in systemic plant resistance to pathogens and pests. *Cell* 70: 899

193. Fisher, R.F. and S.R. Ling. 1992. Rhizobium-plant signal exchange. *Nature* 357: 655

194. Fletcher, J.T., M.J. Hims, F.C. Archer and A. Brown. 1982. Effect of adding calcium and sodium to field soils on the incidence of clubroot. *Ann. Appl. Biol.* 100: 245

195. Flor, H.H. 1955. Host-parasite interaction in flax rust— its genetics and other implications. *Phytopathology* 45: 680

196. Flor, H.H. 1956. Complimentary genic system in flax and flax rust. *Adv. Genet.* 8: 267

197. Flor, H.H. 1971. Current status of the gene-for-gene concept. *Annu. Rev. Phytopathol.* 9: 275

198. Ford, E.J., A.H. Gold and W.C. Snyder. 1970. Induction of chlamydospore formation in *Fusarium solani* by soil bacteria. *Phytopathology* 60: 479

199. Forster, H. and I. Rasched. 1985. Purification and characterization of extracellular pectinesterase from *Phytophthora infestans. Plant Physiol.* 77: 109

200. Friend, J. 1984. The role of lignification in the resistance of plants to attack by pathogens. *Appl. Biochem. Biotechnol.* 9: 325

201. Fry, W.F. 1982. *Principles of Plant Disease Management.* Academic Press

202. Fuchs, J.G., Y. Moenne-Loccuz and G. Dejago. 1997. Non-pathogenic *Fusarium oxysporum* (Fo 47) induces resistance to Fusarium wilt in tomato. *Plant Dis.* 81: 492

203. Fuhr, I., G. Hill and P. Blaise. 1998. Effect of foliar sprays of phosphates and carbonates on grapevine powdery mildew. *Bull. OILB-SROP* 21: 17

204. Gabriel, D.W. 1984. Genetics of plant parasite populations and host-parasite specificity, In: *Ref. No.* 335

205. Gabriel, D.W. and Rolfe, B.G. 1990. Working models of specific recognition in plant-microbe interactions. *Annu. Rev. Phytopathol.* 28: 365

206. Gadoury, D.M., R.C. Seem, D.A. Rosenberger, W.F. Wilcox, W.E. MacHardy

and L.P. Berkett. 1992. Disparity between morphological maturity of ascospores and physiological maturity of asci in *Venturia inaequalis*. *Plant Dis.* 26: 277

207. Gaffney, T., L. Friedricj, B. Vernooij, *et al*, 1993. Requirement of salicylic acid for the induction of systemic acquired resistance. *Science* 261: 754

208. Gangawane, L.V. 1997. Management of fungicide resistance in plant pathogens. *Indian Phytopathol.* 50: 305

209. Garbers, C. and C. Simmons. 1994. Approaches to understanding auxin accumulation. *Trend Cell Biol.* 4: 245

210. Gaumann, E. 1950. *Principles of Plant Infection.* Hafner, New York

211. Geels, F.P. and B. Schippers. 1983. Reduction of yield depression in high frequency potato cropping soil after seed tuber treatment with antagonistic fluroescent *Pseudomonas* spp. *Phytopath. Z.* 108: 207

212. Georgopoulos, S.G. 1995. The genetics of fungicide resistance, pp. 39-52. In: *Modern Selective Fungicides* H. Lyre (ed.) Gustav Fischer Verlag, NY

213. Gianinazzi, S. 1984. Genetic and molecular aspects of resistance induced by infections or chemicals. pp. 321-342. In: *Ref. No. 335*

214. Gilchrist, D.G. 1998. Programmed cell death in plant disease: the purpose and promise of cellular suicide. *Annu. Rev. Phytopathol.* 36: 393

215. Gindrat, D. 1979. Biological soil disinfestation, pp. 251-287. In: *Soil Disinfestation.* D. Mulder (ed.). Elsevier, Amsterdam

216. Gnanamanickam, S.S. and T.W.M. Mew. 1992. Biological control of blast of rice (*Oryza sativa*) with antagonistic bacteria and its mediation by a *Pseudomonas* antibiotic. *Ann. Phytopath. Soc. Japan* 58: 380

217. Goodman, R.N., Z. Kiraly and K.L. Wood. 1986. *The Biochemistry and Physiology of Plant Disease.* Univ. Missouri Press, Columbia

218. Goodman, R.N. and A. Novacky. 1994. *The hypersensitive reaction in plants to pathogens: A resistance phenomenal* Am. Phytopathol. Soc., St. Paul, Minn. USA

219. Gopalan, S. and Shang Yang He. 1996. Bacterial genes involved in the elicitation of hypersensitive response and pathogenesis. *Plant Dis.* 80: 604

220. Gorlach, J., S. Volrath, G. Knauf-Beiter, G. Henry, *et al.* 1996. Benzothiadiazole, a novel class of inducer of systemic acquired resistance, activates gene expression and disease resistance in wheat. *Plant Cell* 8: 629

221. Gottstein, H.D. and J.A. Kuc. 1989. Induction of systemic resistance to anthracnose in cucumber by phosphates. *Phytopathology* 79: 176

222. Gow, N.A.R. 1994. Growth and guidance of the fungal hypha. *Microbiology* 140: 3193

223. Graniti, A. 1972. The evolution of toxin concept in plant pathology. In: *Ref. No. 554*

224. Grant, T.J. 1958. Heat treatment for obtaining sources of virus-free citrus budwood. *Proc. Fl. State Hortic Soc.* 71: 51

225. Greenberg, J.T. 1997. Programmed cell death in plant-pathogen interactions. *Annu. Rev. Plant Physiol. Mol. Biol.* 48: 525

226. Grogan, R.G. and R.N. Campbell. 1966. Fungi as vectors and hosts of viruses. *Annu. Rev. Phytopathol.* 4: 29

227. Gross, D.C. 1991. Molecular and genetic analysis of toxin production by pathovars of *Pseudomonas syringae*. *Annu. Rev. Phytopathol.* 29: 247

228. Gustafson, J.P. (ed). 1984. *Gene Manipulation in Plant Improvement.* Plenum

Press, New York

229. Hahn, M.G. 1996. Microbial elicitors and their receptors in plants. *Annu. Rev. Phytopathol.* 34: 387

230. Halbrock, K. and D. Scheel. 1987. Biochemical responses of plants to pathogens. In: *Ref. No.* 116

231. Hammerschmidt, R. and J. Kuc. 1995. *Induced Resistance to Disease in Plants.* Kluwer, Dorcrecht

232. Hammond-Kosack, K.E. and J.D.G. Jones. 1995. Plant disease resistance genes: unravelling how they work. *Can. J. Bot.* 73 (Suppl. 1): S 495

233. Hammond-Kosack, K.E. and J.D.G Jones. 1996. Resistance gene-dependent plant defense responses. *Plant Cell* 8: 1773

234. Hammond-Kosack, K.E. and J.D.G. Jones. 1997. Plant disease resistance genes. *Annu. Rev. Plant Physiol. Plant Mol. Biol.* 48: 575

235. Hampson, M.C. 1992. Some thoughts on demography of the Great Potato Famine. *Plant Dis.* 76: 1284

236. Hampson, M.C. 1996. A qualitative assessment of wind dispersal of resting spores of *Synchytrium endobioticum*, the causal organism of wart disease of potato. *Plant Dis.* 80: 779-782

237. Harris, K.F. 1981. Arthropods and nematode vectors of plant viruses. *Annu. Rev. Phytopathol.* 19: 391

238. Hashiba, T. 1987. An improved system for biological control of damping off by using plasmids in fungi, pp. 331-351. In: *Ref. No.* 116

239. Havel, L. and D. Durzan. 1996. Apoptosis in plants. *Bot. Acta* 109: 26

240. Deleted

241. Hayward, A.C. 1995. *Pseudomonas solanacearum.* In: *Ref. No.* 481, Vol. 1

242. Heath, M.C. 1981. A generalised concept of host-parasite specificity. *Phytopathology* 71: 1121

243. Heath, M.C. 1982. The absence of active defence mechanism in compatible host-pathogen interactions, pp. 143-156. In: *Ref. No.* 553

244. Heath, M.C. 1991. The role of gene-for-gene interactions in the determination of host-species specificity. *Phytopathology* 81: 127

245. Heitefuss, R. and P.H. Williams. 1976. *Physiological Plant Pathology.* Springer—Verlag, Berlin

246. Helgeson, J.P. and B.J. Deverall (eds.). 1983. *Use of Tissue Culture and Protoplasts in Plant Pathology.* Academic Press, New York

247. Herrera-Estrella, L., L.S. Rosales and R. Rivera-Bustamante. 1996. Transgenic plants for disease resistance, pp. 33-80. In: *Ref. No.* 487

248. Hervas, A., J.L. Trapero-Casas and R.M. Jiminez-Diaz. 1995. Induced resistance against Fusarium wilt of chickpea by non-pathogenic race of *Fusarium oxysporum* f.sp. *ciceris* and non-pathogenic isolates of *Fusarium oxysporum*. *Plan Dis.* 79: 1110

249. Hillocks, R.J. 1986. Cross protection between strains of *Fusarium oxysporum* f.sp. *vasinfectum* and its effect of vascular wilt resistance mechanism. *J. Phytopathol* 117: 216

250. Hoch, H.C. and R.C. Staples. 1991. Signalling for infection structure formation in fungi. In: *Ref. No.* 122

251. Hoitink, H.A.J. and P.C. Fahy. 1986. Basis for the control of soil-borne plant pathogens with compost. *Annu. Rev. Phytopathol.* 24: 93

252. Homma, Y. and K. Kegasawa. 1984. Predation of larvae of plant parasitic nematodes by soil vampyrellid amoebae. *Japanese J. Nematol* 14: 1

253. Homma, Y. and M. Ishii. 1984. Perforation of hyphae and sclerotia of *Rhizoctonia solani* by mycophagous soil amoebae from vegetable field soils in Japan. *Ann. Phytopath Soc. Japan* 50: 229

254. Hopkins, D.L. 1988. Plant pathogenesis and host-parasite specificity in Rickettsia-like bacteria., pp. 235-246 In: *Ref. No. 478*

255. Hornby, D. 1983. Suppressive soils. *Annu. Rev. Phytopathol.* 21: 65

256. Hornby, D., R.J. Cook, Y. Henis, W.R. Ko *et al*, 1993. *Biological Control of Soil-borne Plant Pathogens.* CAB Int., Wallingford

257. Horsfall, J.G. and A.E. Dimond. 1959. The diseased plant, pp. 1-17. In: *Plant Pathology—An Advanced Treatise.* Vol. I. J.G. Horsfall and A.E. Dimond (eds.). Academic Press

258. Horsfall, J.G. and E.B. Cowling. 1977. *Plant Disease— An Advanced Treatise.* Vol. 1. Academic Press

259. Horsfall, J.G. and E.B. Cowling. 1978. *Plant Disease— An Advanced Treatise,* Vols 2-3. Academic Press

260. Horsfall, J.G. and E.B. Cowling. 1980. *Plant Disease— An Advanced Treatise,* Vol 5. Academic Press

261. Huber, D.M. and R.D. Watson. 1970. Effect of organic amendments on soil-borne plant pathogens. *Phytopathology* 60: 22

262. Huber, D.M. and R.D. Watson. 1974. Nitrogen form and plant disease. *Annu. Rev. Phytopathol.* 12: 139

263. Hull, R. 1989. The movement of viruses in plants. *Annu. Rev. Phytopathol.* 27: 213

264. Hulloli, S.S., R.P. Singh and J.P. Verma. 1998. Management of bacterial blight of cotton induced by *Xanthomonas axonopodis* pv. *malvacearum* with the use of neem-based formulations. *Indian Phytopath* 51: 21

265. Hussey, R.S. and R.W. Roncadori. 1982. Vesicular-arbuscular mycorrhizae may limit nematode activity and improve plant growth. *Plant Dis.* 66: 9

266. Hutcheson, S.W. 1998. Current concepts of active defense in plants. *Annu. Rev. Phytopathol.* 36: 59

267. Hwang, S.F. 1992. Effects of vesicular-arbuscular mycorrhizal fungi on development of Verticillium and Fusarium wilts of alfalfa. *Plant Dis.* 76: 239

268. Hyakumochi, M. 1997. Induced systemic resistance against anthracnose in cucumber due to plant growth promoting fungi and studies on mechanisms, pp. 164-169. In: *Ref. No. 403*

269. Hyre, R.A. 1955. Three methods of forecasting late blight of potato and tomato in north eastern United States. *Am. Potato J.* 32: 362

270. Ingham, J.I. 1973. Disease resistance in higher plants: The concept of pre-infectional and post-infectional resistance. *Phytopath. Z.* 78: 314

271. Ishiie, T., Y. Doi, K. Yora and H. Asuyama. 1967. Suppressive effects of antibiotics of tetracycline group on symptom development of mulberry dwarf disease. *Ann. Phytopathol. Soc. Japan* 33: 267

272. Jackson, A.O. and C.B. Taylor. 1996. Plant-microbe interactions: life and death at the interface. *Plant Cell* 8: 1651

273. Jacobsen, B.J. and P.A. Backman. 1993. Biological and cultural plant disease control: Alternatives and supplements to chemicals in IPM strategy. *Plant Dis.* 77: 311

372

274. James, W.C. 1971. An illustrated series of assessment keys for plant diseases, their preparation and usuage. *Can. Plant Dis. Surv.* 51: 39

275. James, W.C. 1971. A manual of assessment keys for plant diseases. *Can. Dep. Agr. Publ.* 1458, 80 p.

276. James, W.C. 1974. Assessment of plant diseases and losses. *Annu. Rev. Phytopathol.* 12: 27

277. Janisiewicz, W.J. 1988. Biological control of diseases of fruits, pp. 153-165, In: *Biocontrol of Plant Diseases*. K.G. Mukerji and K.L. Garg (eds.). CRC Press, Boca Raton, Florida

278. Janisiewicz, W.J., D.L. Peterson and R. Bors. 1994. Control of storage decay of apple with *Sporobolomyces roseus*. *Plant Dis.* 78: 466

279. Jayashree, K., K.B. Pun and S. Doraiswamy. 1999. Effect of plant extracts and derivatives, butter milk and virus inhibitory chemicals on pumpkin yellow mosaic virus transmission. *Indian Phytopath.* 52: 357

280. Johnson, R. 1981. Durable resistance: definition of, genetic control, and attainment in plant breeding. *Phytopathology* 71: 567

281. Jones, A.L., S.L. Lillevik, P.D. Fisher and T.C. Stebbins. 1980. A microcomputer-based instrument to predict primary apple scab infection period. *Plant Dis.* 64: 69

282. Jones, A.L., P.D. Fisher, R.C. Seem, J.C. Kroon and J. Van De Motter. 1984. Development and commercialization of an in-field microcomputer delivery system for weather-driven predictive models. *Plant Dis.* 68: 458

283. Jones, E.B.G. 1994. Fungal adhesions. *Mycol. Res.* 98: 961

284. Jubina, P.A. and V.K. Girija. 1998. Antagonistic rhizobacteria for management of *Phytophthora capsici*, the incitant of foot rot of black pepper. *J. Mycol. Pl. Pathol.* 28: 147

285. Kahl, G. 1988. Wound response and wound repair in plants, pp. 147-174. In: *Ref. No.* 478

286. Kajirawa, T. 1971. Structure and physiology of haustoria of various parasites, pp. 255-277. In: *Morphological and Biochemical Events in Plant-Parasite Interaction*. S. Akai and S. Ouchi (eds.)

287. Kalita, P., L.C. Bora and K.N. Bhagbati. 1996. Phylloplane microorganisms of citrus and their role in management of citrus canker. *Indian Phytopath.* 49: 234

288. Kao, C.C., E. Barlow and L. Sequeria. 1992. Extracellular polysaccharide is required for wild type virulence of *Pseudomonas solanacearum*. *J. Bacteriol.* 174: 1068

289. Kassemeyer, H.H., G. Busam and P. Blaise. 1998. Induced resistance of grapevine-Perspective of biological control of grapevine diseases. *Bull. OILB-SROP* 2: 43

290. Katan, J. 1981. Solar heating (solarization) of soil for control of soil-borne pests. *Annu. Rev. Phytopathol.* 19: 211

291. Katan, J. 1987. Soil solarization, pp. 77-105. In: *Ref. No.* 116

292. Katsui, N., A. Murai, M. Takasugi, K. Imaizumi, T. Masumane and K. Tomiyama. 1968. The structure of rishitin, a new antifungal compound from diseased potato tubers. *Chem. Commun.* 1968: 43

293. Kaur, J. and M. Dhillon. 1988. Pre-infectional anatomical defence mechanisms of groundnut leaf against *Cercosporidium personatum*. *Indian Phytopath.* 41: 376

294. Keck, R.W. and T.K. Hodges. 1973. Membrane permeability in plants: changes induced by host specific pathotoxins. *Phytopathology* 63: 226

295. Keen, N.T. 1981. Evaluation of the role of phytoalexins, pp. 155-177. In: *Ref. No.* 488

296. Keen, N.T. 1981. Mechanisms conferring specific recognition in gene-for-gene plant—parasite systems, pp. 67-84. In: *Specificity in Plant Disease.* Plenum Press

297. Keen, N.T. 1981. Specific recognition in gene-for-gene host parasite system. *Adv. Plant Pathol.* 1: 35

298. Keen, N.T. 1982. Specific recognition in gene-for-gene plant-parasite systems. *Adv. Plant Pathol.* 2: 35

299. Keen, N.T. 1985. Progress in understanding the biochemistry of race-specific interaction, pp. 85-101. In: *Genetic Basis of Biochemical Mechanism of Plant Disease.* J.V. Groth and W.R. Bushnell (eds.). Am. Phytopath. Soc. St. Paul, Minn

300. Keen, N.T. 1990. Gene-for-gene complimentarity in plant-pathogen interactions. *Annu. Rev. Gent.* 24: 447

301. Keen, N.T. 1992. The molecular biology of disease resistance. *Plant Mol. Biol.* 19: 109

302. Keen, N.T. and M. Legrand. 1980. Surface glycoproteins: Evidence that they may function as the race specific phytoalexin elicitors of *Phytophthora megasperma* f.sp. *glycinea. Physiol. Plant Pathol.* 17: 175

303. Keen, N.T. and M.J. Holliday. 1982. Recognition of bacterial pathogens by plants, pp. 179-221. In: *Ref. No.* 392, Vol. 2

304. Keen, N.T. and B.J. Stastkawicz. 1988. Host range determinants in plant pathogens and symbionts. *Annu. Rev. Microbiol.* 42: 421

305. Keen, N.T. and W.O. Dowson. 1992. Pathogen avirulence genes and elicitors of plant defence, pp. 85-114. In: *Ref. No.* 70

306. Kerr, A. 1980. Biological control of crown gall through production of Agrocin 84. *Plant Dis.* 64: 24

307. Kerr, A. 1989. Commercial release of a genetically engineered bacterium for the control of crown gall. *Agric. Sci.* 2: 41

308. Kerry, B.R. 1981. Fungal parasites: A weapon against cyct nematodes. *Plant Dis.* 65: 390

309. Kerry, B.R. 1990. An assessment of progress toward microbial control of plant parasitic nematodes. *J. Nematol.* (Suppl.) 22: 621

310. Kerry, B.R. and K. Evans. 1996. New strategies for the management of plant parasitic nematodes, pp. 134-152. In: *Principles and Practices of Managing Soil-borne Plant Pathogens.* R. Hall, (ed.). APS Press, St. Paul, Minn

311. Kessman, H., T. Staub, C. Hoffman *et al,* 1994. Induction of systemic acquired disease resistance in plants by chemicals. *Annu. Rev. Phytopathol.* 32: 439

312. Khilare, V.C. and L.V. Gangwane. 1997. Application of medicinal plant extracts in the management of thiophanate resistant *Penicillium digitatum* causing green mold of mosambi. *J. Mycol. Pl. Pathol.* 27: 134

313. King, R.R., C.H. Lawrence and M.C. Clark. 1991. Correlation of phytotoxin production with pathogenecity of *Streptomyces scabies* isolates from scab infected potato tubers. *Am. Potato J.* 68: 275

314. King, R.R., C.H. Laverence and L.A. Colhoun. 1992. Chemistry of phytotoxins

associated with *Streptomyces scabies*, the causal organism of potato common scab. *J. Agric. Food Chem.* 40: 834

315. Kiraly, Z. 1980. Defence triggered by the invader: Hypersensitivity, pp. 201-229. In: *Ref. No. 260*

316. Kiraly, Z. 1996. Sustainable agriculture and the use of pesticides. *J. Environ. Sci. Health* B31: 283

317. Klement, Z. 1982. Hypersensitivity, pp. 149-177. In: *Ref. No. 392*. Vol. 2

318. Kloepper, J.W. 1983. Effect of seed piece inoculation with plant growth promoting rhizobacteria on populations of *Erwinia carotovora* on potato roots and daughter tubers. *Phytopathology* 73: 217

319. Kloepper, J.W. 1996. Host specificity in microbe-microbe interactions. *Bioscience* 46: 406

320. Kloepper, J.W. and M.N. Schroth. 1981. Development of a powder formulation of rhizobacteria for inoculation of potato seed pieces. Phytopathology 71: 590

321. Kloepper, J.W., J. Leong, M. Teintze and M.N. Schroth. 1980. Enhanced plant growth by siderophore produced by plant growth promoting rhizobacteria. *Nature* 286: 885

322. Kloepper, J.W., S. Tuzum and J.A. Kuc. 1992. Proposed definitions related to induced disease resistance. *Biocontrol. Sci. Technol.* 2: 349

323. Kloepper, J.W., S. Tuzum, L. Liu and G. Wein. 1993. Plant growth promoting rhizobacteria as inducers of systemic disease resistance, pp. 156-165. In: *Pest Management: Biologically Based Technologies*. R.D. Lumsden and J.L. Vaughan (eds.). Am. Chem. Soc., Washington

324. Kluepfel, D.A. 1993. The behaviour and tracking of bacteria in the rhizosphere. *Annu. Rev. Phytopathol.* 31: 441

325. Knight, S.C., V.M. Anthony, A.M. Brady, A.J. Greenland *et al*, 1997. Rationale and perspectives on the development of fungicides. *Annu. Rev. Phytopathol.* 35: 349

326. Knogg, W. 1996. Fungal infection of plants. *Plant Cell* 8: 1711

327. Knott, D.R. 1989. *The wheat Rusts-Breeding for Resistance*. Springer-Verlag, Berlin

328. Ko, W.H. 1985. Natural suppression of soil-borne plant diseases. *Plant Prot. Bull. Taiwan* 27: 171

329. Kocks, C.G. and J.C. Zadoks. 1996. Cabbage refuge piles as source of inoculum for black rot epidemics. *Plant Dis.* 80: 789

330. Koizumi, M. 1998. Production systems of fruit tree nurseries to control graft-transmissible diseases. *J. Jpn. Soc. Hort. Sci.* 67: 1093

331. Kolattukudy, P.E. 1980. Biopolyster membrane of plants: cutin and suberin. *Science* 208: 990

332. Kolattukudy, P.E. and W. Koller. 1983. Fungal penetration of the first line defensive barriers of plants, pp. 79-100. In: *Biochemical Plant Pathology*. J.A. Callow (ed.). Wiley and Sons, Chichester

333. Koller, W. 1992. *Target Cites of Fungicidal Action*. CRC Press

334. Koller, W., C. Yao, F. Trail and D.M. Parker. 1995. Role of cutinase in the invasion of plants. *Can. J. Bot* 73: S1109

335. Kosuge, T. and E.W. Nester (eds.). 1984. *Plant-Microbe Interactions: Molecular and Genetic Perspectives*. Macmillan New York

336. Kranz, J. and B. Han. 1980. Systems analysis in epidemiology. *Annu. Rev.*

Phytopathol. 18: 67

337. Krause, R.A. and L.B. Massie. 1975. Predictive systems: Modern approaches to disease control. *Annu. Rev. Phytopathol.* 13: 31

338. Karuse, R.A., L.B. Massie and R.A. Hyre. 1975. BLITECAST: A computerised forecast of potato late blight. *Plant Dis. Rep.* 59: 95

339. Kubo, Y. and I. Furusawa. 1991. Melanin biosynthesis: Pre-requisite for successful invasion of the host by appessoria of *Colletotrichum* and *Pyricularia*, pp. 205-218. In: *Ref. No.* 122

340. Kuc, J. 1966. Resistance of plants to infectious agents. *Annu. Rev. Microbiol.* 20: 337

341. Kuc, J. 1972. Phytoalexins. *Annu. Rev. Phytopathol.* 10: 207

342. Kuc, J. 1976. Phytoalexins, pp. 632-652. In: *Ref. No.* 245

343. Kuc, J. 1978. Changes in intermediary metabolism caused by disease, pp. 349-374. In: *Ref. No.* 259, Vol. 3

344. Kuc, J. 1982. Induced immunity to plant disease. *Bioscience* 32: 854

345. Kuc, J. 1987. Plant immunization and its applicability for disease control, pp. 255-274, In: *Ref. No.* 116

346. Kuc, J. 1995. Induced systemic resistance—An overview. pp. 169-175. In: *Ref. No.* 231

347. Kumar, S. and S.K. Sugha. 1999. Role of alpha and beta conidia in pathogenesis of *Phomopsis vexans*. *J. Mycol. Pl. Pathol.* 29: 166

348. Lamb, C.J. 1994. Plant disease resistance genes in signal perception and transduction. *Cell* 75: 419

349. Lamb, C.J. and R.A. Dixon. 1997. The oxidative burst in plant disease resistance. *Annu. Rev. Plant Physiol. Plant Mol. Biol.* 48: 251

350. Lamb, T.G., D.W. Tonky and D.A. Kluepfel. 1996. Movement of *Pseudomonas aureofaciens* from the rhizosphere to aerial plant tissue. *Can. J. Microbiol.* 42: 1112

351. Lamberti, F., J.M. Walter and N.A. van der Graaf (eds.). 1983. *Durable Resistance in Crops.* Plenum Press, New York

352. Langhans, R.W., R.K. Horst and E.D. Earle. 1977. Disease-free plants via tissue culture propagation. *Hort. Science* 12: 149

353. Lapwood, D.H., L.M. Wellings and J.H. Hawkins. 1973. Irrigation as a practical means to control potato common scab (*Streptomyces scabies*). Final experiments and conclusion. *Plant Pathology* 22: 35

354. Large, E.C. 1966. Measuring plant disease. *Annu. Rev. Phytopathol.* 4: 9

355. Latterell, F.M. and A.E. Rossi. 1986. Longevity and pathogenic stability of *Pyricularia oryzae*. *Phytopathology* 76: 231

356. Leach, J.E. and F.F. White. 1996. Bacterial avirulence genes. *Annu. Rev. Phytopathol.* 34: 153-179

357. Leben, C. 1965. Epiphytic microorganisms in relation to plant disease. *Annu. Rev. Phytopathol.* 3: 209

358. Leigh, J.A. and D.L. Coplin. 1992. Exopolysaccharides in plant—bacterial interactions. *Annu. Rev. Microbiol.* 46: 307

359. Leong, J. 1986. Siderophores: Their biochemistry and possible role in the biocontrol of plant pathogens. *Annu. Rev. Phytopathol.* 24: 187

360. Leroux, P. 1996. Recent developments in the mode of action of fungicides.

Pestic. Sci. 47: 191

361. Levine, A., R. Tenhaken, R. Dixon and C. Lamb. 1994. H_2O_2 from the oxidative burst orchestrates the plant hypersensitive disease resistance response. *Cell* 79: 583-593

362. Lin. T.S. and P.E. Kolattukudy. 1978. Induction of a biopolyster hydrolase (cutinase) by low levels of cutin monomers in *Fusarium solani* f.sp. *pisi. J. Bacteriol.* 133: 942

363. Linthorst, H.J.M. 1991. Pathogenesis related proteins of plants. *Crit. Rev. Plant Sci.* 10: 123

364. Liu, L., J.W. Kloepper and S. Tuzum. 1995. Induction of systemic resistance in cucumber against Fusarium wilt by plant growth promoting rhizobacteria. *Phytopathology* 85: 695

365. Liu, L., J.W. Kloepper and S. Tuzum. 1995. Induction of systemic resistance in cucumber angular leaf spot by plant growth promoting rhizobacteria. *Phytopathology* 85: 843

366. Liu, L., J.W. Kloepper and S. Tuzum. 1995. Induction of systemic resistance in cucumber by plant growth promoting rhizobacteria: duration of protection and effect of host resistance on protection and root colonization. *Phytopathology* 85: 1064

367. Loegering, W.Q. 1978. Current concept in inter-organisimal genetics. *Annu. Rev. Phytopathol.* 16: 309

368. Loper, J.E. and J.S. Buyer. 1991. Siderophores in microbial interactions on plant surfaces. *Mol. Plant-Microbe Interact* 4: 5

369. Lucas, J.A. and C. Sherriff. 1988. Pathogenesis and host specificity in downy mildew fungi, pp. 321-349. In: *Ref. No.* 478

370. Luke, H.H. and V.E. Gracen Jr. 1972. Helminthosporium toxins, pp. 139-168. In: *Microbial Toxins, Vol 8. Fungal Toxins.* S. Kadis *et al,* (eds.). Academic Press, New York

371. Lukens, R. 1984. *Antimicrobial Agents in Crop Production.* Lea and Febiger, Philadelphia

372. Lyr, H. (ed.). 1995. *Modern Selective Fungicides.* Gustav Fischer Verlag, New York

373. MacKenzie, D.R. 1981. Scheduling fungicide applications for potato late blight with BLITECAST. *Plant Dis.* 65: 394

374. Mankau, R. 1980. Biological control of nematode pests by natural enemies. *Annu. Rev. Phytopathol.* 18: 415

375. Marx, D.H. 1972. Ectomycorrhizae as biological deterrents to pathogenic root infections. *Annu. Rev. Phytopathol.* 10: 429

376. Matteoni, J.A. and W.A. Sinclair. 1983. Stomatal closure in plants infected with mycoplasma-like organisms. *Phytopathology* 73: 398

377. Mayama, S., T. Tani, Y. Matsumura, T. Ueno and H. Fukami. 1981. The production of phytoalexins by oat in response to crown rust, *Puccinia coronata* f.sp. *avenae. Physiol. Plant Pathol.* 19: 217

378. Mayee, C.D. and V.V. Datar. 1989. Measuring plant disease, pp. 485-496. In: *Perspectives in Plant Pathology.* V.P. Agnihotri *et al,* (eds.). Today and Tomorrow Printers and Publishers, New Delhi

379. Mayee, C.D. and A.P. Suryavanshi. 1995. Structural defence mechanisms in

groundnut to late leaf spot pathogen. *Indian Phytopath.* 48: 160

380. McLaughlin, R.J. and L. Sequeira. 1988. Evaluation of an avirulent strain of *Pseudomonas solanacearum* for biological control of bacterial wilt of potato. *Amer. Potato J.* 65: 244

381. Mehdy, M.C. 1994. Active oxygen species in plant defence against pathogens. *Plant Physiol* 105: 467

382. Mendgen, K., M. Hahn and H. Deising. 1996. Morphogenesis and mechanisms of penetration by plant pathogenic fungi. *Annu. Rev. Phytopathol.* 34: 367

383. Menzies, J.D. 1959. Occurrence and transfer of a biological factor in soil that suppresses potato scab. *Phytopathology* 49: 648

384. Mew, T.W. and W.C. Ho. 1977. Effect of soil temperature on resistance of tomato cultivars to bacterial wilt. *Phytopathology* 67: 909

385. Misaghi, I.J. 1982. *Physiology and Biochemistry of Plant-Pathogen Interactions.* Plenum Press, New York

386. Miller, P.R. and M.O. Brien (eds.). 1957. Prediction of plant disease epidemics. *Annu. Rev. Microbiol.* 11: 77

387. Mills, W.D. 1944. Efficient use of sulfur dusts and sprays during rain to control apple scab. *Cronell Ext. Bull.* 630

388. Mittler, R. and Lam, E. 1996. Sacrifice in the face of foe: pathogen-induced programmed cell death in plants. *Trends Microbiol.* 4: 10

389. Moffett, M.L. and B.A. Wood. 1985. Resident populations of *Xanthomonas campestris* pv. *malvacearum* on cotton leaves: a source of inoculum for bacterial blight. *J. Appl. Bacteriol.* 58: 607

390. Mondal, K.K., R.P. Singh, P. Dureja and J.P. Verma. 2000. Secondary metabolites of cotton rhizobacteria in the suppression of bacterial blight of cotton. *Indian Phytopath.* 53: 22

391. Money, N.P. 1995. Turgor pressure and the mechanics of fungal pressure. *Can. J. Bot.* 73: 596

392. Mount, M.S. and G.S. Lacy (eds.). 1982. *Phytopathogenic Prokaryotes*, Vol. I and II. Academic Press, New York

393. Muller, R. and P.S. Gooch. 1982. Organic amendments in nematode control. An examination of the literature. *Nematropica* 12: 319

394. Mussell, H. 1980. Tolerance to disease, pp. 39-52. In: *Ref. No.* 260

395. Narula, K.L. and R.S. Mehrotra. 1987. Biocontrol potential of Phytophthora leaf blight of colocasia by phyllosphere microflora. *Indian Phytopath.* 40: 384

396. Nicholson, R.L. 1996. Adhesion of fungal propagules, pp. 117-134. In: *Histology, Ultrastructure and Molecular Cytology of Plant-Microbe Interactions.* M. Nicole and V. Gianinazzi-Pearson (eds.). Kliwer, Dordreht

397. Nicholson, R.L. and L. Epstein. 1991. Adhesion of fungi to the plant surface prerequisite for pathogenesis, pp. 3-23. In: *Ref. No.* 122

398. Nicholson, R.L. and R. Hammerschmidt. 1992. Phenolic compounds and their role in disease resistance. *Annu. Rev. Phytopathol.* 30: 369

399. Newman, M.A., M.J. Daniels and J.M. Dow. 1995. Lipopolysaccharide from *Xanthomonas campestris* induces defence related gene expression in *Brassica campestris*. *Mole. Plant-Microbe Interact.* 8: 778

400. Nishimura, S., K. Kohmoto, H. Otani, H. Fukami and T. Ueno. 1976. The involvement of host specific toxin in the early steps of infection by *Alternaria kikuchiana* and *Alternaria mali*, pp. 94-100. In: *Ref. No.* 509

401. Northover, J. and K.E. Schneider. 1993. Activity of plant oils on diseases caused by *Podosphaera leucotricha, Venturia inaequalis* and *Albugo occidentalis. Plant Dis.* 77: 152

402. Oerke, E.C., H.W. Dehne, F. Schonberg and A. Weber. 1994. *Crop Protection and Crop Production.* Elsevier. Amsterda

403. Ogoshi, A., K. Kabayashi, Y. Homma *et al,* (eds.). 1997. *Plant Growth Promoting Rhizobacteria: Present Status and Future Prospects.* Fac. Agric. Hokkaido Univ., Sapporo

404. Osborne, A.E. 1995. Pathogenecity of the black rot bacterium *Xanthomonas campestris* pv. *campestris* to crucifers, pp. 153-165 In: *Ref. No.* 481, Vol. 1

405. Osborne, A.E. 1996. Preformed antimicrobial compounds and plant defence against fungal attack. *Plant Cell* 8: 1821

406. Palti, J. 1981. *Cultural Practices and Infectious Crop Diseases.* Springer-Verlag, Berlin

407. Papavizas, G.C. and R.D. Lumsden. 1980. Biological control of soil-borne fungal propagules. *Annu. Rev. Phytopathol.* 18: 389

408. Park, E.W. and S.M. Lin. 1985. Overwintering of *Pseudomonas syringae* pv. *glycinea* in the field. *Phytopathology* 75: 520

409. Park, P. 1976. Initial changes in the ultrastructure of leaf cells of Japanese pear caused by *Alternaria kikuchiana* toxin, tenuazinic acid and citrinin, pp. 66-69. In: *Ref. No. 509*

410. Parke, J.L., R.E. Rand, A.E. Joy and E.B. King. 1991. Biological control of Pythium damping off and Aphanomyces root rot of peas by application of *Pseudomonas cepacia* or *Pseudomonas fluorescens* to seed. *Plant Dis.* 75: 987

411. Parkinson, J.S. and E.C. Kofold 1992. Communication modules in bacterial signalling proteins. *Annu. Rev. Genet.* 26: 71

412. Pathak, K.N. and R.P. Nath. 1993. Interaction of nematodes with bacterial plant pathogens, pp. 244-253. In: *Handbook of Economic Nematology.* K. Sitaramaiah and R.S. Singh (eds.). Cosmo Publications, New Delhi

413. Patil, S.S. 1974. Toxins produced by phytopathogenic bacteria. *Annu. Rev. Phytopathol.* 12: 259

414. Paxton, J.D. 1988. Phytoalexins in plant-parasite interactions, pp. 537-549. In: *Ref. No. 478*

415. Pearson, R.C., D.G. Riegel and D.M. Gadoury. 1994. Control of powdery mildew in vineyards using single-application vapor action treatments of triazole fungicides. *Plant Dis.* 78: 164

416. Pennel, R.L. and C. Lamb. 1997. Programmed cell death in plants. *Plant Cell.* 9: 1157

417. Perry, R.N. 1996. Chemoreception in plant parasitic nematodes. *Annu. Rev. Phytopathol.* 34: 181-199

418. Petrini, O. and G.B. Ouellette (eds.). 1994. *Host Wall Alterations by Parasitic Fungi.* APS, St. Paul, Minn., USA

419. Plumb, R.T. and J.M. Thresh (eds.). 1983. *Plant Virus Epidemiology: The Spread and Control of Insect-borne Viruses.* Blackwell, Oxford

420. Preece, T.F. and C.H. Dickinson. 1971. *Ecology of Leaf Surface Microorganisms.* Academic Press

421. Prusky, D. 1988. Hypersensitivity: An Overview, pp. 485-522. In: *Ref. No. 478*

422. Prusky, D. 1996. Pathogen quiescence in postharvest diseases. *Annu. Rev.*

Phytopathol. 34: 413
423. Pryor, T. 1987. The origin and structure of fungal disease resistance genes in plants. *Trends Genet.* 3: 157
424. Punn, K.B., S. Doraiswamy and R. Jeyrajan. 1999. Screening of plant species for the presence of antiviral principles against okra yellow vein mosaic virus. *Indian Phytopath.* 52: 221
425. Purcell, A.H. 1990. Homopteran transmission of xylem inhabiting bacteria, pp. 243-266. In: *Advances in Disease Vector Research* K.F. Harris (ed.)
426. Purkayastha, R.P. 1998. Disease resistance and induced immunity in plants. *Indian Phytopath.* 51: 211
427. Raupach, G.S., L. Liu, J.F. Murphy, S. Tuzum and B. Kloepper. 1996. Induced systemic resistance in cucumber and tomato against cucumber mosaic cucumovirus using plant growth promoting rhizobacteria (PGPR). *Plant Dis* 80: 891
428. Rausch, T., G. Kahl and W. Hilgenberg. 1984. Primary action of indole-3-acetic acid in crown gall tumors. *Plant Physiol.* 75: 354
429. Reuvini, R. and M. Reuvini. 1998. Foliar-fertilizer therapy—a concept in integrated pest management. *Crop. Prot.* 17: 111
430. Rhodes-Roberts, M. and F.A. Skinner (eds.). 1982. *Bacteria and Plants.* Academic Press, New York
431. Ride, J.P. 1983. Cell wall and other structural barriers in defence, pp. 215-236. In: *Biochemical Plant Pathology.* J.A. Callow (ed.). John Wiley and Sons, London
432. Rohde, R.A. 1972. Expression of resistance in plants to nematodes. *Annu. Rev. Phytopathol.* 10: 233
433. Rohringer, R., N.K. Hower, W.K. Kim an D.J. Samborski. 1974. Evidence for a gene specific RNA determining resistance in wheat to stem rust, *Nature* 279: 585
434. Romantschuk, M. 1992. Attachment of plant pathogenic bacteria to plant surfaces. *Annu. Rev. Phytopathol.* 30: 225
435. Rovira, A.D. and C.B. Davey. 1974. Ecology of the rhizosphere, pp. 153-204. In: *Plant Root and Its Environment.* E.W. Carson (ed.). Univ. Press, Va, Charlottesville
436. Rudolph, K.W.E. 1976. Non-specific toxins, pp. 270-315. In: *Ref. No. 245*
437. Rudolph, K.W.E. 1995. *Pseudomonas syringae* pathovars, pp. 47-165. In: *Ref. No. 481*, Vol. 1
438. Ryals, J.A., S. Uknes and E. Ward. 1994. Systemic acquired resistance. *Plant Physiol.* 104: 1109
439. Ryals, J.A., U.H. Neuenschwander, M.G. Willies *et al*, 1996. Systemic acquired resistance. *Plant Cell* 8: 1809
440. Sall, M.A. 1980. Epidemiology of grape powdery mildew. A model. *Phytopathology* 70: 338
441. Salmond, G.P.C. 1994. Secretion of extracellular virulence factors by plant pathogenic bacteria. *Annu. Rev. Phytopathol.* 32: 181
442. Schafer, J.F. 1971. Tolerance of plant disease. *Annu. Rev. Phytopathol.* 9: 235
443. Scheffer, R.P. 1976. Host-specific toxins in relation to pathogenesis and disease resistance, pp. 347-369. In: *Ref. No. 245*
444. Scheffer, R.P. and O.C. Yoder. 1972. Host specific toxins and selective toxicity,

380

pp. 251-272. In: *Ref. No. 554*

445. Schippers, B. 1988. Biological control of pathogens with rhizobactria. *Phil. Trans. R. Soc. London* B. 318: 283

446. Schippers, B. and W. Gams (eds.) 1979. *Soil-borne Plant Pathogens.* Academic Press, New York

447. Schippers, B., A.W. Bakker and P.A.H.M. Bakker. 1987. Interactions of deleterious and beneficial rhizosphere microorganisms and the effect of cropping practices. *Annu. Rev. Phytopathol.* 25: 339

448. Schlosser, E. 1980. Preformed internal chemical defence, pp. 161-177. In: *Ref. No. 260*

449. Schlosser, E. 1988. Preformed chemical barriers in host-parasite incompatibility, pp. 465-476. In: *Ref. No. 478*

450. Schlosser, E. 1994. Alternative control of powdery mildew on grapevine. Abstr. in *46th Int. Symp. Plant Prot.* Gent, Belgium, May 3, 1995

451. Schnathorst, W.C. 1965. Environmental relations in powdery mildews. *Annu. Rev. Phytopathol.* 3: 343

452. Schneider, R.W. (ed.). 1982. *Suppressive Soils and Plant Disease.* Am. Phytopathol. Soc., St. Paul, Minn., USA

453. Schoenbeck, F. 1979. Endomycorrhiza in relation to plant disease, pp. 271-280. In: *Ref. No. 446*

454. Schoenbeck, F. and E. Schlosser. 1976. Preformed substances as potential protectants, pp. 653-678. In: *Ref. No. 245*

455. Schottel, J.L. 1995. Streptomyces pathogenesis, pp. 253-271. In: *Ref. No. 481*, Vol. 1

456. Schroth, M.N. and J.G. Hancock. 1982. Disease-suppressive soil and root colonizing bacteria. *Science* 216: 1376

457. Schuster, M.L. and D.C. Coyne. 1974. Survival mechanisms of phytopathogenic bacteria. *Annu. Rev. Phytopathol.* 12: 199

458. Scott, P.R. and A. Bainbridge (eds.). 1978. *Plant Disease Epidemiology.* Blackwell, Oxford

459. Seem, R.C. 1984. Disease incidence and severity relationship. *Annu. Rev. Phytopathol.* 22: 133

460. Sequeira, L. 1963. Growth regulators in plant diseases. *Annu. Rev. Phytopathol.* 1: 5

461. Sequeira, L. 1973. Hormone metabolism in diseased plant. *Annu. Rev. Plant Physiol.* 24: 353

462. Sequeira, L. 1980. Defence triggered by the invader: Recognition and compatibility phenomena, pp. 179-200. In: *Ref. No. 260*

463. Sharma, T.R. and B.M. Singh. 1990. Physiologic races of *Erysiphe graminis tritici* in Himachal Pradesh. *Indian Phytopath.* 43: 33

464. Shaw, P.D. 1988. Plasmids in phytopathogenic bacteria, pp. 221-234. In: *Ref. No. 478*

465. Shaykh, M., C. Soliday and P.E. Kolattukudy. 1977. Proof for the production of cutinase by *Fusarium solani* f.sp. *pisi* during penetration into its host *Pisum sativum. Plant Physiol.* 60: 170

466. Shipton, P.J. 1977. Monoculture and soil-brone plant pathogens. *Annu. Rev. Phytopathol.* 15: 387

467. Sikora, R.A. 1992. Management of the antagonistic potential in agricultural ecosystems for the biological control of plant parasitic nematodes. *Annu. Rev. Phytopathol.* 30: 245-270

468. Sindhan, G.S., I. Hooda and R.D. Parashar. 1999. Evaluation of plant extracts for the control of powdery mildew of pea. *Indian Phytopath.* 52: 257

469. Singh, R.P., S. Lal and K. Singh. 1988. Effect of ambient temperature on red rot of sugarcane. *Indian Phytopath.* 41: 86

470. Singh, R.S. 1982. *Plant Pathogens— The Fungi.* Oxford and IBH

471. Singh, R.S. 1989. *Plant Pathogens— The Prokaryotes.* Oxford and IBH, New Delhi

472. Singh, R.S. 1998. *Plant Diseases.* 7th Ed. Oxford and IBH, New Delhi, India

473. Singh, R.S. 2000. *Diseases of Fruit Crops.* Oxford and IBH, New Delhi

474. Singh, R.S. 2001. *Plant Disease Management.* Oxford and IBH, New Delhi

475. Singh, R.S. and K. Sitaramaiah. 1970. Control of plant parasitic nematodes with organic amendments of soil. *PANS* 16: 287

476. Singh, R.S. and K. Sitaramaiah. 1994. *Plant Pathogens— The Nematodes.* Oxford and IBH, New Delhi

477. Singh, R.S., H.S. Chaube and N. Singh. 1972. Studies on the control of black scurf of potato. *Indian Phytopath.* 25: 343

478. Singh, R.S., U.S. Singh, W.M. Hess and D.J. Weber (eds.). 1988. *Experimental and Conceptual Plant Pathology.* Oxford & IBH, New Delhi

479. Singh, U.P. 2000. Pea powdery mildew—an ideal pathosystem. *Indian Phytopath* 53: 1

480. Singh, U.S. and R.S. Singh. 1988. Philosophy of defence in plants, pp. 459-464. In: *Ref. No. 478*

481. Singh, U.S., R.P. Singh and K. Kohmoto (eds.). 1995. *Pathogenesis and Host Specificity in Plant Diseases: Histopathological, Biochemical and Molecular Bases,* Vol. 1. *Prokaryotes,* Vol. II. *Eukaryotes.* Pergamon/Elsevier, New York

482. Singh, U.S., R.K. Khetarpal and J. Kumar. 1989. Plant Disease Concept—An Overview, pp. 497-503. In: *Ref. No. 5*

483. Smiley, R.W. 1975. Forms of nitrogen and the pH in the root zone and their importance in root infections. pp. 52-62. In: *Ref. No. 88*

484. Smith, D.C. 1994. *Symbiotic Interactions.* Oxford Univ. Press

485. Sneh, B. 1998. Use of non-pathogenic or hypovirulent fungal strains to protect plants against closely related fungal pathogens. *Biotechnol. Adv.* 16: 1

486. Snyder, B.A. and R.L. Nicholson, 1990. *Science* 248: 1637

487. Stacey, G. and N.T. Keen (ed.). 1996. *Plant-Microbe Interactions.* Chapman & Hall, New York

488. Stapples, R.C. (ed.). 1981. *Plant Disease Control: Resistance and Susceptibility.* Wiley-Interscience, New York

489. Starr, M.P. 1984. Landmarks in the development of phytobacteriology. *Annu. Rev. Phytopathol.* 22: 169

490. Steiner, U. and F. Schoenbeck. 1995. Induced disease resistance in monocots, pp. 86-110. In: *Ref. No. 231*

491. Sticher, L., B. Mauch-Mani and J.P. Metraux. 1997. Systemic acquired resistance, *Annu. Rev. Phytopathol.* 35: 235

492. Stirling, G.R. 1991. *Biological Control of Plant-Parasitic Nematodes. CAB Int.,*

382

Wallingford, UK. 292 pp.

493. Stoessel, A. 1980. Phytoalexins—a biogenetic perspective. *Phytopath. Z.* 99: 251

494. Stoessel, A. 1982. Biosynthesis of phytoalexins, pp. 133-180. In: *Ref. No.* 36

495. Stoessel, P. 1988. Pathogenesis and host-parasite specificity in *Phytophthora* pp. 273-300 In: *Ref. No.* 478

496. Strobel, G.A. 1974. Phytotoxins produced by plant parasites. *Annu. Rev. Plant Physiol.* 25: 541

497. Sun, S.K. and J.W. Huang. 1985. Formulated soil amendment for controlling Fusarium wilt and other soil-borne diseases. *Plant Dis.* 69: 917

498. Sutton, J.C., T.J. Gillaspie and P.D. Hildebrand. 1984. Monitoring weather factors in relation to plant disease. *Plant Dis.* 68: 78

499. Sutton, J.C., De-Wei, Li, G. Peng *et al*, 1997. *Gilocladium roseum*. A versatile adversery of *Botrytis cinerea* in crops. *Plant Dis.* 81: 316

500. Swain, T. 1977. Secondary compounds as protective agents. *Annu. Rev. Plant Physiol.* 28: 479

501. Tani, T. 1988. Pathogenesis and host-parasite specficity in rusts, pp. 301-320. In: *Ref. No.* 478

502. Tani, T. and S. Mayama. 1982. Evaluation of phytoalexins and preformed antifungal substances in relation to fungal infection, pp. 301-313. In: *Ref. No.* 27

503. Tarr, S.A.J. 1972. *Principles of Plant Pathology*. Macmillan Press, London, 603 pp.

504. Tenhaken, R., A. Levine, L.F. Brisson, R.A. Dixon and C. Lamb. 1995. Functions of the oxidative burst in hypersensitive resistance. *Proc. Natl. Acad. Sci. USA* 92: 4158

505. Thomas, C.E. and E.L. Jordian. 1992. Host effect on selection of virulence factors affecting sporulation of *Pseudoperonospora cubensis*. *Plant Dis.* 76: 905

506. Thomason, S.V., M.N. Schroth, W.J. Moller and W.O. Reil. 1982. A forecasting model for fire blight of pear. *Plant Dis.* 66: 576

507. Thresh, J.M. (ed.) 1981. *Pests, Pathogens and Vegetation*. Pitman, London

508. Tomiyama, K., T. Sakuma, N. Ishazaka, N. Sato, N. Katsui M. Takasugi and T. Masamura. 1968. A new antifungal substance isolated from resistant potato tuber tissue infected by pathogens. *Phytopathology* 58: 115

509. Tomiyama, K., J.M. Daly, I. Uritani, H. Oku and S. Ouchi (eds.). 1976. *Biochemistry and Cytology of Plant-Parasite Interaction*. Elsevier, New York, 256 pp

510. Toussoun, T.A., R.V. Bega and P.E. Nelson. 1970. *Root Diseases and Soil-borne Pathogens*. California Univ. Press, Berkeley

511. Trione, E.J. 1960. The HCN content of flax in relation to flax wilt resistance. *Phytopathology* 50: 482

512. Trivedi, P.C. and K.R. Barker. 1986. Management of nematodes by cultural practices. *Nematropica* 16: 213

513. Trivedi, N. and A.K. Sinha. 1978. Production of fungitoxic substances in rice in response to *Drechslera* infection. *Trans. Brit. Mycol. Soc.* 70: 57

514. Tu, J.C. 1978. Protection of soybean from sever Phytophthora root rot by *Rhizobium*. *Physiol. Plant Pathol.* 12: 233

515. Tuzum, S. and B. Kloepper. 1995. Practical applications and implementation of induced resistance, pp. 152-168. In: Ref. No. 231

516. Vance, C.P., T.K. Kirk and R.T. Sherwood. 1980. Lignification as a mechanism of disease resistance. Annu. Rev. Phytopathol. 18: 259

517. Van der Molen, G.E., C.H. Beckma and E. Rodenhorst. 1977. Vascular gelation: a general response phenomenon following infection. Physiol. Plant Pathol. 11: 95

518. Van der Plank, J.E. 1963. Plant Disease Epidemics and Control Academic Press

519. Van der Plank, J.E. 1975. Principles of Plant Infection. Academic Press

520. Van der Plank, J.E. 1978. Genetics and Molecular Basis of Plant Pathogenesis. Springer-Verlag, Berlin

521. Van der Plank, J.E. 1984. Disease Resistance in Plants, 2nd Ed. Academic Press

522. Van Loon, L.C. 1997. Induced resistance in plants and the role of pathogenesis related proteins. Eur. J. Plant Pathol. 103: 753

523. Van Loon, L.C., P.A.H.M. Bakker and C.M.J. Pieterse. 1998. Systemic resistance induced by rhizosphere bacteria, Annu. Rev. Phytopathol. 36: 453

524. Vasavan, M.G., K.M. Rajan and M.J. Thomas. 1980. Herbicides in plant disease control. Int. Rice Res. Newsl. 5(4): 18

525. Verma, J.P. 1995. Advances in bacterial blight of cotton. Indian Phytopath. 48: 1

526. Verma, J.P., R.P. Singh, H.D. Chowdhury and P.P. Sinha. 1983. Usefulness of phylloplane bacteria in the control of bacterial blight of cotton. Indian Phytopath. 36: 574

527. Vinette, J.R. 1982. How bacteria find their hosts?, pp. 3-30. In: Ref. No. 392, Vol. 2

528. Volpin, H., Y. Elkind, Y. Ojon and Y. Kapuinik. 1994. A vesicular arbuscular mycorrhizal fungus (Glomus intraradix) induces defence response in alalfa roots. Plant Physiol. 104: 683

529. Waggoner, P.E. 1965. Microclimate and plant disease. Annu. Rev. Phytopathol. 3: 103

530. Waggoner, P.E. and J.G. Horsfall. 1969. EPIDEM, a simulator of plant disease written for a computer. Bull. Conn. Agric. Exp. Sta. 698

531. Walker, J.C. 1965. Use of environmental factors in screening for disease resistance. Annu. Rev. Phytopathol. 3: 197

532. Walton, J.D. and D.G. Panaccione. 1993. Host-selective toxins and disease specificity — perspectives and progress. Annu. Rev. Phytopathol. 31: 275

533. Watson, A.G. and E.G. Ford. 1972. Soil fungistasis— a reappraisal. Annu. Rev. Phytopathol. 10: 327

534. Webster, R.K. 1974. Recent advances in the genetics of plant pathogenic fungi. Annu. Rev. Phytopathol. 12: 331

535. Wei, G., J.W. Kloepper and S. Tuzum. 1996. Induced systemic resistance to cucumber diseases and increased growth by plant growth-promoting Rhizobacteria under field conditions. Phytopathology 86: 221

536. Wei, Z-M, R.J. Laby, C.H. Zumoff et al, 1992. Harpin, elicitor of the hypersensitive response produced by the plant pathogen Erwinia amylovora. Science 257: 85

384

537. Weinhold, A.R., J.W. Oswald, T. Bowman, J. Bishop and D. Wright. 1964. Influence of green manuring and crop rotation on common scab of potato. *Am. Potato J.* 41: 265

538. Weller, D.M. 1988. Biological control of soil-borne plant pathogens in the rhizosphere with bacteria. *Annu. Rev. Phytopathol.* 26: 379

539. Weltzien, H.C. and N. Ketterer. 1986. Control of downy mildew *Plasmopara viticola* on grapevine leaves through water extract of composted organic wastes. *Jour. Phytopathol.* 116: 186

540. Wheeler, H. 1975. *Plant Pathogenesis.* Springer-Verlag, Berlin

541. Wheeler, H. 1976. Permeability alterations in diseased plants, pp. 413-429. In: *Ref. No.* 245

542. Wheeler, H. and H.H. Luke. 1963. Microbial toxins in plant diseases. *Annu. Rev. Microbiol.* 17: 223

543. Whipps, J.M. 1997. Developments in the biological control of soil-borne plant pathogens. *Adv. Bot. Res.* 26: 1

544. White, R.F. 1979. Acetyl salicylic acid (aspirin) induces resistance to tobacco mosaic virus in tobacco. *Virology* 99: 410

545. Whitney, P.J. 1976. *Microbial Plant Pathology.* Hutchinson, London

546. Wilcox, W.F., D.I. Wasson and J. Kovach. 1992. Development and evaluation of integrated reduced spray program using sterol methylation inhibition fungicides for control of primary scab. *Plant Dis.* 76: 669

547. Williamson, V.M. 1998. Root-knot nematode resistance genes in tomato and their potential for future use. *Annu. Rev. Phytopathol.* 36: 277

548. Williamson, V.M. and R.S. Hussey. 1996. Nematode pathogenesis and resistance in plants. *Plant Cell* 8: 1735

549. Windels, C.E. and S.E. Lindow. 1985. *Biological Control in the Phylloplane.* Am. Phytopathol. Soc., St. Paul, Minn., USA

550. Windham, M.T., Y. Elad and R. Baker. 1986. A mechanism for increased plant growth induced by *Trichoderma* spp. *Phytopathology* 76: 518

551. Wood, R.K.S. 1967. *Physiological Plant Pathology.* Blackwell Scientific Publications, Oxford.

552. Wood, R.K.S. 1973. Specificity in plant diseases, pp. 1-16, In: *Ref. No.* 94

553. Wood, R.K.S. 1982. *Active Defence Mechanisms in Plants.* Plenum Press, New York

554. Wood, R.K.S., A. Ballio and A. Graniti (eds.). 1972. *Phytotoxins in Plant Diseases.* Academic Press, New York

555. Wyss, U. 1988. Pathogenesis and host-parasite specificity in nematodes, pp. 417-432. In: *Ref. No.* 478

556. Xu, G;-H. and D.C. Gross. 1986. Selection of fluorescent pseudomonads antagonistic to *Erwinia carotovora* and suppressive to potato seed piece decay. *Phytopathology* 76: 414

557. Yang, Y., J. Shah and D.F. Klessig. 1997. Signal perception and transduction in plant defense responses. *Genes Dev.* 11: 1621

558. Yarham, D.J. 1979. The effect on soil-borne diseases of changes in crop and soil management, pp. 371-383. In: *Ref. No.* 446

559. Yarwood, C.E. 1976. Modification of the host response: Predisposition, pp. 703-718. In: *Ref. No.* 245

560. Yoshikawa, M. and H. Masago. 1982. Biochemical mechanism of glyceollin accumulation in soybean, pp. 265-279. In: *Ref. No. 27*

561. Yoshikawa, M., N. Yamaoka and Y. Takeuchi. 1993. Ellicitors: their significance and primary modes of action with induction of plant defense reactions. *Plant Cell Physiol.* 34: 1163

562. Young, H.C. Jr., J.M. Prescott and E.E. Saari. 1978. Role of disease monitoring in preventing epidemics. *Annu. Rev. Phytopathol.* 16: 253

563. Zadoks, J.C. 1984. A quarter century of disease warning. *Plant Dis.* 68: 352

564. Zadoks, J.C. and R.D. Shein. 1979. *Epidemiology and Plant Disease Management.* Oxford Univ. Press, London

565. Zakaria, M.J. and J.L. Lockwood. 1980. Reduction in *Fusarium* populations in soil by oil seed meal amendments *Phytopathology* 70: 240

566. Zentmyer, G.A. 1961. Chemotaxis of zoospores for root exudates. *Science* 133: 1595

567. Zhang, W., W.A. Dick and H.A.J. Hoitink. 1996. Compost induced systemic acquired resistance in cucumber to Pythium root rot and anthracnose. *Phytopathology* 86: 1066

568. Zuber, M. and K. Manibhushanrao. 1979. Phytoalexins from the germinating seeds of rice (*Oryzae sativa*). *Curr. Sci.* 48: 497

Index